Dynamic Behaviour of Heat Exchangers

Volume 3

International Series on Developments in Heat Transfer

Objectives:

The Developments in Heat Transfer book Series publishes state-of-the-art books and provides valuable contributions to the literature in the field of heat transfer. The overall aim of the Series is to bring to the attention of the international community recent advances in heat transfer by authors in academic research and the engineering industry.

Research and development in heat transfer is of significant importance to many branches of technology, not least in energy technology. Developments include new, efficient heat exchangers, novel heat transfer equipment as well as the introduction of systems of heat exchangers in industrial processes. Application areas include heat recovery in the chemical and process industries, and buildings and dwelling houses where heat transfer plays a major role. Heat exchange combined with heat storage is also a methodology for improving the energy efficiency in industry, while cooling in gas turbine systems and combustion engines is another important area of heat transfer research.

To progress developments within the field both basic and applied research is needed. Advances in numerical solution methods of partial differential equations, high-speed, efficient and cheap computers, advanced experimental methods using LDV (laser-doppler-velocimetry), PIV (particle-image-velocimetry) and image processing of thermal pictures of liquid crystals, have all led to dramatic advances during recent years in the solution and investigation of complex problems within the field.

The aims of the Series are achieved by contributions to the volumes from invited authors only. This is backed by an internationally recognised Editorial Board for the Series who represent much of the active research worldwide. Volumes planned for the series include the following topics: Compact Heat Exchangers, Engineering Heat Transfer Phenomena, Fins and Fin Systems, Condensation, Materials Processing, Gas Turbine Cooling, Electronics Cooling, Combustion-Related Heat Transfer, Heat Transfer in Gas-Solid Flows, Thermal Radiation, the Boundary Element Method in Heat Transfer, Phase Change Problems, Heat Transfer in Micro-Devices, Plate-and-Frame Heat Exchangers, Turbulent Convective Heat Transfer in Ducts, Enhancement of Heat Transfer and other selected topics.

S del Guidice
Department of Energy Technology
and Mechanical Engineering
University of Udine
Viale Ungheria 43
33100 Udine
Italy

D B Ingham
Department of Applied Mathematical
Studies
The University of Leeds
School of Mathematics
Leeds
West Yorkshire LS2 9JT
UK

M Faghri
Department of Mechanical
Engineering and Applied Mechanics
The University of Rhode Island
Kingston
RI 02881-0805
USA

Y Jaluria
Mechanical Engineering Department
Rutgers University
PO Box 909
Piscataway
NJ 08855-0909
USA

T Fusegi
Tokyo Gas Col. Ltd.,
Energy Technology Research Institute
Shibarua 1-Chome
Minato-Ku
Tokyo 105
Japan

S Kotake
Department of Mechanical
Engineering
University of Tokyo
Bunkyo-Ku
Tokyo 113
Japan

P J Heggs
Department of Chemical Engineering
UMIST
PO Box 88
Manchester
M60 1QD
UK

M Lamvik
Department of Mechanics
Norwegian University of Science and
Technology
Thermo and Fluid Dynamics
K. Hejes veg I B
N-7034 Trondheim
Norway

C Herman
Department of Mechanical
Engineering
John Hopkins University
3400 N. Charles Street
Baltimore
Maryland 21218-2686
USA

P S Larsen
Department of Energy Engineering
Fluid Mechanics Section
Technical University of Denmark
DK-2800 Lyngby
Denmark

D B Murray
Department of Mechanical and
Manufacturing Engineering
Trinity College Dublin
Parsons Building
Trinity College
Dublin 2
Ireland

A J Nowak
Institute of Thermal Technology
Technical University of Silesia
44-101
Gliwice
Konarksiego 22
Poland

P H Oosthuizen
Department of Mechanical
Engineering
Heat Transfer Laboratory
Queen's University
Kingston
Ontario
Canada K7L 3N6

W Roetzel
Institut für Thermodynamik
Universtät der Bundeswehr Hamburg
D-22039 Hamburg
Germany

B Sarler
Faculty of Mechanical Engineering
University of Ljubljana
Askerceva 6
1000 Ljubljana
Slovenia

A C M Sousa
Department of Mechanical
Engineering
University of New Brunswick
PO Box 4400
Fredericton
New Brunswick
Canada E3B 5A3

D B Spalding
CHAM
Bakery House
40 High Street
Wimbledon Village
London SW19 5AU
UK

J Szmyd
St Staszic
University of Mining and Metallurgy
Mickiewicza 30
30059 Krakow
Poland

E Van den Bulck
Department of Mechanical
Engineering
Katholieke Universiteit Leuven
Celestijnenlaan 300 A
B-3001 Heverlee
Belgium

K H Winters
AEA Technology plc
Product Development Division
424.4 Harwell
Didcot
Oxfordshire, OX11 ORA
UK

S Yanniotis
Department of Food Industries
Agricultrual University of Athens
IERA ODOS 75
11855 Athens
Greece

M Zerroukat
Wessex Institute of Technology
Ashurst Lodge
Ashurst
Southampton
Hampshire SO40 7AA
UK

Dynamic Behaviour of Heat Exchangers

Volume 3

W. Roetzel
Universität der Bundeswehr Hamburg
Germany

Y. Xuan
Nanjing University of Science and Technology
China

WIT PRESS Boston, Southampton
Computational Mechanics Publications

Wilfried Roetzel
Yimin Xuan

Published by

WIT Press/Computational Mechanics Publications
Ashurst Lodge, Ashurst, Southampton, SO40 7AA, UK
Tel: 44(0)1703 293223;
Fax: 44(0)1703 292853
Email: witpress@witpress.com
http://www.witpress.com

For USA, Canada and Mexico

Computational Mechanics Inc
25 Bridge Street, Billerica, MA 01821, USA
Tel: 978 667 5841;
Fax: 978 667 7582
Email: cmina@ix.netcom.com

British Library Cataloguing-in-Publication Data

A Catalogue record for this book is available from the British Library

ISBN: 1 85312 5067 WIT Press/Computational Mechanics Publications
ISSN: 1369-7331

Library of Congress Catalog Card Number: 97-81196

Printed in Great Britain by Bookcraft Ltd., Bath.

Contents

Preface

Many books on heat exchangers have already appeared. They all deal with steady-state operating conditions, which are usually and necessarily considered for heat exchanger design and rating. This book deals exclusively with transient processes which are of growing interest in research and industry particularly as the steady-state thermal calculation methods have already been highly developed.

Considering the heat exchanger as part of a system reveals that the dynamic behaviour is as important as the steady-state performance. However, the mathematical treatment is far more complicated. Simplified mathematical models such as the traditional or the dispersed plug flow model have to be applied for the prediction of outlet temperature responses to changes in inlet temperatures or flow rates of an exchanger.

This book proposes and describes suitable mathematical models and methods for the analytical and numerical treatment of transient processes in shell-and-tube, cross-flow, plate, and other types of heat exchangers. All necessary information is provided to the research engineer in industry or the graduate student and teacher in the university to enable them to write and apply programmes for the fast calculation of transient processes in heat exchangers. Examples are presented to explain the influence of various design parameters on the dynamic behaviour.

It is also shown, how the knowledge and understanding of dynamic behaviour can be used to measure indirectly the heat transfer coefficients and other parameters. Finally some new developments in heat exchanger modeling are discussed.

The nomenclature is mainly explained in the text and, with respect to the cited literature, not completely uniform in all chapters. At the end of the book a common nomenclature is given with the frequently used symbols, definitions and units.

The authors know that the book is incomplete. Nevertheless they hope that it might be a basis for further research in this important and interesting field. To facilitate the application of the methods presented it is planned to develop suitable software for selected methods in the years to come.

This book is the result of a research cooperation between the Universität der Bundeswehr Hamburg (UniBw H) and the Nanjing University of Science and Technology (NUST). The authors wish to express their thanks to the presidents of both universities, Prof. Dr. Gerhard Strunk (UniBw H) and Prof. Dr. Hongzhi Li (NUST), who supported the cooperation and mutual research visits of the authors to Hamburg and Nanjing.

Sincere thanks are especially due to Dr.-Ing. Xing Luo for his valuable help in completing the manuscript. Dr. Luo, the former doctoral student of the first author in Hamburg contributed the results of his very recent doctoral thesis to the chapters on cross-flow and transient test techniques. He also calculated and recalculated the numerous examples presented in the book and he did the unpleasant final editorial work for the manuscript.

Finally the authors would like to thank the Series Editor Prof. Dr. Bengt Sunden, Lund Institute of Technology, as well as the Publishing Director of Computational Mechanics Publications, Mr. Lance Sucharov, for the fruitful and effective cooperation.

Wilfried Roetzel
Yimin Xuan

Chapter 1

Introduction

1.1 Chronological development

Since the first industrial revolution, various types of heat exchangers have been invented, innovated and put into operation for different applications, such as chemical processes, power engineering, cryogenic techniques, automobiles, aero- and space-vehicles. For a long time, the design and rating procedures of heat exchangers have been based on the steady-state simulation of heat transfer and flow phenomena occurring in the heat exchangers. The mean temperature difference approach, as well as the effectiveness and number of transfer units approach, are two methods widely used for design and rating purposes. The steady-state procedure reveals the energy balance relationship among fluid streams involved in heat exchange and the time-independent temperature profiles of these streams, but it provides no information about the dynamic behaviour of heat exchangers.

Besides being in steady-state operation, heat exchangers experience transient processes which are induced for different reasons. During start-up, shutdown, and deliberate variations in flow rates and temperatures of incoming fluid streams, heat exchangers are in a transient operation phase. In addition, the scale deposit on heat transfer surfaces also induces time-dependent operation of heat exchangers. Strictly speaking, heat exchangers are always in dynamic operation, although they are usually assumed to be in the steady-state operation in most cases. After the transient operation, a heat exchanger may either approach a new steady state or return to the original steady state. It is also possible that a heat exchanger is in a transient operation all the time, for example, in the case of periodic variations of flow rates and temperatures of incoming streams. Gupta (1986) listed the six types of changes concerned with the dynamics of a heat exchanger as follows:

(1) Impulse change in the inlet flow rates and/or temperatures of the fluid streams, which may be applied to dislodging the scale deposit on the heat transfer surface.

(2) Step change in the inlet flow rates and/or temperatures of the fluid streams, which may happen due to process requirements or during start-up and shutdown.

(3) Periodic change in the inlet flow rates and/or temperatures of the fluid streams.

(4) Gradual fouling up of the heat transfer surface due to scale deposit which increases the overall heat transfer resistance.

(5) Gradual reduction in the heat transfer area due to the plugging of the tubes as and when they start to leak.

(6) Gradual blanketing of the heat transfer surface due to noncondensable gases present in a condensing stream or released from liquids due to change in temperature and/or pressure or if the venting is inadequate.

Dynamic analysis of heat exchanger is usually concerned with the first three items, although a good analysis may cover the rest of the above-mentioned items as well.

Optimum operation, treatment of accidents, dynamic design and real-time control demand more accurate description of the time-domain behaviour of heat exchangers, although the steady-state design and rating procedure is still used. Dynamic analysis of heat exchangers provides information about transient responses subject to various disturbances. There are many cases in which the system dynamic behaviour is of prime design consideration. For example, the dynamic analysis is very important in nuclear plants where severe transients may cause accidents and one may have to estimate the possibility of accidents and may be able to prevent them by means of accurate knowledge of dynamic behaviour. Since the 1950s more and more attention has been paid to the transient behaviour of a variety of heat exchangers with or without phase change and there are plenty of published papers on the dynamics of heat exchangers. Compared with the steady-state procedure, dynamic simulation gives more information about the operational situation and effects of exchanger structures. Some of the advantages of dynamic analysis of an exchanger were assessed by Gupta (1986). He concluded that the dynamic information is useful in evaluating the following:

(1) the effects on the fluid streams (heat transfer rate, pressure drop, freezing, polymerization fouling, reduction in scale deposit, vibration damage);

(2) the effects on the processes downstream;

(3) the effects on any auxiliary controls required;

(4) the new state and the time required to reach it;

(5) the time required to return to the previous steady state;

(6) the changes required in the steady-state operating conditions so that the operation is more stable or the lag time is minimal;

(7) whether the system is statistically stable (i.e., after a disturbance from a steady state, the system tends to the original steady state) or is dynamically stable (i.e., after a disturbance from a steady state, the system will ultimately attain the original steady state, the convergence procedure may be periodic or aperiodic in nature);

(8) the thermal stresses within an exchanger and loads on the bolts making a flanged joint, etc.

In all, dynamic simulation can serve as an aid to exchanger operation (i.e., control and regulation) and a design tool. Increasing safety and environmental concerns and stricter regulation have provided strong incentives for the study of the dynamics of heat exchangers.

There is a number of factors affecting the transient behaviour of heat exchangers. Among them are flow rates and temperatures at inlets, heat transfer coefficients, the nature and magnitude of disturbances of incoming fluid streams, the size and construction features of the exchanger, flow arrangement (i.e., cocurrent, countercurrent, cross-flow), flow distribution (i.e., even or maldistribution), thermal properties of fluid streams and exchanger materials as well as phase change. Some of them may not appear in the steady-state simulation procedure. A comprehensive algorithm of dynamic simulation should contain as many factors as possible in order to provide an accurate description of the exchanger dynamics. The idealizations involved in dynamic simulation are generally similar to those for the steady-state analysis of heat exchangers. In some cases, however, the assumptions of time-independent flow rates and heat transfer coefficients may be violated.

The dynamic analysis of heat exchangers can be performed either experimentally or theoretically. With respect to complexity of dynamic simulation, both the lumped parameter and distributed parameter approaches can be used for equation-oriented dynamic simulation. The first assumes that temperatures of fluid streams and solid materials (if they are involved) are only functions of the time variable, i.e., the fluid and solid material have infinite thermal conductivity, which greatly simplifies the mathematical model describing the transient behaviour of exchangers and is convenient for controller design. The latter takes variations of the temperatures with both time and space variables into account, which may more closely coincide with the real operation than the first. As pointed out by Pantelides and Barton (1992), the modelling of distributed parameter systems is of strategic importance for the future evolution of general-purpose modelling environments.

For both the lumped and distributed parameter systems, the transient behaviour can generally be described mathematically by mixed sets of differential and algebraic equations. To obtain the transient responses of heat exchangers, the differential equation systems based on the above-mentioned approaches are solved by means of different numerical methods such as the finite-difference method, the finite-element method, numerical inversion of the Laplace transform, and numerical inversion of the Fourier transform (Margan, 1991). Furthermore, one may derive transient functions to analyse the response characteristics in the frequency domain.

There are various types of heat exchangers with different constructions and flow arrangements. Since many variables are involved and a variety of disturbances are possible, no general solution can be obtained for the dynamic simulation of heat exchangers. For some specific types of heat exchangers, one can find published papers. The transient responses to step disturbances are often reported. For deriving differential equations of transient behaviour, different flow patterns may be used except for the lumped parameter approach which pays no attention to the flow pattern of the fluid streams in the exchanger. For application of the distributed parameter approach, the most common case corresponds to assuming the plug-flow pattern, which is widely used in the steady-state procedure and the early phase of dynamic simulation. To account for the possible flow maldistribution, the conventional plug-flow model is modified and the dispersed plug-flow model (Roetzel and Xuan, 1992; Xuan and Roetzel, 1993) may be applied to the dynamic simulation of heat exchangers.

1.2 Fundamental equations of dynamic simulation

The dynamic simulation of all types of heat exchangers is based on the continuity equation, momentum and energy balance equations. These conservation equations are the starting point for the derivation of a set of differential equations for the transient analysis. Depending upon the specific case, the general forms of these equations may be simplified.

The thermal energy equation for general flow processes is written as (Whitaker, 1977)

$$\rho c_p \frac{\mathrm{D}\theta}{\mathrm{D}\tau} = \nabla(k\nabla\theta) + \beta\theta\frac{\mathrm{D}p}{\mathrm{D}\tau} + \Phi + \dot{s} \tag{1.1}$$

with the substantial derivative of temperature θ

$$\frac{\mathrm{D}\theta}{\mathrm{D}\tau} = \frac{\partial\theta}{\partial\tau} + \mathbf{v}\cdot\nabla\theta, \tag{1.2}$$

and correspondingly of pressure p. The second term with the expansion coefficient β on the right-hand side of eqn (1.1) represents compression and expansion work and may be either positive (compression) or negative (expansion). If this term is included in the energy equation, the temperature θ has to be the thermodynamic (absolute) temperature. The third term accounts for the viscous dissipation and is always positive. Both these terms are generally neglected in modelling the dynamic process in an exchanger. $\dot{s}$ represents the energy source or sink. Hence, eqn (1.1) can be simplified to

$$\rho c_p\left(\frac{\partial \theta}{\partial \tau} + \mathbf{v} \cdot \nabla \theta\right) = k\nabla^2\theta + \dot{s}, \tag{1.3}$$

with a constant thermal conductivity. A detailed derivation and discussion of the energy equation has been given by Baehr and Stephan (1996).

By assuming a one-dimensional flow pattern in the exchanger, the positive source term takes heat transfer from the wall to the fluid into account

$$\dot{s} = \frac{hP}{A_c}(\theta_w - \theta), \tag{1.4}$$

where A_c is the cross-sectional area for the plug-flow and P is the wetted perimeter of the flow channel. The temperature θ is the adiabatic mixing temperature of the flow stream yielding its correct enthalpy flow.

Thus, the transient energy equation of a fluid becomes

$$\rho c_p\left(\frac{\partial \theta}{\partial \tau} + u\frac{\partial \theta}{\partial l}\right) = k\frac{\partial^2 \theta}{\partial l^2} + \frac{hP}{A_c}(\theta_w - \theta). \tag{1.5}$$

This equation is valid for each stream in the exchanger. For special cases under consideration, it may further be simplified. For example, the axial conduction term on the right-hand side is omitted ($k = 0$) for the traditional plug-flow model:

$$\rho c_p\left(\frac{\partial\theta}{\partial\tau}+u\frac{\partial\theta}{\partial l}\right)=\frac{hP}{A_c}(\theta_w-\theta)\,. \tag{1.6}$$

In addition, the derivative of temperature against time may be neglected for gas flow. If flow maldistribution occurs, the dispersed plug-flow model may be used. In this case, the thermal conductivity k in eqn (1.5) should be replaced by the apparent or dispersive conductivity $\mathcal{D}$ to form the so-called dispersed plug-flow model which is a half empirical model and treats the multi-dimensional flow pattern approximately by a one-dimensional formulation.

By introducing the thermal flow rate $\dot{W}$ and the capacitance C of fluid residing in the exchanger, respectively,

$$\dot{W}=uA_c\rho c_p\,,\ C=A_cL\rho c_p\,,$$

eqn (1.5) can be rewritten as

$$\frac{C}{L}\frac{\partial\theta}{\partial\tau}+\dot{W}\frac{\partial\theta}{\partial l}-\mathcal{D}A_c\frac{\partial^2\theta}{\partial l^2}+\frac{hP}{A_c}(\theta-\theta_w)=0\,. \tag{1.7}$$

This differential equation is frequently used in the following chapters.

For the temperature distribution in the wall, the Fourier conduction equation corresponding to eqn (1.3) is used with $\mathbf{v}=0$:

$$(\rho c_p)_w\frac{\partial\theta_w}{\partial\tau}=\nabla(k\nabla\theta)_w+\dot{s}_w\,. \tag{1.8}$$

The conduction term on the right-hand side accounts for axial and/or transverse heat conduction. If the lumped parameter approach is adopted, the transverse conduction resistance is neglected while the axial conduction resistance may be assumed to be infinite. Thus,

$$(\rho c_p)_w \frac{\partial \theta_w}{\partial \tau} = \dot{s}_w . \tag{1.9}$$

The source term $\dot{s}_w$ depends upon the heat exchange between fluid streams and the wall, which is related to the type of heat exchangers and flow arrangement. In two-fluid tubular exchangers, for example, the term $\dot{s}_w$ in eqn (1.9) can be expressed as

$$\dot{s}_w = \sum_{i=1}^{2} \left[\frac{hP}{A_{cw}} (\theta - \theta_w) \right]_i . \tag{1.10}$$

In some particular cases, the transverse resistance in the wall has to be taken into account. Neglecting the axial conduction, one can write the energy equation for the tube wall as follows:

$$(\rho c_p)_w \frac{\partial \theta_w}{\partial \tau} = k \left(\frac{\partial^2 \theta_w}{\partial r^2} + \frac{1}{r} \frac{\partial \theta_w}{\partial r} \right) . \tag{1.11}$$

There is no heat source term in eqn (1.11) and the boundary conditions are of the third type. If axial heat conduction in the wall (along the flow direction) is taken into account and transverse conduction resistance is neglected, eqn (1.8) becomes

$$(\rho c_p)_w \frac{\partial \theta_w}{\partial \tau} = k_w \frac{\partial^2 \theta_w}{\partial l^2} + \sum_{i=1}^{2} \left[\frac{hP}{A_{cw}} (\theta - \theta_w) \right]_i . \tag{1.12}$$

All these differential equations constitute the basis of the dynamic analysis of heat exchangers. In cases in which phase change takes place, however, it may be more convenient to use the enthalpy as the dependent variable rather than the temperature and the energy equations for fluid streams may assume different forms. Roetzel (1996) has reviewed research development of transient analysis of heat exchangers.

1.3 Application of dynamic theory of heat exchangers

For a long time steady-state design and rating procedures have played the dominant role in the optimal design and performance estimation of heat exchangers. However, these methods cannot deal with all of the problems occurring in exchangers. As mentioned before, the dynamic analysis provides more information about operation of heat exchangers than the steady-state investigation alone. Such information may help engineers to improve design, operation and control of heat exchangers to avoid over-design and operational accidents. Improved design and control will lead to a reduction in capital and operation cost and better performance. Application of the dynamic theory of heat exchangers is mainly in the following aspects.

1. Analysis of transient responses. Dynamic analysis reveals the effects of various factors such as structures, dimensions, flow rates and fouling resistance on transient responses subject to all possible disturbances and provides reference information for operation decision and possible design modification.

2. Controller design. Most control algorithms are equation-oriented, except the fuzzy control. From dynamic analysis, one can obtain input–output dynamic relationships of the system under consideration. Controllers are designed on the basis of these relationships which may be linearized for nonlinear cases. From dynamic feature equations, one can determine transfer functions of heat exchangers, whether the apparatus are considered as one input–one output, one input–multi output or multi input–multi output systems. These transfer functions primarily define the controller structure and regulation strategies.

3. Dynamic design. The conventional steady-state design procedure mainly provides structural data and heat transfer surface area to satisfy the given heat duty under certain constraints, but pays little attention to operability and controllability of apparatus. Since dynamic analysis can determine the characteristics of both the transient behaviour and the steady-state performance, the dynamic design procedure developed from the dynamic analysis may determine structural and surface data by synthesising constraints proposed according to the heat duty, capital and operational cost, and regulation performance.

4. Operation imitation. The dynamic analysis can predict transient responses to any disturbances and reveal the effects of various factors on transient features of heat exchangers. The dynamic analysis programme opens up possibilities for operation imitation, accident diagnosis, analysis of preventive measures against different accidents, and training of operation engineers. All these functions may be very important for steam generators in nuclear power stations where safety is the first consideration.

5. Efficiency improvement. Since the dynamic analysis can simulate both the transient and the steady-state operation features of thermal apparatus, it can be applied to the optimization of the operating conditions in order to improve efficiency, increase product quality, decrease consumption and environmental pollution. The improvement effect may be more remarkable for heat exchanger networks, furnaces, kilns and ovens. There are different furnaces, kilns and ovens in a variety of industrial applications. To some extent, they can be considered as heat exchangers from the view point of heat exchange. Because dynamic simulation involves a set of partial differential equations describing the chemical reactions, mass balance and energy balance in these units, dynamic analysis may yield the optimal design and multi-variable optimal control for the objectives such as fuel saving, reduction in pollutant emissions, acceptable product quality and unit lifetime.

In brief, the urgent demand for optimal operation and regulation, controller design, dynamic design, safety and environmental concerns gives an impetus to research into the transient behaviour of a variety of heat exchangers, which is also the incentive to produce this monograph. Knowledge of dynamic simulation of exchangers has been finding more and more applications.

1.4 References

Baehr, H. D. and Stephan, K. (1996) Wärme- und Stoffübertragung, 2. Auflage, Springer, Berlin.

Gupta, J. P. (1986) *Fundamentals of Heat Exchanger and Pressure Vessel Technology*, Hemisphere Publishing Corporation, New York.

Margan, E. (1991) Calculating the Transient Response, *Elektrotehuiski Vestnik*, **58**, 11–22.

Pantelides, C. C. and Barton, P. I. (1992) Equation-Oriented Dynamic Simulation Current Status and Future Perspectives, *Computers and Chemical Engineering*, **17**, s262–s285.

Roetzel, W. (1996) Transient Analysis in Heat Exchangers, in *New Development in Heat Exchangers*, (ed. by Afgan, N. *et al.*), pp. 547–575, Gordon and Breach Publishers, OPA Amsterdam, B. V.

Roetzel, W. and Xuan, Y. (1992) Analysis of Transient Behaviour of Multipass Shell and Tube Heat Exchangers with the Dispersion Model, *International Journal of Heat and Mass Transfer*, **35**, 2953–2962.

Xuan, Y, and Roetzel, W. (1993) Stationary and Dynamic Simulation of Multipass Shell and Tube Heat Exchangers with the Dispersion Model for Both Fluids, *International Journal of Heat and Mass Transfer*, **36**, 4221–4231.

Whitaker, S. (1977) *Fundamental Principles of Heat Transfer*, Pergamon Press, New York.

Chapter 2

Geometry and structure of heat exchangers

2.1 Description and types of heat exchangers

The heat exchanger is an apparatus that facilitates heat exchange among fluids with different temperatures. In a heat exchanger two or more fluid streams are involved in the heat transfer process. Heat exchangers are used in almost every industrial application, e.g., in the space shuttle, aircraft, automobile, electronic technology, power engineering, chemical engineering, the food industry and metallurgical industry. In addition, heat exchangers are widely used in our domestic life. In fact, every living organism itself is a system of heat exchangers with a simple or complicated geometry. It may be stated that heat exchangers are found wherever heat flow appears.

Industrial heat exchangers cover a very broad range of temperature and pressure: from cryogenic temperatures (4 K) to very high temperatures (above 1700 K) and from vacuum states to hyperpressures (above 3000 bar). An enormous range of heat transfer capacities are performed in heat exchangers: the smallest (< 1 W) are used in miniature cryocoolers for infrared thermal imaging, heat seeking missile guidance or superconducting electronic applications; the largest (> 1 GW) are the boilers, condensers and air-coolers for recooling water in power stations (Walker, 1982).

For various service aspects and objects there are different types of heat exchangers, some are simpler and some are more complicated. Naturally, they are usually not so complicated as the heat exchange systems in living organisms. The fluids participating in the heat transfer process within heat exchangers may be single-phase streams such as gases and liquids or two-phase streams such as vapours and gases or liquids with particles or slurries. The sort of heat-exchange fluids, the range of applied temperature and pressure and the application site make particular demands upon the material, the geometry, the compactness and the weight of heat exchangers. Consequently, heat exchangers are manufactured from metal and nonmetal materials in different structures. Therefore, a diversity of types, shapes, arrangements and combinations of heat exchangers are put into industrial operation.

According to the geometrical and operating characteristics, a few criteria are used for sorting a great diversity of types of heat exchangers. The words *recuperative* and *regenerative* classify all industrial heat exchangers into two groups: (1) recuperative heat exchangers; (2) regenerative heat exchangers. Recuperative heat exchangers refer to apparatus in which fluids are separated from each other by a solid wall and flow continuously in the particular conduit. Heat is transferred from hot fluids to cold ones across the solid wall. If all operating conditions remain stable, a steady state can be reached in recuperative heat exchangers. Regenerative heat exchangers have quite different structures and are composed of solid matrices. The heat exchange fluids possess a common conduit and flow over the same surface in a regenerator during separate time spans. While the hot fluid is flowing through the solid matrices (the hot period), the solid matrices absorb and store heat; during the cold fluid flow (the cold period) heat is transferred from the solid matrices to the cold fluid. In this way heat exchange between fluids in a regenerator takes place indirectly. Evidently, the temperatures of fluids and solid matrices in regenerative heat exchangers are functions of time and space.

Besides the above-mentioned two types of heat exchangers, another type of devices is known as the direct-contact heat exchanger (Kreith and Boehm, 1988). As the name implies, a hot stream and a cold stream are brought into contact in such an apparatus, instead of being separated by a solid wall. Heat and mass transfer, and consequently both sensible and latent heat exchange processes may simultaneously be involved. Some typical examples of direct-contact heat exchangers are air-cooling towers, distillation units and fluidized beds. Recently, they are also used for waste heat recovery from the effluents. Many unsolved problems are still to be faced for simulation and design of this type of heat exchange devices.

Recuperative heat exchangers exist in many different variations, depending upon their geometry, the manufacturing technique and the flow arrangement. A common methodology is to classify recuperative exchangers into the tubular and plate types. In tubular exchangers tubes form the heat transfer surface and flow channels. Double-pipe heat exchangers and shell-and-tube heat exchangers are two typical types of the tubular exchanger. Plate exchangers are composed of thin plates with various geometrical structures assembled by different methods. The stacks of die-formed plates may be brazed or welded or sealed with gaskets. Plate heat exchangers, spiral heat exchangers and compact heat exchangers (plate-fin types) belong to this category. With development of heat transfer enhancement techniques, some structural improvements are introduced in both tubular and plate exchangers. New materials give an impulse to the invention of new heat exchangers.

Because of the geometrical advantage, the tubular exchanger can endure higher pressure and serve as a pressure vessel as well as a heat transfer apparatus in some cases. The tubular exchangers can virtually be applied to such broad ranges of pressures and temperatures that there is almost no limitation. The standards or codes for design, manufacture, maintenance and inspection of tubular exchangers are familiar to many engineers. Among the various standards, the TEMA standard proposed by the Tubular Exchanger Manufacturers Association (TEMA) in the United States of America and the API Standard 660 by the American Petroleum Institute (API) are extensively accepted. The newest versions of these standards are TEMA Standards (1991) and API Standard 660 (1993).

Although the tubular exchangers find wider use, more and more plate exchangers are used in various units or systems, especially in cases in which the cost, weight and the compactness are important, for example, in vehicles, aeroplanes, space shuttles and electronic equipment. Up-to-date, research achievements of heat exchangers have mainly been related to plate heat exchangers. Compared with tubular exchangers, different flow arrangements can conveniently be accomplished in plate exchangers to meet the specific demand. They are especially suitable for heat exchange among more than two fluid streams which are not allowed to be mixed together.

2.2 Shell-and-tube heat exchangers

Many industrial applications, for example, chemical, petrochemical, food, refrigeration and power stations, utilise shell-and-tube heat exchangers. They serve as reboilers, evaporators, condensers, heaters and coolers. To some extent tubular exchangers imply shell-and-tube heat exchangers with a shell and tubes. As pointed out by Singh and Soler (1984), there are two domains in a shell-and-tube heat exchanger: the "tubeside" domain commonly refers to the inside of the tubes and the inlet/outlet nozzles as well as the headers; the "shellside" refers to the remaining space between the shell and the tubes in the heat exchanger.

A lot of types of shell-and-tube heat exchangers are constructed and put into operation. According to the construction styles, they are divided into three groups: (i) fixed tubesheet, (ii) U-tubesheet and (iii) floating tubesheet. In view of the shell construction different shell types (E, F, G, H, J, K, and X) may be designated. TEMA standards provide a basic nomenclature and methodology scheme for shell-and-tube heat exchangers, which are broadly accepted to distinguish the diversity of shell-and-tube heat exchangers.

Generally, two fluid streams flow through the shell-and-tube heat exchanger in an industrial operation. The shellside fluid flows through the space between the shell and tubes, and the tubeside fluid flows inside the tubes. The tubeside stream may flow through all tubes in a parallel connection and the total tubeside flow length equals the tube length. This flow arrangement is referred to as a single tubeside pass arrangement.

In a multipass flow arrangement the tubeside fluid passes subsequently through separate parts of the bundle (connected in series) and the total tubeside flow length equals the tube length multiplied by the number of passes.

The shellside fluid may also flow in one or more than one pass. The shellside passes may be in parallel connection or in series. Shellside passes in parallel connection are denoted as different passes if they undergo different temperature changes.

A variety of shellside and tubeside flow arrangements come into operation. The arrangement chosen for any particular application may result from a compromise among a number of factors, such as thermal effectiveness, manufacturing techniques and costs, allowable pressure drops, vibration and fouling of the heat transfer surface. Evidently, one cannot expect to have a flow arrangement to satisfy all these demands because some of them conflict inherently. Some demands are satisfied at the price of neglecting the other. With respect to thermal effectiveness the countercurrent arrangement in a single-pass shell-and-tube heat exchanger is the ideal configuration, in which the mean driving temperature difference between the fluids is a maximum for given inlet and outlct temperatures and heat exchange between them is most effective.

Multipass flow arrangements are possible subject to the demands for the large heat transfer surface and compact installation. Because of manufacturing technique and costs, heat exchangers are seldom constructed with more than two shell passes, although apparatus with as many as six passes have been successfully designed and fabricated (Singh and Soler, 1984). With regard to the tubeside arrangement, one-tube pass, two-, four-, six-, eight-tube passes are found in industrial application. Apart from the one-tube pass, exchangers are seldom designed with odd numbers of tube passes for convenient arrangement, although heat exchangers with odd tube passes have higher thermal effectiveness, provided the number of countercurrent passes exceeds the number of cocurrent passes (Roetzel, 1988). Generally, thermal effectiveness of multi-tubeside-pass heat exchangers with more than one shell pass is higher than that of single shell pass exchangers. It is emphasized that reverse heat transfer may take place in multipass heat exchangers, which means that in some tube passes heat may be transferred from the originally cold stream to the originally hot one. This phenomenon should be avoided by designers and operating engineers.

In shell-and-tube heat exchangers various forms of baffles, such as segmental, strip, disk and doughnut, and orifice baffles, may be used to enforce the shellside fluid to move perpendicular to the tubes in processing along the shell. The shellside flow related to the particular baffles are briefly described by Walker (1982). It is due to the geometrical structure, manufacturing clearances and tolerances, and operating conditions that some types of flow maldistribution may appear in exchangers. On the shell side the following main types of flow maldistribution may occur: (i) the stagnant flow near baffles; (ii) bypass-flow around the tube bundle; (iii) leakage-flow through the gap between the baffle and the shell as well as the orifice between baffles and tubes. The classical description on the shellside flow is made by Tinker (1951). The flow maldistribution on the tube side mainly results from uneven distribution of the stream among the tubes, especially for the laminar flow. The flow maldistribution prevents effective heat transfer between fluid streams and induces deterioration. For some types of shell-and-tube heat exchangers, Roetzel and Spang (1997) proposed a set of rating and design methods.

A variation of shell-and-tube heat exchangers is the lamella heat exchanger consisting of flattened tubes which is designed for longitudinal flow on both sides. The flattened tubes–lamellas are made up of two strips of plates, formed and spot- or seam-welded together in a continuous operation. The lamellas are welded together at both ends either directly at bent-out end edges or by joining the ends with steel bars in between, depending on the space required between the lamellas, instead of using tube sheets. This type of exchanger has a lower weight than conventional tubular heat exchangers. Another type of shell-and-tube heat exchanger is the spiral tube heat exchanger. Because of its simplicity, compactness and the effective way of solving the thermal expansion problem, the spiral tube heat exchanger is used in process industries and cryogenic systems.

According to the operating temperature and pressure ranges, many metallic and nonmetallic materials can be used to manufacture shell-and-tube heat exchangers. For example, Teflon is used as a tube material to handle acids or other aggressive fluids. Graphite heat exchangers are also available for highly corrosive liquids.

2.3 Plate heat exchangers (PHE)

The plate heat exchanger was patented in the second half of the nineteenth century. As implied by the name, this type of exchanger is composed of plates. The pressed stainless steel plate heat exchanger was introduced in the 1930s. Since then stainless steel has been the standard material for the plate heat exchanger, although a wide range of alternative materials including composites are available. The plates are assembled together by different methods: gasketed, brazed or welded. Plate heat exchangers are commercially available and in use in a wide spectrum of industrial branches. Although plate heat exchangers cannot tolerate as high temperatures and pressures as tubular exchangers do, they are preferred in some applications because of the advantage of greater compactness, adjustable heat transfer surface area, easy inspection and cleaning as well as convenient flow arrangement. Heat exchange among more than two streams and their flow arrangements are easily accomplished in plate heat exchangers. They are used as liquid–liquid heat exchangers, evaporators and condensers, particularly in the process industry. With advances in enhanced heat transfer techniques, plate heat exchangers tend towards wider application, higher compactness and effectiveness. For example, the typical plate with form "∧" coated with a boiling enhancement (porous surface) is put into operation for ammonia evaporator applications (Panchal and Rabas, 1993). Reay (1994) briefly describes some types of plate heat exchangers which serve under different temperature and pressure ranges, such as plate-and-frame heat exchangers, partially welded plate heat exchangers, brazed plate heat exchangers and the Bavex hybrid welded plate heat exchangers.

In plate heat exchangers the fluid streams flow through the fine and narrow conduits between alternate plates. Corrugated plate surfaces and discrete contact points between two adjacent plates induce disturbance of the fluid flow and destruction of the laminar sublayer. The artificial flow turbulence enhances heat transfer between fluids, so that the surface area requirements of plate heat exchangers are typically 25% of those of equivalent shell-and-tube apparatus for liquid–liquid duties. Theoretically, the thermal effectiveness of a plate heat exchanger operation in a heat recovery system could approach 100% by increasing the number of plates, but this number would be infinite. Taking the cost of energy, the capital cost of the plates and the operating cost of pumping power, Reay (1979) thought it was worth the increased capital expenditure to install additional plates to give recovery efficiencies in excess of 90%.

A great diversity of flow arrangements for two or more than two streams can be realized for a particular demand. It is impossible to propose a versatile system of equations to describe temperature profiles in plate heat exchangers. The main factors that degrade the thermal performance of plate heat exchangers are as follows: (i) the uneven distribution of a stream among all channels in one pass which may result from the nonidentical pressure drop characteristics among the channels and conduits; (ii) the edge-effect which occurs because the streams flowing in the end channels of the stacked plates have thermal contact with the other streams only on one side, although the streams in the other channels receive or give off heat on two opposite sides of a channel; (iii) even or uneven number of plates which may cause unsymmetrical distribution of the stream among channels. Application of the plate-and-frame heat exchanger is mainly limited to 300°C and 20 bar by the mechanical characteristics of gaskets.

Another type of plate heat exchanger is the spiral-plate heat exchanger, which was firstly developed in Sweden about 10 years after the invention of the plate heat exchanger, to recover waste heat from contaminated water effluents in pulp mills. Since then its application has been extended to other industrial branches in a wide variety of sizes and duties. The elementary construction of a spiral-plate heat exchanger is described by Walker (1982) and Auth and Loiacono (1978). It may be made of any metals, including carbon steel, stainless steel, titanium, nickel and the high-nickel alloys, depending upon its application. According to the type, the edges of the plate strips can be either bent or seam-welded on one or both sides of the channel, or only machined against the covers to prevent bypassing and leakage.

The curved flow paths are formed by the rolled strips. The strips are spaced at a uniform distance apart by means of studs welded to the sheets. The channel width for a stream remains constant in the whole exchanger, but may be different for each stream. Flexible flow arrangements are accomplished in a spiral-plate heat exchanger, although two fluid streams are commonly involved in the heat transfer process in a spiral-plate apparatus. Either of the hot and cold streams enters at the centre and flows outwards or enters at the periphery and flows to the centre. To some extent the countercurrent flow is realized, but it is less thermally effective than the pure countercurrent flow in the tubular exchanger. The deterioration is caused by the fact that each of the two streams face different temperatures at the inner and outer walls of the flow channel. Although both fluids flow in opposite directions, the effectiveness of this flow arrangement is less than that of the pure countercurrent flow. Interaction between rotational flow and secondary flow induced by centrifugal force in the spiral channel contributes to heat transfer enhancement. The theory of the spiral plate exchanger is rather complicated because the heat transfer process is affected by the curved channels. One cannot expect to apply the thermal simulation method of a shell-and-tube heat exchanger with countercurrent flow to the spiral exchanger. Bes and Roetzel (1993) propose a rating and design method for the spiral heat exchanger.

Compared with the pressed plate heat exchangers, the spiral plate apparatus is relatively expensive because of the complicated fabrication. The most remarkable characteristic of this type of exchanger is its anti-fouling mechanism, which makes it especially suitable for applications related to contaminated fluids and fluids with slurries or solid particles.

2.4 Compact heat exchangers

Compact heat exchangers are characterized as apparatus with a high specific surface area – a high ratio of heat transfer surface to heat exchanger volume. In most literature heat exchangers having over 700 m^2/m^3 of the specific surface area are classified as compact heat exchangers. Generally, plate-fin and tube-fin exchangers and regenerators belong to this category. A tubular exchanger with a tube diameter less than 5 mm may be considered as a compact heat exchanger.

Compact heat exchangers played a major historically technological role in the development of lightweight, minimum volume, and high-effectiveness exchangers for aircraft, vehicular and marine transportation systems (Shah *et al.*, 1990). Now its application is extended to many high-technology fields and some industrial branches, such as spacecraft, cryogenic equipment, electronic technology and the process industry. Chronological development of compact heat exchangers may date back to the time after World War I when the urgent need for small-size, lightweight, and high-effectiveness heat exchangers gave an impetus to the research and manufacture of compact heat exchangers. The classical book about compact exchangers was written by Kays and London (1955, 1964, 1984).

The main merits of compact heat exchangers are as follows:

(1) High effectiveness. Compared with shell-and-tube heat exchangers, most of compact exchangers can operate under small temperature differences between hot and cold streams. Therefore, heat exchange between streams can more thoroughly performed and a high effectiveness is reached.

(2) Smaller volume. High compactness of exchangers leads to a smaller volume and the benefits of reduced material (especially for expensive nickel or titanium) and installation costs.

(3) Convenient multistream and multipass arrangements.

Compact heat exchangers, for example the plate-fin apparatus, are readily configured for multipass and multistream applications which are common in the process of air liquefaction.

To achieve the highest compactness and effectiveness possible, a variety of high-performance extended or secondary surfaces are used on the stream side with the smaller heat transfer coefficient or simultaneously on all sides, for example, rectangular, offset strip, triangular, perforated, wavy, and louvered fins as well as matrix surfaces. With these auxiliary surfaces, the specific surface area on the gas side reaches up to 6000 m^2/m^3. These extended surfaces increase the heat transfer areas and promote flow turbulence which produces high heat transfer coefficients even for low flow rates. Because of their highly compact structure, the heat transfer surfaces in this type of heat exchangers are likely to foul and the fouling layer is difficult to clean once it has been deposited on the surface (except for plate-and-frame exchangers). This is one of the major limitations that affect their adoption.

In general, high compact surfaces result in compact exchangers with a large frontal area and a short flow length. Thus, the thermal performance of compact heat exchangers is sensitive to flow distribution among the large number of fine flow channels. It is very important to design the inlet and outlet headers to maintain a uniform flow distribution. For gas-to-gas applications (for example, an intercooler in a gas-turbine plant), the front areas on both sides are quite different, and then the uniform flow distribution on each side is vital to the highly thermal performance of compact heat exchangers. Thermal contact resistance between the preliminary and secondary surfaces plays a key role in the heat transfer process of compact heat exchangers. The ideal case is zero contact resistance. Because of the light weight, high performance and compactness, a transient process in compact apparatus takes a short time compared with shell-and-tube heat exchangers. A more accurate dynamic simulation is necessary for start-up and control of compact heat exchangers.

A new development of compact heat exchangers is the so-called microstructure heat exchanger which is mainly applied to electronic and spaceflight equipment at low temperatures. The distinctive feature of the microstructure heat exchangers is the miniature size of the channels and extended surfaces (or fins). Microfabrication techniques, for example the laser technology and chemical manufacturing methods, are used for the construction of this type of exchanger. By using these techniques, dimensions of fins can be down to the order of 20-400 μm. These exchangers have very high surface area densities and can operate under very small temperature differences between the hot and cold fluids. Typical apparatus are perforated-plate exchangers with the surface area density of 12 000 m^2/m^3 and the printed circuit heat exchangers with the surface area densities of 1000–5000 m^2/m^3. The thermal simulation theory of compact heat exchangers is applicable for the microstructure heat exchangers. Wild *et al.* (1994) gave an example of the micro heat exchanger installed in a closed loop cycle where heat was rejected to a pool of superfluid helium (He II). In their experimental research the micro heat exchanger operated in the cryogenic temperature range of 1.8 K to 10 K.

2.5 Thermal regenerators

As pointed out previously, in regenerative heat exchangers the hot and cold gases flow alternately over the same surface through porous matrices. These compact matrices may be constructed of plates, particles or wire meshes of metal, nonmetal and alloys with high heat capacities. While the hot stream is flowing over, the matrices absorb heat; the absorbed heat is transferred to the cold stream during the cold stream flow. The performance characteristics of regenerators lie in the facts of their periodic-flow and time-dependent temperature distributions of both streams and the solid matrices. The typical regenerators are the rotary and fixed (or valved) types (Hausen, 1976; Kays and London, 1984).

In the rotary type of exchanger, there is only one porous matrix wheel which rotates around its axis and the rotation movement makes the matrix pass alternately through the hot and cold stream regions. In this way heat is indirectly transferred from the hot liquid to the cold. The fixed (or valved) type of exchanger has two identical matrices. The hot and cold streams alternately flow through each of the two matrices by means of a periodic switching of the quick operating valves. Evidently, the mass of these two matrices divided by the period of valve operations in this type corresponds to the mass of the single matrix in the rotary type times the rotation speed. Thus, the heat exchanger design theory will apply equally to both types once this correspondence on the matrix mass rate is recognised (Kays and London, 1984). Thermal performance of regenerators is a function of such variables and parameters as the surface area, heat transfer coefficient, heat capacity and thermal conductivity of the matrix, and the rotation speed or switching period of valves. Generally, thermal simulation of regenerators is more complicated than that of recuperators because a transient process is involved in the first. The classical and substantive theory for regenerators was developed by Hausen and others in the 1930s (Hausen, 1976).

The main advantages of regenerative exchangers are their high-compactness, high-performance and self-cleaning ability. They are used for waste heat recovery in gas turbines and power plants and as preheaters and precoolers in cryogenic engineering. Besides stagnant and uneven maldistribution of the fluid, there are some other types of flow maldistribution in regenerators, for example, carryover and leakage, which may not occur in recuperators. The carryover refers to the phenomenon that there is always some mixing of the hot and cold fluids because the voids in the matrix are filled with one fluid at the instant when another fluid enters the matrix and it is impossible to separate the hot and cold fluids completely. The effect of the carryover becomes very significant in a large, highly porous matrix with short-duration blow times. The leakage is fluid mixing when the fluid on the high-pressure side penetrates into the fluid on the low-pressure side through unsealed parts, which may be common in regenerative inlet-air preheater application for gas-turbine units. This leakage appears in the rotary type of exchangers, which offers a severe challenge in the mechanical design of the rotating drives and the fluid seals separating the two streams.

If the inlet volumetric gas flow rate is different from that of the exhaust gas in a rotary regenerator, the free flow areas of both gases are usually also different from each other. For optimum heat transfer, rotational speed is typically 20 r.p.m. (Reay, 1979). Regenerators with stainless-steel matrices can operate in exhaust gases at temperatures in excess of 1000 K. For application at even higher temperatures ceramic balls or matrices are used. Besides gas-to-gas heat recovery applications, the rotary type of regenerators are applicable in heating, ventilating and air-conditioning. For this purpose, nonmetallic matrices, called hygroscopic wheels, are used to transfer moisture as well as sensible heat.

2.6 Heat pipe heat exchangers

The first high-performance element of heat transfer by means of phase-change of liquids was invented in the 1940s (Gaugler, 1942). The first heat pipe appeared in the 1960s for temperature control of space vehicles and the phrase "heat pipe" was introduced at that time (Grover, 1963). The device invented by Grover is principally identical to that invented by Gaugler, but Grover included a primitive analysis and presented results of experiments carried out on stainless steel heat pipes incorporating a wire mesh wick and sodium as the working fluid. Since then the remarkable merits of its high-performance heat transfer and isothermal feature attract more and more attention. A typical heat pipe consists of a sealed pipe, a working liquid and a wick layer. The wick layer is not necessary if a heat pipe operates under the influence of gravity. It can be divided into three sections: evaporation, condensation and insulation. In the evaporation section heat is transferred to the working liquid from the fluid flowing over the heat pipe and the working liquid is evaporated. The vapour moves axially towards the condensation section where heat is transferred to the cold fluid flowing over the outside of the heat pipe. The vapour condenses and the condensate flows back to the evaporation section under gravitational and/or capillary forces. In this way heat is transferred from the hot fluid to the cold fluid.

In principle, many kinds of metal materials (for example, carbon steel, stainless steel, aluminium, copper) can be used as pipes which can be filled with one of many kinds of liquids (for example, water, alcohol, refrigerants and liquid metals). But the assembly selection of a pipe material and a working liquid is not arbitrary. The operating temperature range of the heat pipe and the saturated temperature–pressure relationship of a working fluid are two important factors to be considered for selecting the pipe material and the working fluid. Besides some factors like ratio of strength to weight, thermal conductivity, ease of fabrication including weldability, machineability and ductility, porosity, and wettability, compatibility between the working fluid and the pipe as well as wick materials is subject to prior consideration. It is not true that any of the pipe materials agrees well with any of the working fluids in a closed container. For example, carbon steel and water cannot be coupled to assemble a heat pipe without any pre-treatment, otherwise a noncondensable gas, nitrogen, is produced and performance of the heat pipe degrades, although carbon steel and water are ideal for waste heat recovery below 600 K in view of economy and heat transfer performance. Dunn and Reay (1982) gave a brief discussion about the compatible problem of the pipe materials and the working fluids.

A heat pipe works in a two-phase (liquid and vapour) region and has good isothermal feature because the pressure drop within the pipe is negligibly small. With heat transfer process related to phase change, the heat pipe is generally considered as an excellent heat transfer element whose apparent axial thermal conductivity exceeds over pure silver, copper and aluminium. However, thermal performance of a heat pipe is constrained by some limits (Dunn and Reay, 1982). Clearly, a heat pipe itself is a heat exchanger. In order to satisfy the demand for a given heat load in most cases, a great number of heat pipes are assembled. The statement that a heat pipe has excellent performance of heat transfer does not imply that heat can be transferred from a hot fluid to a cold fluid almost without any thermal resistance by means of a heat pipe; it just means that the thermal resistance within a heat pipe is negligibly small. In fact, there are a series of thermal resistances during the heat transfer process between hot and cold fluids in a heat pipe exchanger and heat pipe heat exchangers have lower efficiencies than some other gas-to-gas heat recovery systems. These types of resistance are listed as follows: the convective resistance R_1 between the hot fluid and the outside surface of the evaporation section of the heat pipe; the conductive resistance R_2 of the pipe wall at the evaporation section; the conductive resistance R_3 of the composite layer of the wick and working liquid; the evaporative resistance R_4 at the evaporation section; the convective resistance R_5 of the vapour from the evaporation section to the condensation section caused by pressure drop of flowing vapour; the condensation resistance R_6 at the condensation section; the conductive resistance R_7 of the composite layer of the wick and working fluid; the conductive resistance R_8 of the pipe material, and the convective resistance R_9 between the outside surface of the condensation section and the cold fluid. Therefore, the total thermal resistance between the hot and cold fluids in a heat pipe exchanger is approximately expressed as

$$R = \sum_{i=1}^{9} R_i . \tag{2.1}$$

Among these types of thermal resistances, the convective resistances R_1 and R_9 are usually controlling. Since their magnitudes are greater than the others, the thermal performance of the heat pipe exchanger is determined mainly by them. A number of extended surfaces are attached to the outside surface of a heat pipe to compensate for the deficiency due to greater convective resistance, especially for heat pipe exchangers in gas-to-gas applications. Typically, the ratio of extended surface area to that of the bare tube is between 20 : 1 and 30 : 1. Although the heat pipe originated from military and spaceflight applications, it has found wide use in many waste heat recovery systems since the 1970s. Reay (1979) listed some application examples of heat pipe heat exchangers. The thermal effectiveness of a heat pipe tends to increase with increasing pipe length between 1 m and 6 m, and when the heat pipe diameter increases above 12 mm (up to 19 mm) a 9% increase in effectiveness is predicted. It should be pointed out that various heat transfer enhancement techniques are being introduced to improve the thermal performance of the heat pipe and a variety of new heat pipes are being invented and put into applications for different cases.

2.7 Multi-fluid heat exchangers

Generally, heat exchange occurs between two fluids in a single heat exchanger. In the cryogenic gas processing and petrochemical industries, however, thermal energy transfer often takes place among more than two fluids in a single exchanger. According to some processing demands in cryogenic and chemical processes such as air and noble gases separation systems, purification and liquefaction of hydrogen, and ammonia gas synthesis, up to six or even more streams are involved in an apparatus. The heat transfer process is far more complicated than that in a two-fluid exchanger.

In view of the exchanger types, multi-fluid heat exchangers may be tubular, plate-and-frame, plate-fin, or printed circuit heat exchangers. But multi-stream flow provides challenges for thermal and mechanical design. The geometrical structure of such exchangers varies with the number of participating streams. Some types of three-fluid heat exchangers are schematically illustrated by Sekulic and Shah (1995). Although all of the above-mentioned types of exchangers can serve as multi-fluid heat exchangers, the types of exchangers with high compactness and performance are more suitable because multi-fluid exchangers commonly operate under small temperature differences among fluids (for example, in the cryogenic temperature range) in a limited space. On the other hand, high compactness and performance are two merits of multi-fluid units that incorporate heat transfer process among more than two fluids in a single apparatus. To some extent, plate-and-frame and plate-fin heat exchangers are more frequently used for heat exchange among three or more fluids.

The parallel-flow and cross-flow arrangements are two major types of flow arrangements in a multi-fluid exchanger. The first mainly refers to cocurrent and countercurrent arrangements as well as any combinations of both cocurrent and countercurrent arrangements; the second includes a number of cross-, cocurrent and countercurrent flow combinations such as cocurrent cross-flow and countercurrent cross-flow as well as other flow arrangements. Besides the geometrical structure, surface area, heat transfer coefficient, and the flow arrangement may remarkably affect the thermal performance of a multi-fluid heat exchanger because it controls the utilization of temperature differences among the fluids. Since the number of fluid streams, flow arrangements and types of apparatus vary from case to case, thermal design theory of multi-fluid exchangers is more complicated than that of two-fluid exchangers and one cannot find a general theory.

2.8 Cross-flow heat exchangers

Cross-flow heat exchangers are not a form of apparatus with any new structure. All operating heat exchangers through which fluids flow in cross directions are classified as cross-flow exchangers. They may be tubular, plate-fin or other types of exchangers. Typical cross-flow heat exchangers are compact units, for example, tube-fin and plate-fin exchangers. Although the thermal effectiveness of a heat exchanger in a cross-flow arrangement is somewhat lower than that of an exchanger in a countercurrent-flow arrangement, cross-flow arrangements are widely applied. The reasons are different in various applications, but two of them may be common: (i) the cross-flow arrangement is more convenient and economical than other flow arrangements in some cases; (ii) for most of the compact heat exchangers with a dense installation of fins the cross-flow arrangement is preferable and other arrangements are impracticable with respect to effective heat transfer and low flow friction. In addition to the pure cross-flow arrangement, a number of combinations among cross- cocurrent and countercurrent flow arrangements are also used, especially in multipass and multi-fluid heat exchangers. The overall cocurrent and countercurrent cross-flow patterns are two typical arrangements in multipass cross-flow exchangers. According to the tubeside fluid flow direction, there exists the so-called codirected or counterdirected flow patterns. If a fluid enters each pass at the same end in a multipass cocurrent or countercurrent cross-flow exchanger, the flow arrangement is called the codirected cocurrent or countercurrent cross-flow; if a fluid enters passes at alternative ends, the flow is called counterdirected cocurrent or countercurrent cross-flow. As pointed out by Roetzel and Neubert (1979), the codirected cross-flow gives a higher thermal effectiveness than the counterdirected cross-flow. Furthermore, a fluid can be either mixed or unmixed between passes, although it is independent of the flow pattern whether it is mixed or unmixed within passes. Theoretically, there are a variety of flow arrangements in multipass cross-flow heat exchangers. Practically, however, the countercurrent cross-flow is most frequently used.

For the one-pass cross-flow arrangement, there are a few possible patterns of flow distribution. In the cross-flow heat exchanger with both fluids unmixed each fluid stream is assumed to be broken up into a large number of separate flow channels through the apparatus without transverse mixing. This occurs in a cross-flow exchanger in which each fluid stream flows transversely over dense fins through channels formed by the fins. In the cross-flow heat exchanger with one fluid mixed and the other unmixed, one fluid is considered to flow through separate channels which may be formed by fins and no transverse mixing takes place, while the other fluid is perfectly transversely mixed, e.g., inside the tubes of one row. The cross-flow exchanger with both fluids mixed refers to an apparatus through which each fluid flows with perfect transverse mixing. It is an uncommon flow arrangement. However, it can serve as a good approximation for a multipass shell-and-tube heat exchanger with one shell pass and a high even number of tube passes. The temperature of a mixed fluid stream is a function of only one spatial coordinate, and the temperature of an unmixed fluid is a two-dimensional function. Therefore, thermal simulation of cross-flow with both fluids unmixed is more complicated and no analytical solution can be found in closed form. Recently Luo (1998) considered a modification of the traditional pure cross-flow in which axial dispersion is taken into account in one fluid (One fluid unmixed, the other fluid transversally unmixed and axially partially mixed). He obtained the transient responses to arbitrary inlet temperature variations after uniform initial temperature distributions using numerical inverse Laplace transform.

As pointed out previously, a cross-flow arrangement is frequently used in compact heat exchangers, for example, radiators in vehicles and intercoolers in gas-turbines. They have a common geometrical feature of great ratio of the frontal area to the flow length. Fluid distribution has a significant effect on thermal performance of these types of cross-flow heat exchangers. Therefore, much attention is paid to the design of fluid distributors. Fox and Haseler (1994) discussed some types of distributors.

2.9 References

API, (1993) *Shell-and-Tube Heat Exchangers for General Refinery Services,* Washington, DC.

Auth, W. J. and Loiacono, J. (1978) Plate and Spiral Plate Heat Exchangers, in Practical Aspects of Heat Transfer, *Proceedings 1976 Fall Lecture Series, N.U. AIChE,* American Institute of Chemical Engineers, New York.

Bes, Th. and Roetzel, W. (1993) Thermal Theory of the Spiral Heat Exchanger, *International Journal of Heat and Mass Transfer*, **36**, 765–773.

Dunn, P. D. and Reay, D.A. (1982) *Heat Pipes*, Pergamon Press, New York.

Fox, T. A. and Haseler, L. E. (1994) Heat Transfer in the Distributors of Plate-fin Heat Exchangers, in *Proceedings of the Tenth International Heat Transfer Conference*, Brighton, UK, (ed. by Hewitt, G. F.), **4**, 369–374.

Gaugler, R. S. (1942) *Heat Transfer Device*, US Patent 2350348, published on June 6, 1944.

Grover, G. M. (1963) *Evaporation-Condensation Heat Transfer Device*, US Patent 3229759, published on January 18, 1966.

Hausen, H. (1976) *Wärmeübertragung im Gegenstrom, Gleichstrom und Kreuzstrom,* Springer-Verlag, Berlin.

Kays, W. M. and London, A. L. *Compact Heat Exchangers*, 1st edn, 1955; 2nd edn, 1964; 3rd edn, 1984, McGraw-Hill Book Company, New York.

Kreith, F. and Boehm, R. F. (1988) *Direct-Contact Heat Exchangers,* Hemisphere Publishing Corporation, Washington.

Luo, X. (1998) *Das axiale Dispersionsmodell für Kreuzstromwärmeübertrager*, Fortschritt-Berichte VDI, Reihe 19, Wärmetechnik/Kältetechnik, Nr. 109, VDI-Verlag GmbH, Düsseldorf.

Panchal, C. B. and Rabas, T. J. (1993) Thermal Performance of Advanced Heat Exchangers for Ammonia Refrigeration Systems, *Heat Transfer Engineering*, **14**(4), 42–57.

Reay, D. A. (1979) *Heat Recovery Systems-A Directory of Equipment and Techniques*, E. & F. N. Spon, London.

Reay, D. A. (1994) Compact Heat Exchangers: A Review of Current Equipment and R & D in the Field, *Heat Recovery Systems and CHP*, **14**, 459–574.

Roetzel, W. (1988) Thermische Berechnung von dreigängigen Rohrbündelwärmeübertragern mit zwei Gegenstromdurchgängen gleicher Größe, *Wärme- und Stoffübertragung*, **22**, 3–11.

Roetzel, W. and Neubert, J. (1979) Calculation of Mean Temperature Difference in Air-Cooled Cross-Flow Heat Exchangers, *Journal of Heat Transfer*, **101**, 511–513.

Roetzel, W. and Spang, B. (1997) Berechnung von Wärmeübertragern, *Abschnitt C des VDI-Wärmeatlas* (8. erw. Auflage), VDI-Verlag GmbH, Düsseldorf.

Sekulic, D. P. and Shah, R. K. (1995) Thermal Design Theory of Three-Fluid Heat Exchangers, in *Advances in Heat Transfer*, **26**, (ed. by Hartnett, J. P., Irvine, T. F., Cho, Y. I., Greene, G. A.), Academic Press Inc., San Diego.

Shah, R. K. and Robertson, J. M. (1993) Compact Heat Exchangers for the Process Industry, in *Proceedings of International Conference on Energy Efficiency in the Process Industries,* Athens, October 1992. Elsevier, Amsterdam.

Shah, R. K., Kraus, A. D. and Metzger, D. (ed.) (1990) *Compact Heat Exchangers,* Hemisphere Publishing Corporation, New York.

Singh, K. P. and Soler, A. I. (1984) *Mechanical Design of Heat Exchangers and Pressure Vessel Components,* Cherry Hill, New York.

TEMA, (1991) *Standards of the Tubular Exchanger Manufacturers Association,* New York.

Tinker, T. (1951) Shell Side Characteristics of Segmentally Baffled Shell-and-Tube Heat Exchangers, Part I, II and III, *Proceedings of General Discussion on Heat Transfer*, Institution of Mechanical Engineers, London, 89–116.

Walker, G. (1982) *Industrial Heat Exchangers,* McGraw-Hill Publications Co., New York.

Wild, S., Oellrich, L. R., Hofmann, A. and Schubert, K. (1994) Comparison of Experimental and Computed Performance of Micro Heat Exchangers in the Ranges of LHe and LN_2 Temperatures, in *Proceedings of the Tenth International Heat Transfer Conference* (ed. by Hewitt, G. F.), Brighton, UK (1994), **4**, 441–445.

Chapter 3

Transient analysis of shell-and-tube heat exchangers

Among a variety of types of heat exchangers, shell-and-tube heat exchangers are widely used in almost all industrial branches. A number of thermal simulation and design methods have been developed for shell-and-tube exchangers operating in the steady state. These theories promote application and improvement of shell-and-tube exchangers. But they cannot provide any information about dynamic features of exchangers. Because of various reasons such as start-up, shut-down and accidents, heat exchangers often endure transient processes. Variations of external loads that leads to disturbances of inlet temperatures and flow rates of fluid streams and changes of heat transfer conditions also induce a transitory process in heat exchangers. Fouling — the accumulation of deposits on the heat transfer surface, which itself varies with time, continuously alters the performance of exchangers. To some extent, heat exchangers always operate in a transitory state rather than in a steady state.

Knowledge of transient performance of shell-and-tube heat exchangers is required for process control, optimum operation, and prevention of accidents: Transient analysis of a shell-and-tube exchanger can be described as follows: according to the transient process occurring in an exchanger, theoretical and/or experimental methods are applied to developing a set of the related governing equations with respect to various influence factors. Then, transient temperature profiles of fluid streams or responses are determined and the dynamic characteristics of exchangers are analysed. For transient analysis, some of the assumptions used for the steady-state analysis are involved to simplify mathematical derivation, which will be mentioned in the following sections.

3.1 Transient analysis of cocurrent flow heat exchangers

The cocurrent or countercurrent flow arrangements are the simplest flow patterns in shell-and-tube heat exchangers. This type of exchanger has one shell pass and one tube pass. Although a cocurrent flow heat exchanger has poor thermal performance compared with the countercurrent arrangement, it is occasionally used in the process industry. A remarkable advantage of this apparatus is that the temperature of the tube wall does not sharply change between the inlet and the outlet of the streams and no high thermal stress in the tube material is produced, which may be important for heat exchangers operating under great temperature differences between fluid streams. Even if the inlet temperature of a fluid is very high, the temperature of the wall near the inlet lies in a tolerable region because the wall is cooled by the fluid with a lower temperature on the other side. By neglecting the effect of the heat capacity of the core wall, Romie (1985) analysed the response of a cocurrent flow heat exchanger. Tan (1994) applied the Laplace transform to the transient mass or heat transfer in the continuous cocurrent process and proposed compact analytical solutions.

3.1.1 The governing equations

A cocurrent exchanger is schematically illustrated in Fig. 3.1. Fluid 1 and fluid 2 enter the exchanger at the same end and flow along either side of the heat transfer surface. The inlet temperatures of both fluids may vary with time in any arbitrary forms, and then a transient process occurs in the apparatus.

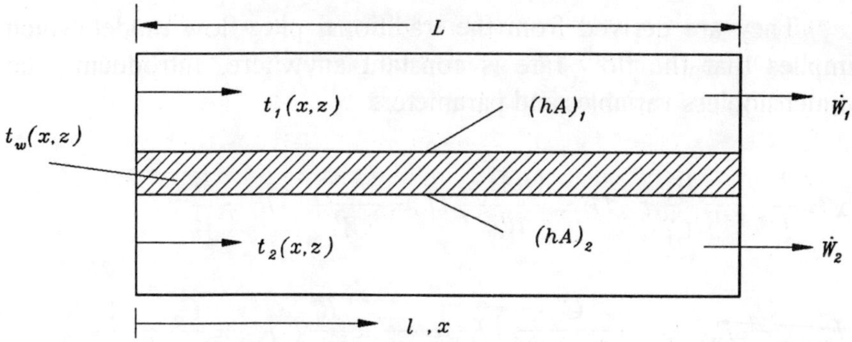

Figure 3.1: Schematic representation of a cocurrent flow heat exchanger

The following assumptions are made to enable the derivation of the governing equations pertinent to transient operation in the heat exchanger (Roetzel and Xuan, 1992a):

(1) The thermal flow rates $\dot{W}_1$ and $\dot{W}_2$ of the fluids and the heat transfer coefficients are constant. The heat transfer surface area is uniformly distributed.

(2) All thermal properties are constant.

(3) Longitudinal heat conduction within the wall and the fluids are neglected and the thermal resistance of the wall to heat transfer from one fluid to the other is negligible.

(4) The inside surface of the shell is adiabatic.

According to these assumptions, one can derive the governing differential equations for the transient temperature profiles of both fluids and the core wall of an exchanger as follows:

$$\dot{W}_1 \frac{\partial \theta_1}{\partial l} + \frac{C_1}{L}\frac{\partial \theta_1}{\partial \tau} + \frac{(hA)_1}{L}\left(\theta_1 - \theta_w\right) = 0 \tag{3.1}$$

$$\dot{W}_2 \frac{\partial \theta_2}{\partial l} + \frac{C_2}{L}\frac{\partial \theta_2}{\partial \tau} + \frac{(hA)_2}{L}\left(\theta_2 - \theta_w\right) = 0 \tag{3.2}$$

$$\frac{C_w}{L}\frac{\partial \theta_w}{\partial \tau} - \frac{(hA)_1}{L}\left(\theta_1 - \theta_w\right) - \frac{(hA)_2}{L}\left(\theta_2 - \theta_w\right) = 0 . \tag{3.3}$$

They are derived from the traditional plug-flow model which implies that the flow rate is constant anywhere. Introducing the dimensionless variables and parameters

$$x = \frac{l}{L},\ z = \frac{\dot{W}_1}{C_1}\tau,\ U_1 = \frac{(hA)_1}{\dot{W}_1},\ U_2 = \frac{(hA)_2}{\dot{W}_2},\ R_1 = \frac{\dot{W}_1}{\dot{W}_2},$$

$$R_\tau = \frac{C_2}{C_1}R_1,\ R_w = \frac{C_w}{C_1 + C_2},\ R_{c1} = \frac{C_1}{C_2} = \frac{R_1}{R_\tau},\ R_{c2} = \frac{C_2}{C_1} = \frac{1}{R_{c1}},$$

$$\alpha_1 = \frac{U_1}{1 + R_{c2}},\ \alpha_2 = \frac{U_2}{R_\tau\left(1 + R_{c1}\right)}$$

into eqns (3.1)–(3.3) yields:

$$\frac{\partial t_1}{\partial z} + \frac{\partial t_1}{\partial x} + U_1\left(t_1 - t_w\right) = 0 \tag{3.4}$$

$$R_\tau \frac{\partial t_2}{\partial z} + \frac{\partial t_2}{\partial x} + U_2\left(t_2 - t_w\right) = 0 \tag{3.5}$$

$$R_w \frac{\partial t_w}{\partial z} - \alpha_1\left(t_1 - t_w\right) - \alpha_2\left(t_2 - t_w\right) = 0. \tag{3.6}$$

The corresponding initial and boundary conditions are given as

$$t_1(x, 0) = g_1(x) \tag{3.7a}$$

$$t_2(x, 0) = g_2(x) \tag{3.7b}$$

$$t_w(x, 0) = g_w(x). \tag{3.7c}$$

$$t_1(0, z) = f_1(z) \tag{3.8a}$$

$$t_2(0, z) = f_2(z). \tag{3.8b}$$

Here the initial conditions take general forms. $g_i(x)$ may be equal to zero, which refers to the uniform temperature distributions at the instant of start-up. Also, $g_i(x)$ may be either the previous steady-state distributions or the temperature distributions at the instant of new transient process. The temperature disturbances $f_i(z)$ are allowed to have arbitrary forms. The simplest case is that only a step variation occurs at the inlet of fluid 1: $f_1(z)=1$ and $f_2(z)=0$. There are many methods suitable for determining transient responses of heat exchangers. Among them, the finite-difference method and the Laplace-transform method are more frequently applied. In the following both these methods are used.

3.1.2 The finite-difference solution

The essential of the finite-difference method is to substitute the finite-difference equation for the differential equation. According to the finite-difference forms of the spatial derivatives, the finite-difference method is further divided into the explicit and implicit representations as well as other combinations of these two preliminary representations (Minkowycz *et al.*, 1988). The first step of this method is to discretize the solution space, i.e., to divide the spatial and time spaces into M $\left(x_0 < x_1 < \ldots < x_M\right)$ and N $\left(z_0 < z_1 < \ldots < z_N\right)$ spans. The grid step may be Δx_i and Δz_n. If a forward finite-difference scheme is applied to the time derivative, one obtains the following explicit finite-difference equations from the previously derived differential equations:

$$t_{1,i}^n = \frac{\Delta z_n}{\Delta x_i} t_{1,i-1}^{n-1} + \left(1 - \frac{\Delta z_n}{\Delta x_i} - U_1 \Delta z_n\right) t_{1,i}^{n-1} + U_1 \Delta z_n t_{w,i}^{n-1} \tag{3.9}$$

$$t_{2,i}^n = \frac{\Delta z_n}{R_\tau \Delta x_i} t_{2,i-1}^{n-1} + \left(1 - \frac{\Delta z_n}{R_\tau \Delta x_i} - \frac{U_2}{R_\tau} \Delta z_n\right) t_{2,i}^{n-1} + \frac{U_2}{R_\tau} t_{w,i}^{n-1} \tag{3.10}$$

$$t_{w,i}^n = \left(1 - \frac{\alpha_1 \Delta z_n}{R_w} - \frac{\alpha_2 \Delta z_n}{R_w}\right) t_{w,i}^{n-1} + \frac{\alpha_1 \Delta z_n}{R_w} t_{1,i}^{n-1} + \frac{\alpha_2 \Delta z_n}{R_w} t_{2,i}^{n-1} \, . \tag{3.11}$$

For numerical stability, all the coefficients on the right-hand side of each of these explicit finite-difference equations must have the same sign. Thus, there exist stability constraints:

$$1-\frac{\Delta z_n}{\Delta x_i}-U_1\Delta z_n \geq 0 \qquad \text{or} \qquad \frac{\Delta z_n}{\Delta x_i}(1+U_1\Delta x_i) \leq 1 \tag{3.12a}$$

$$1-\frac{\Delta z_n}{R_\tau \Delta x_i}-\frac{U_2}{R_\tau}\Delta z_n \geq 0 \quad \text{or} \quad \frac{\Delta z_n}{R_\tau \Delta x_i}(1+U_2\Delta x_i) \leq 1 \tag{3.12b}$$

$$1-\frac{\alpha_1\Delta z_n}{R_w}-\frac{\alpha_1\Delta z_n}{R_w} \geq 0 \quad \text{or} \quad \frac{\alpha_1\Delta z_n}{R_w}\left(1+\frac{\alpha_2}{\alpha_1}\right) \leq 1. \tag{3.12c}$$

These constraints imply that the spatial increment Δx_i cannot be arbitrarily chosen, once the marching increment Δz_n has been chosen according to the constraint (3.12c). One can also take uniform increments Δx and Δz for all spatial and temporal grids. Since the Dirichlet boundary conditions (i.e., the first type of boundary conditions) at the inlets are given, no extra constraints are to be satisfied. With the given initial and boundary conditions, eqns (3.9)–(3.11) can be used to compute the solution at all internal and temporal points and to analyse the transient responses of a cocurrent heat exchanger subject to arbitrary temperature and/or flow disturbances at the inlet. For a cocurrent exchanger, it is convenient to use the implicit-difference algorithm

$$-\frac{\Delta z_n}{\Delta x_i}t_{1,i-1}^n+\left(1+\frac{\Delta z_n}{\Delta x_i}+U_1\Delta z_n\right)t_{1,i}^n-U_1\Delta z_n t_{w,i}^n=t_{1,i}^{n-1} \tag{3.13}$$

$$-\frac{\Delta z_n}{R_\tau\Delta x_i}t_{2,i-1}^n+\left(1+\frac{\Delta z_n}{R_\tau\Delta x_i}+\frac{U_2\Delta z_n}{R_\tau}\right)t_{2,i}^n-\frac{U_2\Delta z_n}{R_\tau}t_{w,i}^n=t_{2,i}^{n-1} \tag{3.14}$$

$$-\frac{\alpha_1\Delta z_n}{R_w}t_{1,i}^n+\left(1+\frac{\alpha_1\Delta z_n}{R_w}+\frac{\alpha_2\Delta z_n}{R_w}\right)t_{w,i}^n-\frac{\alpha_2\Delta z_n}{R_w}t_{2,i}^n=t_{w,i}^{n-1}. \tag{3.15}$$

These implicit equations are always numerically stable. However, they are conjugated and numerical computation becomes somewhat more complicated. Substituting eqns (3.13) and (3.14) into eqn (3.15), one obtains for the elimination of $t^n_{1,i}$ and $t^n_{2,i}$

$$t^n_{w,i} = \frac{1}{A_P}\left(t^{n-1}_{w,i} + A_S t^{n-1}_{1,i} + A_W t^n_{1,i-1} + B_S t^{n-1}_{2,i} + B_W t^n_{2,i-1}\right) \tag{3.15a}$$

in which

$$A_S = \frac{\dfrac{\alpha_1}{R_w}}{\dfrac{1}{\Delta z_n} + \dfrac{1}{\Delta x_i} + U_1}, \qquad B_S = \frac{\dfrac{\alpha_2}{R_w}}{\dfrac{1}{\Delta z_n} + \dfrac{1}{R_\tau \Delta x_i} + \dfrac{U_2}{R_\tau}},$$

$$A_W = \frac{\Delta z_n}{\Delta x_i} A_S, \quad B_W = \frac{\Delta z_n}{R_\tau \Delta x_i} B_S, \qquad A_P = 1 + A_S + A_W + B_S + B_W.$$

Beginning the calculation with $i = 1$, one has known values of $t^n_{1,i-1}$ and $t^n_{2,i-1}$ and the algorithm becomes explicit.

It should be emphasized that if the numerical method such as the finite-difference method is used to determine transient behaviour of a heat exchanger, some of the above-mentioned assumptions can be relaxed. For example, the assumption of constant thermal properties, flow rates and heat transfer coefficients is no longer necessary.

Besides the aforementioned finite-difference method, the other numerical methods such as the finite-element and the method of characteristics are used for dynamic simulation of heat exchangers. Under some assumptions, Heggs and Render (1983) used the method of characteristic curves for transient response of heat exchangers with one fluid of infinite thermal flow rate, which include evaporators, condensers, reboilers, precoolers, and intercoolers.

3.1.3 The Laplace transform solution

Under the above-mentioned assumptions the differential equations (3.4)–(3.6) are linear. Thus, the Laplace transform can be applied. Using s as the Laplace transform variable with respect to the dimensionless time variable z and the given initial conditions, one obtains the following equations in the transformed domain

$$\frac{\mathrm{d}T_1}{\mathrm{d}x} = a_{11}T_1 + a_{12}T_2 + G_1(x,s) \tag{3.16}$$

$$\frac{\mathrm{d}T_2}{\mathrm{d}x} = a_{21}T_1 + a_{22}T_2 + G_2(x,s) \tag{3.17}$$

$$T_w = \frac{\alpha_1 T_1 + \alpha_2 T_2}{\alpha_1 + \alpha_2 + R_w s} + G_w(x,s) \tag{3.18}$$

where

$$a_{11} = \frac{\alpha_1 U_1}{\alpha_1 + \alpha_2 + R_w s} - s - U_1, \quad a_{12} = \frac{\alpha_2 U_1}{\alpha_1 + \alpha_2 + R_w s},$$

$$a_{21} = \frac{\alpha_1 U_2}{\alpha_1 + \alpha_2 + R_w s}, \quad a_{22} = \frac{\alpha_2 U_2}{\alpha_1 + \alpha_2 + R_w s} - R_\tau s - U_2,$$

$$G_1(x,s) = g_1(x) + U_1 G_w(x,s), \quad G_2(x,s) = R_\tau g_2(x) + U_2 G_w(x,s),$$

$$G_w(x,s) = \frac{R_w g_w(x)}{\alpha_1 + \alpha_2 + R_w s}.$$

The ordinary differential equation system (3.16)–(3.18) is readily solved. For the uniform initial temperature distributions, the solutions are as follows:

$$T_1(x,s) = B_1\,\mathrm{e}^{\lambda_1 x} + B_2\,\mathrm{e}^{\lambda_2 x} \tag{3.19}$$

$$T_2(x,s) = \frac{\lambda_1 - a_{11}}{a_{12}} B_1 \,\mathrm{e}^{\lambda_1 x} + \frac{\lambda_2 - a_{11}}{a_{12}} B_2 \,\mathrm{e}^{\lambda_2 x}, \tag{3.20}$$

where B_1 and B_2 are two unknown coefficients to be determined according to the inlet boundary conditions. The distinct eigenvalues λ_1 and λ_2 are represented by

$$\lambda_{1,2} = \frac{(a_{11} + a_{22}) \pm \sqrt{(a_{11} - a_{22})^2 + 4a_{12}a_{21}}}{2}. \tag{3.21}$$

With the corresponding transformed forms of the given boundary conditions (3.8), both the coefficients are derived as

$$B_1 = \frac{1}{\lambda_2 - \lambda_1}\left[(\lambda_2 - a_{11})F_1(s) - a_{12}F_2(s)\right],$$

$$B_2 = \frac{1}{\lambda_1 - \lambda_2}\left[(\lambda_1 - a_{11})F_1(s) - a_{12}F_2(s)\right].$$

So far the temperature distributions in the transformed domain have been found. The next task is to carry out inversion of the Laplace transform to obtain temperature distributions in the time-domain. Generally, it is difficult to analytically perform the inverse transform. Even if the analytical inversion is possible for some cases, infinite series are involved and only finite numbers of series terms are taken for numerical computation. For transient analysis of heat exchangers, the analytical solution to a step temperature disturbance is available subject to some assumptions. But it is almost impossible to find any analytical forms of thermal responses to other types of inlet disturbances. Therefore, numerical inversion is a reasonable choice. Various numerical inversion methods have been proposed. Among them, the algorithm proposed by Gaver (1966) and Stehfest (1970) and the algorithm based on Fourier series (Ichikawa and Kishima, 1972; Crump, 1976) are often applied. If $f(z)$ and $F(s)$

are a Laplace transform pair, these two algorithms are respectively represented as follows:

(1) Numerical inversion using Gaver–Stehfest algorithm (Stehfest, 1970; Jacquot *et al.*, 1983)

$$f(z)=\frac{\ln 2}{z}\sum_{n=1}^{N}K_n F\left(n\frac{\ln 2}{z}\right) \quad \text{(even } N\text{)}$$

$$K_n=(-1)^{n+N/2}\sum_{k=[(n+1)/2]}^{\min(n,N/2)}\frac{k^{N/2}(2k)!}{(N/2-k)!k!(k-1)!(n-k)!(2k-n)!}. \tag{3.22}$$

In the Gaver–Stehfest algorithm the value of N is suggested to be 10 for 8-digit arithmetic and 18 for 16-digit arithmetic. This algorithm cannot be used if there are oscillatory components or very rapid changes in $f(z)$ $(z>0)$.

(2) Numerical inversion using FFT algorithm (Ichikawa and Kishima, 1972)

$$f(z_n)=\frac{\exp(az_n)}{z}\left[\operatorname{Re}\sum_{k=0}^{M-1}F\left(a+\frac{ik\pi}{z}\right)\exp\left(i\frac{2\pi nk}{M}\right)-\frac{1}{2}F(a)\right] \tag{3.23}$$

in which $z_n=nz/M$ $(n\le M/2)$, a is a constant $(4<az<5)$ and the series

$$f_n=\sum_{k=0}^{M-1}F\left(a+\frac{ik\pi}{z}\right)\exp\left(i\frac{2\pi nk}{M}\right)$$

is calculated by means of fast Fourier transform (FFT). If the original function $f(z)$ has discontinuity points, one should pay attention to the additional oscillation (Gibbs' phenomenon) near these points.

The advantages and disadvantages of both these algorithms are discussed in the Appendix. They have successfully been used for dynamic analysis of heat exchangers (Roetzel and Xuan, 1992a). Compared with the finite-difference procedure, the Laplace transform together with its numerical inversion can be used to compute directly transient responses at arbitrary values of z and x, without performing many-step computation pertinent to the selected time and space steps. This is a major advantage of the Laplace method over the finite-difference method. The Laplace method is only suitable for the linear problem.

3.1.4 Dynamic behaviour of cocurrent flow heat exchangers

With the above-derived methods, one can obtain temperature responses to any arbitrary inlet temperature variations whose forms depend upon the specific process. Some typical variations, such as step, ramp, exponential and periodical variations, have been listed by Roetzel and Xuan (1992a). Other inlet temperature changes may be described by super-position (or combination) of these primary forms. If one can obtain the transient responses to these primary temperature disturbances, superposition (or combination) of their results responds to the inlet temperature change with the same superposition (or combination) of the elementary inlet temperature changes.

The transient temperature distributions of both the fluids are functions of the five dimensionless parameters U_1, U_2, R_1, R_w and R_τ and two dimensionless variables x and z. Here U_1 and U_2 are similar to the definition of *number of transfer units* by Kays and London (1984); R_1 represents the ratio of thermal flow rate of fluid 1 to thermal flow rate of fluid 2; R_w ratio of heat capacity of tube wall to the total heat capacities of both fluids, R_τ ratio of the residence time of fluid 1 to that of fluid 2 in the exchanger. For a given exchanger and specific operating conditions, these parameters are known. Assigning different values to them, one can analyse effects of structural parameters and operating conditions on dynamics of exchangers. By assuming $f_2(z)=0$, transient responses to $f_1(z)=1$ and $f_1(z)=z$ are representatively illustrated in Fig. 3.2 and Fig. 3.3 subject to the following parameters: $R_1 = R_\tau = R_w = 1$, and $U_1 = U_2 = 2NTU_1 = 2NTU_2 = 2NTU$.

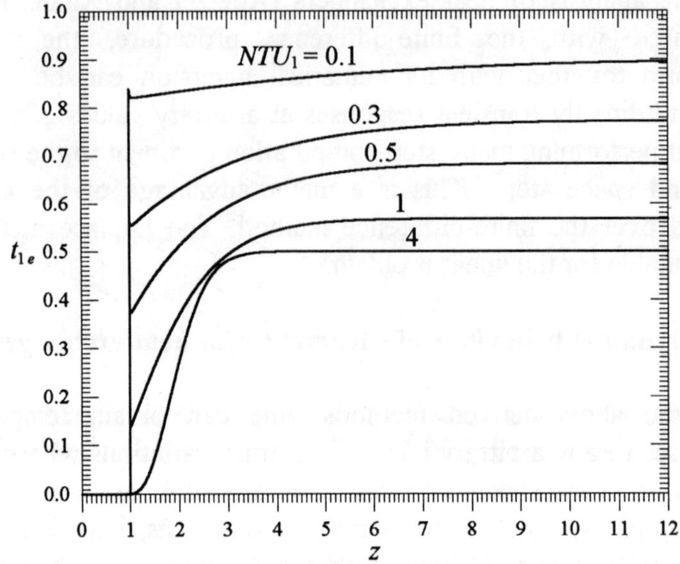

(a) shellside fluid

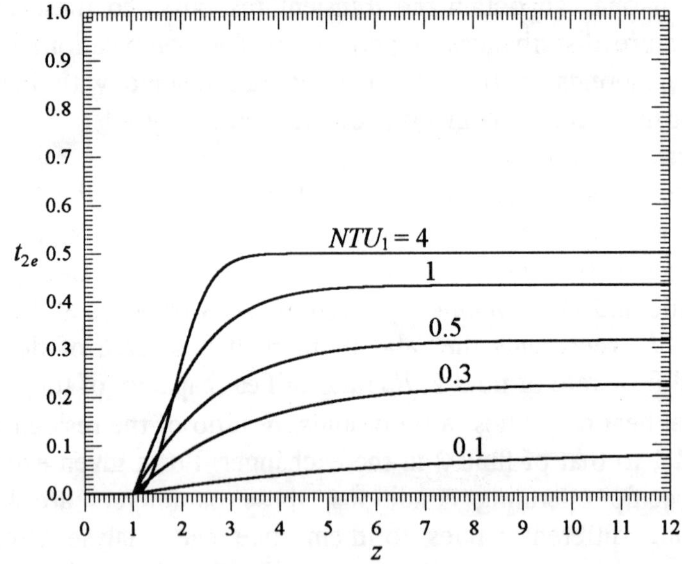

(b) tubeside fluid

Figure 3.2: Exit temperature responses to a shellside step inlet temperature change, cocurrent flow, $R_1 = R_\tau = R_w = 1$, $U_1 = U_2$, $f_1(z) = 1$, $f_2(z) = 0$

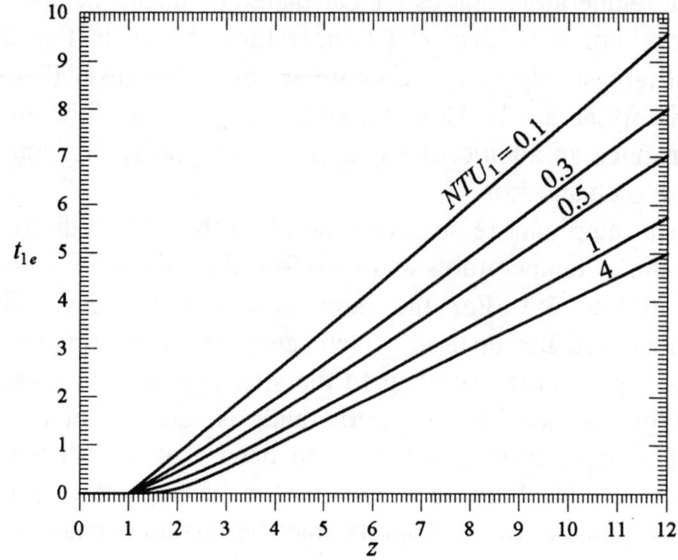

(a) shellside fluid

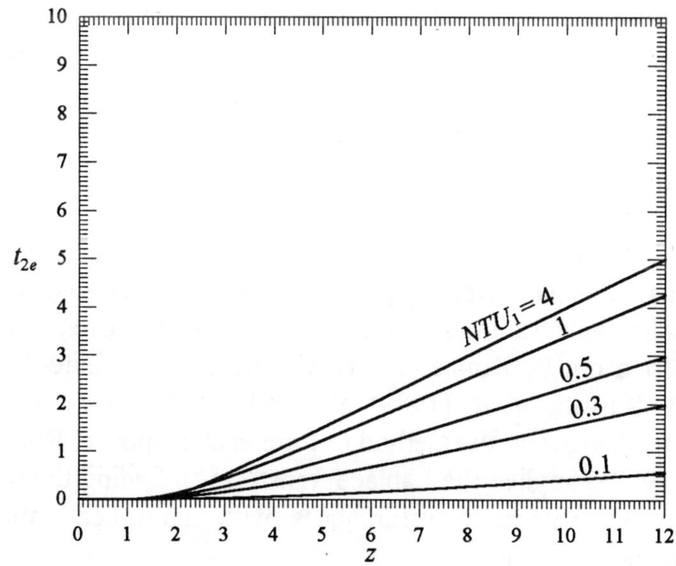

(b) tubeside fluid

Figure 3.3: Exit temperature responses to a shellside ramp inlet temperature change, cocurrent flow, $R_1 = R_\tau = R_w = 1$, $U_1 = U_2$, $f_1(z) = z$, $f_2(z) = 0$

The temperature curves are calculated by means of fast Fourier transform. For a sudden inlet temperature change in Fig. 3.2 the Gaver–Stehfest algorithm cannot be used because there is a discontinuity at $z = 1$. Although near this point the FFT algorithm will introduce an additional impulse, one can easily distinguish and eliminate such an error.

For a ramp inlet temperature variation there is no discontinuity in both outlet temperatures and therefore the curves are smooth, as shown in Fig. 3.3. For the ramp variation the Gaver–Stehfest algorithm could also be used, which needs less computing time.

The results corresponding to the step change have shown that the energy balance between both fluids is satisfied in the steady state. This demonstrates on the one hand that the algorithm for numerical inverse Laplace transform can be applied to predicting transient response to an arbitrary inlet temperature change and on the other hand that a transient simulation method can also be extended to the prediction of steady-state performance if an inlet temperature step change is introduced.

3.2 Transient analysis of countercurrent flow heat exchangers

Since a countercurrent flow heat exchanger has the best thermal performance, it is preferable in most cases. The transient response to an inlet temperature step change in a countercurrent apparatus was carried out by Romie (1984) with the explicit finite-difference method. Kawata *et al.* (1989) obtained the dynamic response to flow rate changes by the method of frequency response. Roetzel and Xuan (1992a) applied the Laplace transform to finding the response to arbitrary temperature disturbances, with taking heat capacity of the tube into account.

3.2.1 The governing equations

The countercurrent flow exchanger is schematically shown in Fig. 3.4. By means of a similar procedure used for the cocurrent flow heat exchanger, one can readily derive a set of differential equations governing transient behaviour of a countercurrent flow exchanger. In fact, the equations for fluid 1 and the tube wall remain the same as eqns (3.4) and (3.6), only the equation for fluid 2 is altered. Thus, the equations are expressed as

$$\frac{\partial t_1}{\partial z} + \frac{\partial t_1}{\partial x} + U_1(t_1 - t_w) = 0 \tag{3.4}$$

$$R_\tau \frac{\partial t_2}{\partial z} - \frac{\partial t_2}{\partial x} + U_2(t_2 - t_w) = 0 \tag{3.24}$$

$$R_w \frac{\partial t_w}{\partial z} - \alpha_1(t_1 - t_w) - \alpha_2(t_2 - t_w) = 0. \tag{3.6}$$

The initial and boundary conditions are given in the same general forms as in (3.7)–(3.8). These equations can be solved with the finite-difference method. The numerical schemes for countercurrent flow are similar to those for cocurrent flow. Only the grid temperature $t_{2,i-1}^{n-1}$ in eqns (3.10), (3.15) and (3.15a) should be replaced by $t_{2,i+1}^{n-1}$. However, eqn (3.15a) is no longer explicit and therefore an iteration beginning from an assumed temperature field of fluid 2 is needed.

If the above equations are linear they can also be solved with

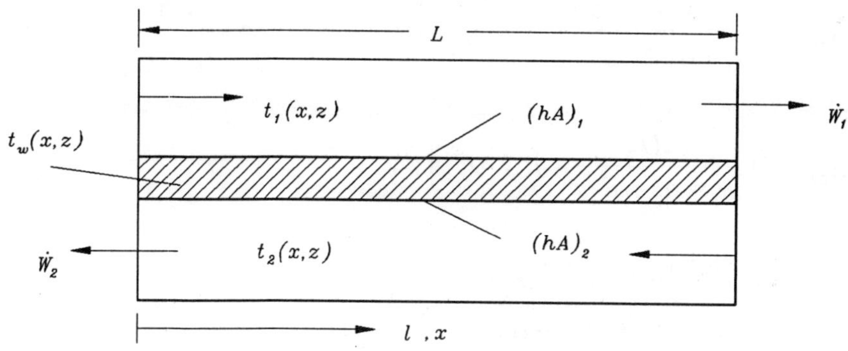

Figure 3.4: Schematic representation of a countercurrent flow heat exchanger

Laplace transform method. For the uniform initial conditions the solution in the image-domain of the Laplace transform are as follows:

$$T_1(x,s) = B_1\,e^{\lambda_1 x} + B_2\,e^{\lambda_2 x} \tag{3.25}$$

$$T_2(x,s) = \frac{\lambda_1 - a_{11}}{a_{12}} B_1\,e^{\lambda_1 x} + \frac{\lambda_2 - a_{11}}{a_{12}} B_2\,e^{\lambda_2 x} \tag{3.26}$$

$$T_w(x,s) = \frac{\alpha_1 T_1 + \alpha_2 T_2}{\alpha_1 + \alpha_2 + R_w s}, \tag{3.27}$$

where

$$\lambda_{1,2} = \frac{(a_{11} + a_{22}) \pm \sqrt{(a_{11} - a_{22})^2 + 4a_{12}a_{21}}}{2},$$

$$a_{11} = \frac{\alpha_1 U_1}{\alpha_1 + \alpha_2 + R_w s} - s - U_1,$$

$$a_{12} = \frac{\alpha_2 U_1}{\alpha_1 + \alpha_2 + R_w s},$$

$$a_{21} = -\frac{\alpha_1 U_2}{\alpha_1 + \alpha_2 + R_w s},$$

$$a_{22} = -\left(\frac{\alpha_2 U_2}{\alpha_1 + \alpha_2 + R_w s} - R_\tau s - U_2\right),$$

$$B_1 = \frac{(\lambda_2 - a_{11})e^{\lambda_2} F_1(s) - a_{12}F_2(s)}{(\lambda_2 - a_{11})e^{\lambda_2} - (\lambda_1 - a_{11})e^{\lambda_1}},$$

$$B_2 = \frac{(\lambda_1 - a_{11})e^{\lambda_1} F_1(s) - a_{12}F_2(s)}{(\lambda_1 - a_{11})e^{\lambda_1} - (\lambda_2 - a_{11})e^{\lambda_2}}.$$

3.2.2 Dynamic behaviour of countercurrent flow heat exchangers

The numerical inverse transform of the solutions (3.25)–(3.27) provides transient characteristics of countercurrent flow heat exchangers. Similar to the cocurrent flow apparatus, the exit temporal temperature distributions $t_1(1,z)$ and $t_2(0,z)$ in a countercurrent flow heat exchanger are controlled by the parameters U_1, U_2, R_w, R_τ and R_1. But its dynamic behaviour is different from that of the cocurrent flow exchanger, even when the operating conditions and geometrical parameters are the same in both types of exchangers. An inlet disturbance of one fluid is immediately transported to the exit of the other fluid because of heat transfer between these two fluids across the tube wall. Evidently, the greater the value of number of transfer units *NTU* and the smaller the capacitance of the wall, the earlier a disturbance at the inlet of one fluid causes a significant temperature change at the exit of the other fluid. An example of transient responses to a step variation of fluid 1 in a countercurrent flow exchanger is illustrated in Fig. 3.5.

3.3 Dynamic model of multipass heat exchangers

Besides one-pass heat exchangers, multipass shell-and-tube exchangers are widely used in the process industry to satisfy demands for heavy loads and great surface areas. Since heat exchangers are seldom designed with more than two shell passes, here the phrase *multipass* mainly refers to a number of passes on the tube side. Such apparatus are designated as 1-*N* shell-and-tube heat exchangers, in which "1" refers to one shellside pass and *N* to *N* tubeside passes, which is the so-called E-type tubular heat exchanger.

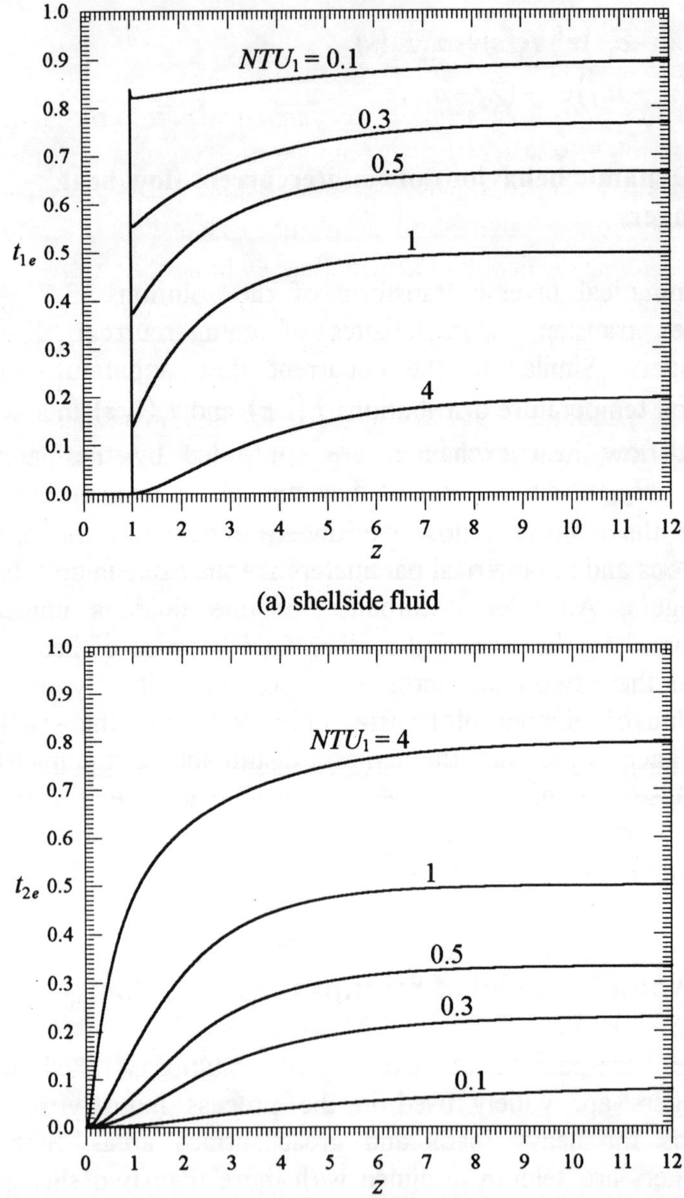

Figure 3.5: Exit temperature responses to a shellside step inlet temperature change, countercurrent flow, $R_1 = R_\tau = R_w = 1$, $U_1 = U_2$, $f_1(z) = 1$, $f_2(z) = 0$

3.3.1 The governing equations

The multipass shell-and-tube heat exchanger under consideration is shown in Fig. 3.6 in which two possible tubeside flow arrangements are presented. The number N of tube passes is arbitrary (even or odd), although an odd number of tube passes is an uncommon choice because of manufacturing and assembling difficulties.

On the basis of the assumptions proposed in the previous context, the transient behaviour of shell-and-tube heat exchangers is modelled. The tubeside and shellside heat transfer coefficients as well as the surface areas are allowed to vary from pass to pass, although they are assumed to be constant within one pass. To develop a general coordinate system, the origin point of the spatial co-ordinate is always set at the location where the shellside fluid enters the exchanger. Thus, a set of transient energy balance equations of both fluids and the tube wall takes the form (Roetzel and Xuan, 1992b).

$$\dot{W}_1 \frac{\partial \theta_1}{\partial l} + \frac{C_1}{L} \frac{\partial \theta_1}{\partial \tau} + \sum_{i=1}^{N} \frac{(hA)_{1i}}{L} \left(\theta_1 - \theta_{wi} \right) = 0 , \tag{3.28}$$

$$(-1)^{i+I} \dot{W}_{2i} \frac{\partial \theta_{2i}}{\partial l} + \frac{C_{2i}}{L} \frac{\partial \theta_{2i}}{\partial \tau} + \frac{(hA)_{2i}}{L} \left(\theta_{2i} - \theta_{wi} \right) = 0 \qquad (i = 1, 2, \ldots , N), \tag{3.29}$$

$$\frac{C_{wi}}{L} \frac{\partial \theta_{wi}}{\partial \tau} - \frac{(hA)_{1i}}{L} \left(\theta_1 - \theta_{wi} \right) - \frac{(hA)_{2i}}{L} \left(\theta_{2i} - \theta_{wi} \right) = 0 \qquad (i = 1, 2, \ldots , N), \tag{3.30}$$

where the value of I in eqn (3.29) indicates the tubeside flow arrangement. $I = 0$ and $I = 1$ correspond to the tubeside flow arrangements I and II (as shown in Fig. 3.6), respectively. For an incompressible fluid with constant specific heat capacity, it must be satisfied that $\dot{W}_{2i} = \dot{W}_2$, i.e., the tubeside thermal flow rate does not vary from pass to pass. However, the tubeside heat capacities C_{2i}

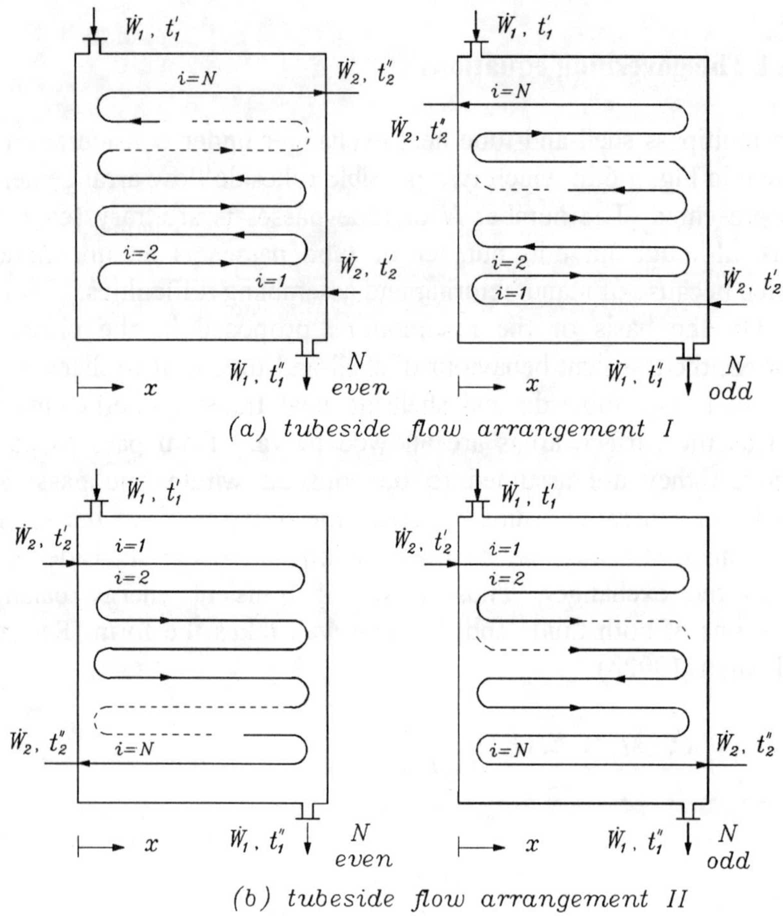

Figure 3.6: Schematic representation of multipass shell-and-tube heat exchangers

may be different for each pass. With the suitable dimensionless variables and parameters, these governing equations take the form:

$$\frac{\partial t_1}{\partial x}+\frac{\partial t_1}{\partial z}+\sum_{i=1}^{N} U_{1i}\left(t_1-t_{wi}\right)=0 \qquad (3.31)$$

$$(-1)^{i+1}\frac{\partial t_{2i}}{\partial x}+\varepsilon_{ci}R_\tau\frac{\partial t_{2i}}{\partial z}+U_{2i}\left(t_{2i}-t_{wi}\right)=0$$

$$\left(i=1,\ 2,\ \dots,\ N\right) \qquad (3.32)$$

$$R_{wi}\frac{\partial t_{wi}}{\partial z}-\alpha_{1i}\left(t_1-t_{wi}\right)-\alpha_{2i}\left(t_{2i}-t_{wi}\right)=0$$

$$\left(i=1,\ 2,\ \ldots,\ N\right) \tag{3.33}$$

where $U_{1i}=\dfrac{(hA)_{1i}}{\dot{W}_1}$, $U_{2i}=\dfrac{(hA)_{2i}}{\dot{W}_2}$, $\alpha_{1i}=\dfrac{U_{1i}}{1+R_{c2}}$, $\alpha_{2i}=\dfrac{U_{2i}}{R_\tau(1+R_{c1})}$, $R_{wi}=\dfrac{C_{wi}}{C_1+C_2}$. Obviously, the following relationships exist between the parameters:

$$(hA)_1=\sum_{i=1}^{N}(hA)_{1i},\ (hA)_2=\sum_{i=1}^{N}(hA)_{2i},\ C_2=\sum_{i=1}^{N}C_{2i},\ C_w=\sum_{i=1}^{N}C_{wi},$$

$$\varepsilon_{1i}=\frac{(hA)_{1i}}{(hA)_1}=\frac{U_{1i}}{U_1},\ \varepsilon_{2i}=\frac{(hA)_{2i}}{(hA)_2}=\frac{U_{2i}}{U_2},\ \varepsilon_{ci}=\frac{C_{2i}}{C_2},\ \varepsilon_{wi}=\frac{C_{wi}}{C_w},$$

$$\sum_{i=1}^{N}\varepsilon_{1i}=1,\ \sum_{i=1}^{N}\varepsilon_{2i}=1,\ \sum_{i=1}^{N}\varepsilon_{ci}=1,\ \sum_{i=1}^{N}\varepsilon_{wi}=1,\ \tau_{r2i}=\frac{C_{2i}}{\dot{W}_{2i}}=\varepsilon_{ci}\tau_{r2}.$$

As mentioned before, the ratios ε_{ji} may vary from pass to pass, which means that the heat transfer coefficient, surface area and geometrical structure are allowed to be different among the tube passes. Such unbalanced multipass exchangers may have some advantages over the conventional balanced exchangers (Roetzel, 1988). The general initial conditions are

$$t_1(x,0)=g_1(x) \tag{3.34}$$

$$t_{2i}(x,0)=g_{2i}(x) \qquad \left(i=1,\ 2,\ \ldots,\ N\right) \tag{3.35}$$

$$t_{wi}(x,0)=g_{wi}(x) \qquad \left(i=1,\ 2,\ \ldots,\ N\right). \tag{3.36}$$

The shellside and tubeside arbitrary inlet temperature changes can be described as

$$t_1(1,z) = f_1(z) \tag{3.37a}$$

$$t_{21}(1-I,z) = f_2(z) \tag{3.37b}$$

respectively. The other N boundary conditions are still needed to form a closed system of differential equations. They are the interfacial conditions between two adjacent tube passes at both ends, which may vary with the number of tubeside passes and flow arrangements and can be expressed as

$$t_{2i}(x_{ci},z) = t_{2,i-1}(x_{ci},z), \qquad (i = 2,\ 3,\ \ldots\ ,\ N) \tag{3.38}$$

where $x_{ci} = \frac{1}{2}\left[1-(-1)^{i+I}\right]$.

Functions $f_1(z)$ and $f_2(z)$ describe any possible temperature disturbances at both inlets of multipass shell-and-tube heat exchangers, which occur either separately or simultaneously. The most common forms of such variations may assume step, ramp, exponential or periodic expressions.

3.3.2 Transient Responses

To obtain the transient responses to the given inlet temperature changes, the solution to the above-proposed equations must be derived. For this purpose, one can use the Laplace transform. According to the given initial conditions, the transformed equations are as follows

$$\frac{\mathrm{d}\,T_1}{\mathrm{d}\,x} = \left(\sum_{i=1}^{N}\frac{\varepsilon_{1i}\alpha_{1i}U_1}{\alpha_{1i}+\alpha_{2i}+R_{wi}s} - s - U_1\right)T_1 + \sum_{i=1}^{N}\frac{\varepsilon_{2i}\alpha_{2i}U_1}{\alpha_{1i}+\alpha_{2i}+R_{wi}s}T_{2i}$$

$$+h_1(x) \tag{3.39}$$

$$(-1)^{i+1}\frac{\mathrm{d}\,T_{2i}}{\mathrm{d}\,x}=\left(\frac{\varepsilon_{2i}\alpha_{2i}U_2}{\alpha_{1i}+\alpha_{2i}+R_{wi}s}-\varepsilon_{ci}R_\tau s-U_2\varepsilon_{2i}\right)T_{2i}$$

$$+\frac{\varepsilon_{2i}\alpha_{1i}U_2}{\alpha_{1i}+\alpha_{2i}+R_{wi}s}T_1+h_{2i}(x)\qquad(i=1,\ 2,\ \ldots\ ,N)\tag{3.40}$$

$$T_{wi}=\frac{\alpha_{1i}T_1+\alpha_{2i}T_{2i}}{\alpha_{1i}+\alpha_{2i}+R_{wi}s}+h_{wi}(x)\qquad(i=1,\ 2,\ \ldots\ ,N)\tag{3.41}$$

where

$$h_1=g_1(x)+\sum_{i=1}^{N}\frac{U_1R_{wi}}{\alpha_{1i}+\alpha_{2i}+R_{wi}s}g_{wi}(x),$$

$$h_{2i}=\varepsilon_{ci}R_\tau g_{2i}(x)+\frac{U_{2i}R_{wi}}{\alpha_{1i}+\alpha_{2i}+R_{wi}s}g_{wi}(x),$$

$$h_{wi}(x)=\frac{R_{wi}}{\alpha_{1i}+\alpha_{2i}+R_{wi}s}g_{wi}(x).$$

With the appropriate boundary and interfacial conditions, eqns (3.39) and (3.40) compose a closed system consisting of $(N+1)$ nonhomogeneous ordinary differential equations of the first order in the imaginary domain. In matrix notation, this system can be expressed as

$$\frac{\mathrm{d}\mathbf{T}}{\mathrm{d}x}=\mathbf{AT}+\mathbf{H}(x)\tag{3.42}$$

where

$$\mathbf{T}=\{T_{21},\ T_{22},\ \ldots\ ,\ T_{2N},\ T_1\}^T$$

and

$$\mathbf{H} = \{h_{21}(x),\ h_{22}(x),\ \dots,\ h_{2N}(x),\ h_1(x)\}^T$$

are two vectors. **A** is a $(N+1)\times(N+1)$ matrix whose elements are given as

$$a_{ij} = \begin{cases} 0, & i \neq j \\ (-1)^{i+I}\left(\dfrac{U_{2i}\alpha_{2i}}{\alpha_{1i}+\alpha_{2i}+R_{wi}s} - \varepsilon_{ci}R_\tau s - U_{2i}\right), & i = j \end{cases} \qquad (i, j \leq N)$$

$$a_{i,N+1} = \frac{(-1)^{i+I}U_{2i}\alpha_{1i}}{\alpha_{1i}+\alpha_{2i}+R_{wi}s} \qquad (i = 1,\ 2,\ \dots,\ N)$$

$$a_{N+1,j} = \frac{(-1)^{i+I}U_{2j}\alpha_{2j}}{\alpha_{1j}+\alpha_{2j}+R_{wj}s} \qquad (j = 1,\ 2,\ \dots,\ N)$$

$$a_{N+1,N+1} = \sum_{i=1}^{N}\frac{U_{1i}\alpha_{1i}}{\alpha_{1i}+\alpha_{2i}+R_{wi}s} - s - U_1\ .$$

In the light of the similar procedure described by Xuan *et al.* (1991), a general solution to the system (3.42) can be derived. With eigenvalues $\{\lambda_i\}$ of **A** and the corresponding eigenvectors $\{\mathbf{u}_i\}$ $(i = 1,\ 2,\ \dots,\ N+1)$, one can build the solution to eqn (3.42) as

$$\mathbf{T} = \mathrm{e}^{\mathbf{A}x}\,\mathbf{D}^* + \mathrm{e}^{\mathbf{A}x}\int \mathrm{e}^{-\mathbf{A}x'}\mathbf{H}(x')dx'\ . \qquad (3.43)$$

Since $\mathrm{e}^{\mathbf{A}x} = \mathbf{U}\,\mathrm{e}^{\mathbf{L}x}\,\mathbf{U}^{-1}$ and $\mathrm{e}^{\mathbf{A}x}\,\mathbf{D}^*$ can be always converted into $\mathbf{U}\,\mathrm{e}^{\mathbf{L}x}\,\mathbf{D}$, one has

$$\mathbf{T} = \mathbf{U}\,\mathrm{e}^{\mathbf{L}x}\,\mathbf{D} + \int \mathbf{U}\,\mathrm{e}^{\mathbf{L}(x-x')}\mathbf{U}^{-1}\mathbf{H}(x')dx' \qquad (3.44)$$

where $e^{\mathbf{L}x}$ is a diagonal matrix, i.e., $e^{\mathbf{L}x} = \mathrm{diag}\{e^{\lambda_1 x}, e^{\lambda_2 x}, \ldots, e^{\lambda_{N+1} x}\}$, $\mathbf{U}$ is a $(N+1) \times (N+1)$ matrix whose columns are the corresponding eigenvectors, i.e., $\mathbf{U} = \{\mathbf{u}_1, \mathbf{u}_2, \ldots, \mathbf{u}_{N+1}\}$ and $\mathbf{D}$ (or $\mathbf{D}^*$) is a coefficient vector which must be determined according to the given interfacial and boundary conditions. The necessary condition of the solution (3.43) or (3.44) is that all eigenvalues are distinct. In general, the eigenvalues of $\mathbf{A}$ in eqn (3.42) describing multipass shell-and-tube heat exchangers are different from each other.

The simplest case corresponds to the uniform initial conditions. In this case, the solution vector is expressed as

$$\mathbf{T} = \sum_{j=1}^{N+1} d_j \mathbf{u}_j \, e^{\lambda_j x}. \tag{3.45}$$

To obtain the particular solution subject to the given interfacial and boundary conditions, $(N+1)$ unknown coefficients d_j $(j = 1, 2, \ldots, N+1)$ must be determined. From eqn (3.45) and the given boundary conditions (3.37) and (3.38), one can find a matrix equation which defines these coefficients

$$\mathbf{WD} = \mathbf{G} \tag{3.46}$$

where $\mathbf{D} = \{d_1, d_2, \ldots, d_{N+1}\}^T$ and $\mathbf{W} = \{w_{ij}\}$ is a $(N+1) \times (N+1)$ matrix whose elements depend upon such factors as the number of tubeside passes, the tube flow arrangement and multiplicity of eigenvalues λ_j. Subtracting the $(i-1)$th line of the matrix equation (3.45) from the ith line $(i = 2, 3, \ldots, N)$ and substituting the interfacial co-ordinate values into them, one obtains

$$w_{1j} = u_{1j}\varphi_j^{(1+I)} \qquad (j = 1, 2, \ldots, N+1) \tag{3.47a}$$

$$w_{ij} = (u_{ij} - u_{i-1,j})\varphi_j^{(i+I)} \qquad (i = 2, 3, \ldots, N;\ j = 1, 2, \ldots, N+1) \tag{3.47b}$$

$$w_{N+1,j} = u_{N+1,j} \qquad (j = 1,\ 2,\ \ldots\ ,\ N+1) \tag{3.47c}$$

where $\varphi_j^{(i)}$ is defined as $\varphi_j^{(i)} = \frac{1}{2}\left[1+(-1)^i\right] + \frac{1}{2}\left[1-(-1)^i\right]e^{\lambda_j}$.

$$\mathbf{G} = \{F_2(s),\ 0,\ 0,\ \ldots\ ,\ 0,\ F_1(s)\}^T \tag{3.48}$$

Consequently, the coefficient vector is obtained as

$$\mathbf{D} = \mathbf{W}^{-1}\mathbf{G} \tag{3.49}$$

and the temperature matrix **T** can be calculated.

Obviously, d_j, $\mathbf{u}_j$ and λ_j in eqn (3.45) may all be functions of the Laplace parameter s. It is impossible to find the inverse transform of these expression analytically. The numerical inverse Laplace transform (3.23) or (3.22) can be used to derive transient responses to the inlet temperatures in the time domain, according to the forms of the inlet disturbances and the desired accuracy.

In order to examine the previously developed simulation method for the transient behaviour of multipass shell-and-tube heat exchangers, the transient responses of an apparatus with one pass are computed and the same results as those in the Sections 3.1 and 3.2 for cocurrent and countercurrent flow exchangers are obtained. Under the uniform initial conditions and the following parameters that $R_1 = R_w = R_\tau = 1_1$, $U_1 = U_2 = 2NTU_1$ and $f_2(z) = 0$, other examples of the exit behaviours are computed for more tubeside passes, different tubeside flow arrangements and different distribution ratios of $(hA)_1$ as well as $(hA)_2$.

The computed results are illustrated in Figs 3.7–3.10. The results show that the energy balance between both fluids is precisely satisfied in the steady-state for the step inlet temperature change, which demonstrates that the algorithm can also be applied to predicting the steady-state behaviour of multipass heat exchangers. The computation shows that for both tubeside flow arrangements, almost the same exit responses appear if the number N of tubeside passes is greater than or equal to 6 ($N \geq 6$). This phenomenon is similar to the relationship between shell-and-tube units with a higher number of tubeside passes and transversely mixed cross-flow in the steady state. As shown in these figures, no immediate shellside exit behaviours respond to any sudden variations taking place at the inlet and there exist time lag between the inlet and exit changes to some extent. Any disturbances in inlet shellside temperature will be carried to the shellside exit by both shellside and the tubeside fluids flowing in the direction of the shellside flow. The time delay of the shellside exit response is equal to the shortest time needed for the shellside and tubeside fluids to reach the shellside exit. However, if the heat transfer resistances between the shellside and tubeside fluids or the thermal capacity of the core wall are very large, the disturbances carried by the tubeside fluid to the shellside exit are weak, the main part of the disturbances will reach the shellside exit at $z = 1$. Obviously, the time delay of the tubeside exit response is the same as that of the shellside exit response if the shellside and tubeside exits are located at the same end. There is no time delay of the tubeside exit response if the tubeside exit and the shellside entrance are at the same end.

Unlike in cocurrent and countercurrent flow heat exchangers, in multipass heat exchangers the temperature variations of shellside and tubeside fluids at the time when the main disturbances reach their exits are always continuous. Meanwhile, at the discontinuity point, the value of which is equal to the time delay, the temperature responses to a sudden inlet temperature change is relatively weak. Therefore no Gibbs' phenomenon occurs if the FFT algorithm is used for the numerical inverse Laplace transform. The Gaver–Stehfest algorithm cannot be used if the outlet temperatures have sudden or very rapid changes. For example, the Gaver–Stehfest algorithm will yield significant errors if it is used to calculate temperature responses in Figs 3.7(a), 3.9(a) and 3.10(a). However, it can be used for Figs 3.7(b), 3.8, 3.9(b) and 3.10(b).

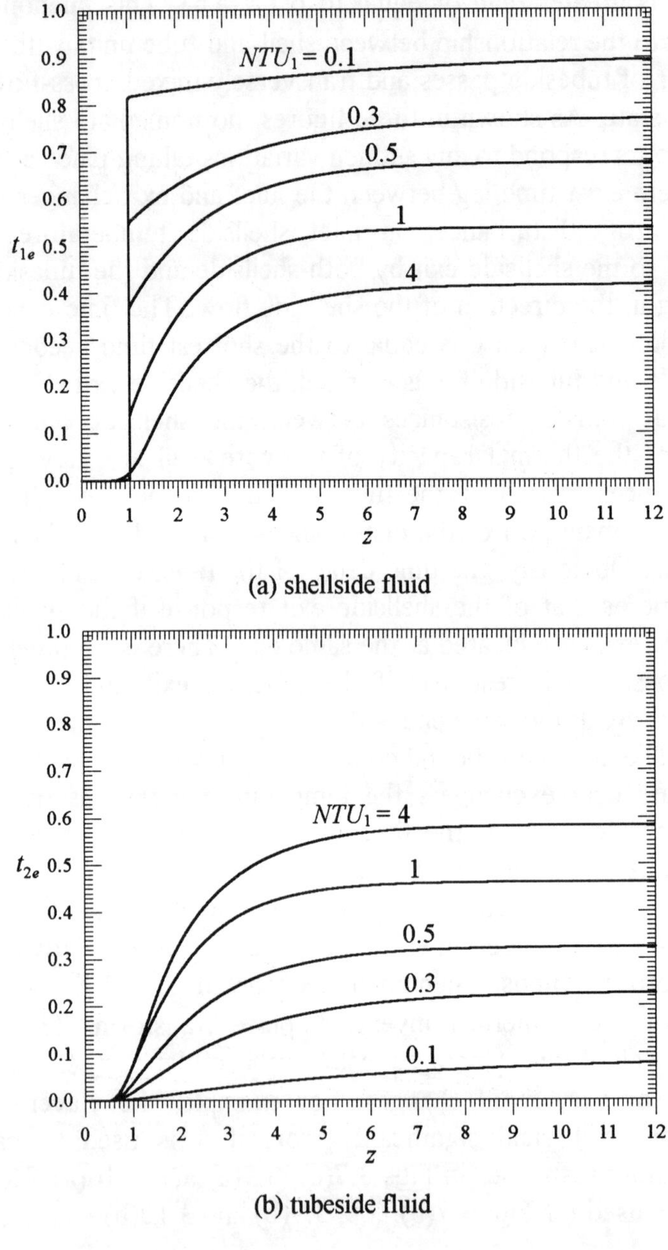

(a) shellside fluid

(b) tubeside fluid

Figure 3.7: Exit temperature responses to a shellside step inlet temperature change of 1–2 heat exchanger with flow arrangement I, $R_1 = R_\tau = R_w = 1$, $U_1 = U_2$, $\varepsilon_{11} = \varepsilon_{12} = 0.5$, $\varepsilon_{21} = \varepsilon_{22} = 0.5$, $\varepsilon_{c1} = \varepsilon_{c2} = 0.5$, $\varepsilon_{w1} = \varepsilon_{w2} = 0.5$, $f_1(z) = 1$, $f_2(z) = 0$

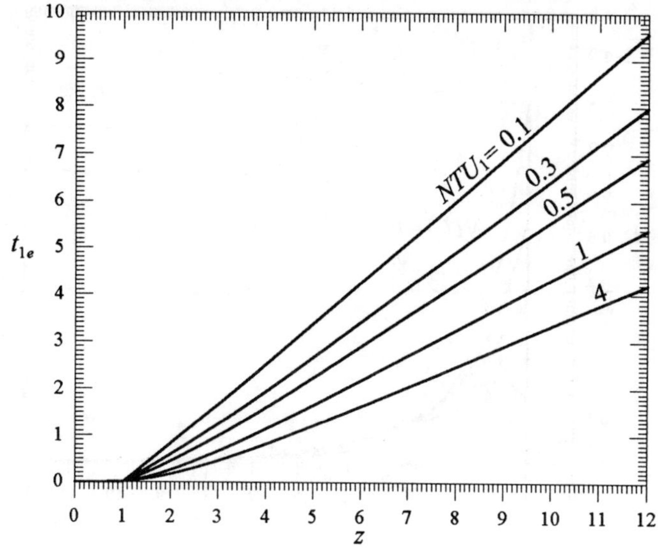

(a) shellside fluid

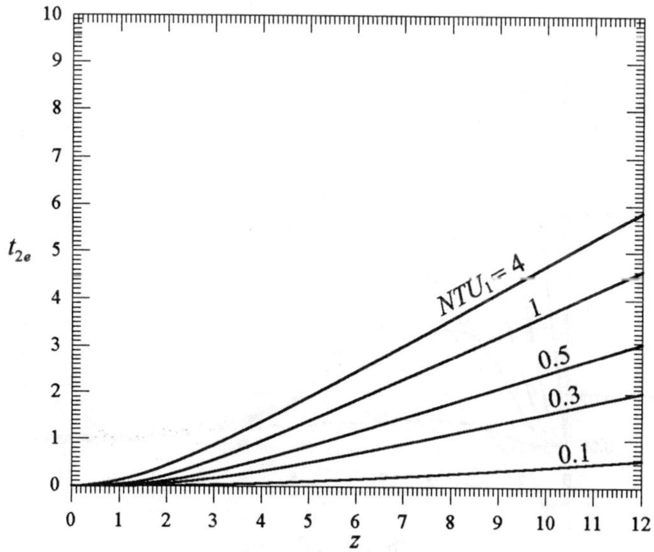

(b) tubeside fluid

Figure 3.8: Exit temperature responses to a shellside ramp inlet temperature change of 1–4 heat exchanger with flow arrangement I, $\varepsilon_{1i} = \varepsilon_{2i} = \varepsilon_{ci} = \varepsilon_{wi} = 0.25 \ (i = 1,\ 2,\ 3,\ 4)$, $R_1 = R_\tau = R_w = 1$, $U_1 = U_2$, $f_1(z) = z$, $f_2(z) = 0$

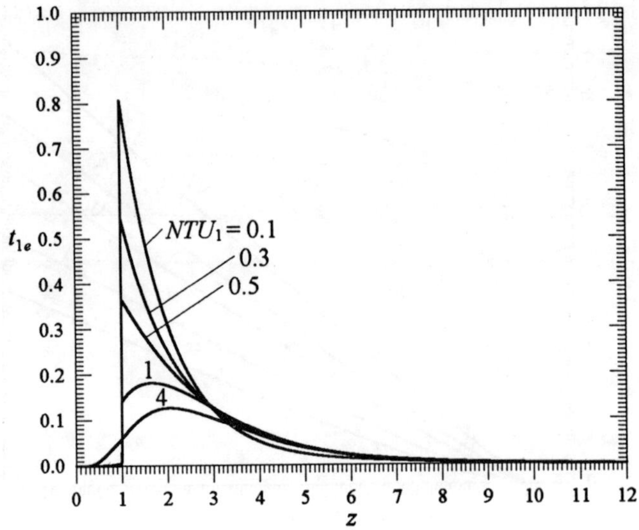

(a) shellside fluid

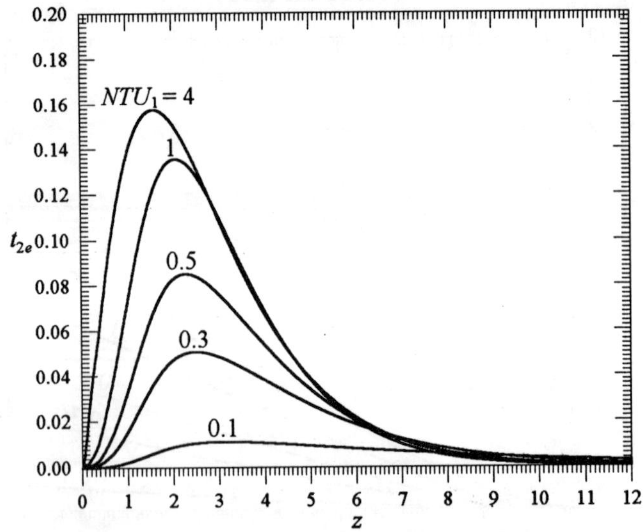

(b) tubeside fluid

Figure 3.9: Exit temperature responses to a shellside exponential inlet temperature change of 1–6 heat exchanger with flow arrangement II, $\varepsilon_{11} = \varepsilon_{14} = \varepsilon_{21} = \varepsilon_{34} = \varepsilon_{c1} = \varepsilon_{c4} = \varepsilon_{w1} = \varepsilon_{w4} = 0.2$, $\varepsilon_{12} = \varepsilon_{13} = \varepsilon_{15} = \varepsilon_{16} = \varepsilon_{22} = \varepsilon_{23} = \varepsilon_{25} = \varepsilon_{26} = \varepsilon_{c2} = \varepsilon_{c3} = \varepsilon_{c5} = \varepsilon_{c6} = \varepsilon_{w2} = \varepsilon_{w3} = \varepsilon_{w5} = \varepsilon_{w6} = 0.15$, $R_1 = R_\tau = R_w = 1$, $U_1 = U_2$, $f_1(z) = e^{-z}$, $f_2(z) = 0$

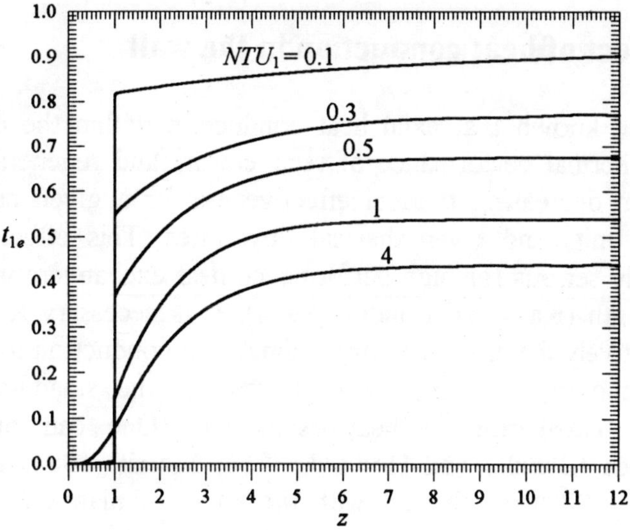

(a) shellside fluid

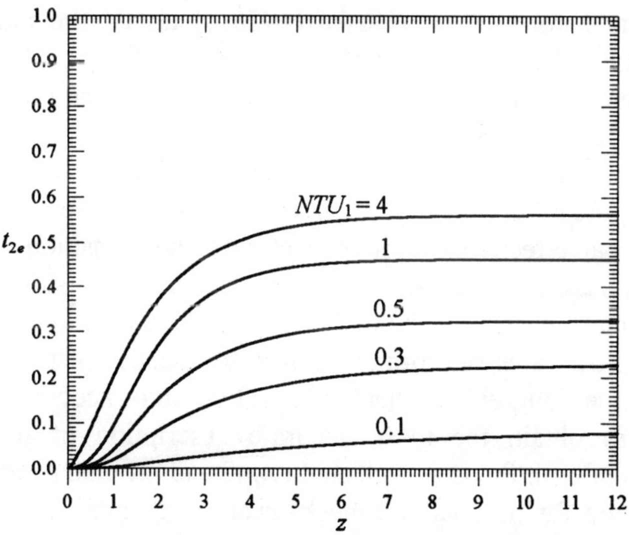

(b) tubeside fluid

Figure 3.10: Exit temperature responses to a shellside step inlet temperature change of 1–8 heat exchanger with flow arrangement II,
$R_1 = R_\tau = R_w = 1,\ U_1 = U_2,\ \varepsilon_{11} = \varepsilon_{12} = \cdots = \varepsilon_{18} = 0.125,$
$\varepsilon_{21} = \varepsilon_{22} = \cdots = \varepsilon_{28} = 0.125,\ \varepsilon_{c1} = \varepsilon_{c2} = \cdots = \varepsilon_{c8} = 0.125,$
$\varepsilon_{w1} = \varepsilon_{w2} = \cdots = \varepsilon_{w8} = 0.125,\ f_1(z) = 1,\ f_2(z) = 0$

3.4 Effect of heat conduction in the wall

It is well known that axial heat conduction within the core wall affects thermal performance of recuperators and regenerators and results in decreasing thermal effectiveness for a given number of transfer units and given thermal flow rates. This effect may be extremely serious for high-performance heat exchangers with short flow length (Kays and London, 1984). It is necessary to estimate quantitatively the effects of longitudinal heat conduction in the solid wall, although it is usually neglected for the steady-state and transient calculation of heat exchangers. One can find some references (Bahnke and Howard, 1964; Mondt, 1964; Li, 1983; Skiepko, 1988) which deal with the effect of matrix longitudinal heat conduction on the performance of periodic-flow (or rotary) heat exchangers under some necessary assumptions. Under the steady-state conduction, Hausen (1976) analysed the influence of core longitudinal heat conduction in a counterflow heat exchanger and pointed out that this influence depends upon the ratio between the hot and cold thermal flow rates.

The effect of core longitudinal heat conduction in the transient state is quite different from that in the steady state. Besides reducing the thermal effectiveness, it may change the transition time from transient state to steady state, which is dependent on such factors as the thermal properties of the core material, the ratio between the thermal flow rates on both sides and the number of transfer units. Taking the longitudinal heat conduction into account, one can expect to obtain the more accurate description of temperature histories of both fluids as well as the wall and provide more accurate information for thermal and mechanical design and control of heat exchangers. With considering the axial heat conduction in the wall, Roetzel and Xuan (1993a) provided a simulation model for transient behaviour of multipass shell-and-tube heat exchangers and analysed the effect of axial heat conduction.

3.4.1 The governing equations

The schematic representation of shell-and-tube heat exchangers with arbitrary tube passes and flow arrangements is shown in Fig. 3.6. A set of differential equations similar to eqns (3.28)–(3.30) is as follows:

$$\dot{W}_1 \frac{\partial \theta_1}{\partial l} + \frac{C_1}{L}\frac{\partial \theta_1}{\partial \tau} + \sum_{i=1}^{N} \frac{(hA)_{1i}}{L}\left(\theta_1 - \theta_{wi}\right) = 0 \tag{3.50}$$

$$(-1)^{i+1} \dot{W}_{2i} \frac{\partial \theta_{2i}}{\partial l} + \frac{C_{2i}}{L}\frac{\partial \theta_{2i}}{\partial \tau} + \frac{(hA)_{2i}}{L}\left(\theta_{2i} - \theta_{wi}\right) = 0 \quad (i = 1,\ 2,\ \ldots,\ N) \tag{3.51}$$

$$\frac{C_{wi}}{L}\frac{\partial \theta_{wi}}{\partial \tau} - k_w A_{cwi} \frac{\partial^2 \theta_{wi}}{\partial l^2} - \frac{(hA)_{wi}}{L}\left(\theta_1 - \theta_{wi}\right) - \frac{(hA)_{2i}}{L}\left(\theta_{2i} - \theta_{wi}\right) = 0 \quad (i = 1,\ 2,\ \ldots,\ N). \tag{3.52}$$

The second term on the left-hand side of eqn (3.52) accounts for the effect of axial heat conduction in the tube wall. With regard to the metallic mass of the baffles and the end plates, one can approximately distribute the heat capacities of these components among all tube passes. In this case, the heat capacity of the wall of tube pass i is given by

$$C^*_{wi} = C_{wi}(1 + \gamma_i) \tag{3.53}$$

where C^*_{wi} is the apparent heat capacity of tube pass i and γ_i a distribution parameter for the metallic components such as the baffles and end plates. The value of γ_i may vary from tube pass to tube pass. In this way, the influence of metallic components on transient behaviour of heat exchangers can be taken into account. For simplicity, the asterisk "∗" is omitted in the following analysis.

With the product of Fourier number and its factors for each pass $Fo\varepsilon_{ai} = \dfrac{a_w \tau_{r1}}{L^2}\varepsilon_{ai} = \dfrac{k_{wi} A_{cwi}}{L C_1 \varepsilon_{wi}}$ and other defined dimensionless variables as well as parameters in Section 3.3, eqns (3.50)–(3.52) are rewritten in the dimensionless form:

$$\frac{\partial t_1}{\partial x} + \frac{\partial t_1}{\partial z} + \sum_{i=1}^{N} U_{1i}\left(t_1 - t_{wi}\right) = 0 \tag{3.54}$$

$$(-1)^{i+1}\frac{\partial t_{2i}}{\partial x} + \varepsilon_{ci} R_{\tau}\frac{\partial t_{2i}}{\partial z} + U_{2i}\left(t_{2i} - t_{wi}\right) = 0 \qquad (i = 1,\ 2,\ \ldots,\ N) \tag{3.55}$$

$$R_{wi}\frac{\partial t_{wi}}{\partial z} - Fo\varepsilon_{ai} R_{wi}\frac{\partial^2 t_{wi}}{\partial x^2} - \alpha_{1i}\left(t_1 - t_{wi}\right) - \alpha_{2i}\left(t_{2i} - t_{wi}\right) = 0 \qquad (i = 1,\ 2,\ \ldots,\ N) \tag{3.56}$$

The definition of Fourier number *Fo* reveals that its value depends upon the characteristic of the tube material, the residence time of fluid 1 in the apparatus, and the geometric dimensions of the apparatus. It indicates the effect of the core longitudinal heat conduction on dynamic behaviour of multipass shell-and-tube exchangers. $Fo = 0$ corresponds to the ideal case that there occurs no longitudinal heat conduction. With increasing *Fo*, the effect of axial heat conduction on the transient process becomes more significant. The initial conditions, boundary and interfacial conditions for fluid 1 and 2 remain the same as those given in Section 3.3.1. What are to be supplemented are 2*N* boundary and interfacial conditions for the tube wall, which depend on the construction of the exchanger. However, the adiabatic boundary conditions

$$\frac{\partial t_{wi}(0,\tau)}{\partial x} = \frac{\partial t_{wi}(1,\tau)}{\partial x} = 0 \qquad (i = 1,\ 2,\ \ldots,\ N) \tag{3.57}$$

can be applied to the most cases with satisfaction. The end effect of the tube bundles can be taken into account by adding the products of heat transfer coefficient and heat transfer surface area $(hA)_{end_i}$ at the tube end to $(hA)_{2i}$ in tubesides.

Application of the Laplace-transform to eqns (3.54)–(3.56) leads to the following forms:

$$\frac{\mathrm{d}T_1}{\mathrm{d}x}+(s+U_1)T_1-U_1\sum_{i=1}^{N}\varepsilon_{1i}T_{wi}=0 \tag{3.58}$$

$$(-1)^{i+1}\frac{\mathrm{d}T_{2i}}{\mathrm{d}x}+(\varepsilon_{ci}R_\tau s+\varepsilon_{2i}U_2)T_{2i}-\varepsilon_{2i}U_2T_{wi}=0 \qquad (i=1,\ 2,\ \ldots,\ N) \tag{3.59}$$

$$Fo\varepsilon_{ai}R_{wi}\frac{\mathrm{d}^2T_{wi}}{\mathrm{d}x^2}-(\alpha_{1i}+\alpha_{2i}+R_{wi}s)T_{wi}+\alpha_{1i}T_1+\alpha_{2i}T_{2i}=0 \qquad (i=1,\ 2,\ \cdots,\ N) \tag{3.60}$$

where the uniform initial conditions are assumed. By means of the expression $Y_i=\frac{\mathrm{d}T_{wi}}{\mathrm{d}x}$ $(i=1,\ 2,\ \cdots,\ N)$, an equivalent system of eqns (3.58)–(3.60) can be derived as

$$\frac{\mathrm{d}\mathbf{T}}{\mathrm{d}x}=\mathbf{AT} \tag{3.61}$$

where $\mathbf{T}=(T_{21},\ T_{22},\ \ldots,\ T_{2N},\ T_1,\ T_{w1},\ T_{w2},\ \ldots,\ T_{wN},\ Y_1,\ Y_2,\ \ldots,\ Y_N)^{\mathrm{T}}$ and $\mathbf{A}$ is a $(3N+1)\times(3N+1)$ matrix whose elements are given by eqns (3.58)–(3.60). Using the similar algorithm to that described in Section 3.3, one is able to find the solution to eqn (3.61). The detailed solution procedure can be found in the paper of Roetzel and Xuan (1993a).

$$\mathbf{T}=\sum_{j=1}^{3N+1}d_j\mathbf{u}_je^{\lambda_j x} \tag{3.62}$$

where λ_j and $\mathbf{u}_j$ $(j = 1, 2, \ldots, 3N+1)$ are distinct eigenvalues and corresponding eigenvectors of matrix **A**. $\mathbf{D} = \{d_1, d_2, \ldots, d_{3N+1}\}^T$ is a coefficient matrix determined by the given boundary conditions, i.e., $\mathbf{D} = \mathbf{W}^{-1}\mathbf{G}$. Substituting the boundary conditions into eqn (3.62) one obtains the nonzero elements of **W** and **G** as

$$w_{1j} = u_{1j}\varphi_j^{(1+I)} \qquad (j = 1, 2, \ldots, 3N+1) \tag{3.63a}$$

$$w_{ij} = \left(u_{ij} - u_{i-1,j}\right)\varphi_j^{(i+I)}$$
$$(i = 2, 3, \ldots, N;\ j = 1, 2, \ldots, 3N+1) \tag{3.63b}$$

$$w_{N+1,j} = u_{N+1,j} \qquad (j = 1, 2, \ldots, 3N+1) \tag{3.63c}$$

$$w_{ij} = u_{i+N,j} \ (i = N+2, N+2, \ldots, 2N+1;\ j = 1, 2, \ldots, 3N+1) \tag{3.63d}$$

$$w_{ij} = u_{ij}e^{\lambda_j} \ (i = 2N+2, 2N+3, \ldots, 3N+1;\ j = 1, 2, \ldots, 3N+1) \tag{3.63e}$$

$$g_1 = F_2(s) \tag{3.64a}$$

$$g_{N+1} = F_1(s) \tag{3.64b}$$

where $\varphi_j^{(i)}$ is defined as

$$\varphi_j^{(i)} = \frac{1}{2}\left[1 + (-1)^i\right] + \frac{1}{2}\left[1 - (-1)^i\right]e^{\lambda_j}.$$

The numerical inversion of the Laplace transform is used to obtain the final temperature profiles and exit temperature responses in the time-domain. Because of the axial heat conduction in tubes there are no discontinuities in the outlet fluid temperatures.

Therefore no Gibbs' impulses would appear by using FFT algorithm if the time interval is small enough.

3.4.2 Effect of core longitudinal heat conduction

To analyse the effect on the temperature profiles, the following parameters $\delta_1 = \left|t_{1e}^* - t_{1e}\right|$ and $\delta_2 = \left|t_{2e}^* - t_{2e}\right|$ are introduced, where the asterisk "*" represents the effect of the core longitudinal heat conduction. The exit temperatures of fluid 1 and 2 are used to discuss the heat conduction effect. Under the condition $U_1 = U_2$, some examples are computed and plotted in Figs 3.11–3.14 to show how the longitudinal heat conduction affects transient temperature profiles and exit temperature responses in shell-and-tube heat exchangers. Comparison between the results with and without considering the axial heat conduction shows that the longitudinal heat conduction leads to a slower temperature change in both fluids. Such an effect becomes more evident with increasing *NTU*.

Figures 3.12 and 3.13 demonstrate the influence of the core longitudinal heat conduction. The influence on the shellside temperature variation increases with the time. However, if the values of NTU_1 are small, a peak in the deviation occurs at $z = 1$, which would be higher than the deviation in steady state. The deviation in the tubeside temperature variation will reach the maximum near $z = 1$ and then decrease with time. The examples illustrated in Fig. 3.14 show that the influence is weaker in a 1–2 heat exchanger than in countercurrent flow.

If the tube ends between the passes are perfectly connected, e.g., the tubes are connected from one pass to the next with U-tube bends, the following boundary conditions could be used

$$t_{wi} = t_{w,i-1}, \ \varepsilon_{k_i} \frac{\partial t_{wi}}{\partial x} = -\varepsilon_{k_{i-1}} \frac{\partial t_{w,i-1}}{\partial x} \quad \text{at } x = \frac{1}{2}\left[1 - (-1)^{i+I}\right]$$
$$(i = 2,\ 3,\ \ldots,\ N) \qquad (3.65)$$

in which $\varepsilon_{k_i} = \left(k_w A_{cw}\right)_i / \left(k_w A_{cw}\right)_r$. The temperature variations according to the boundary conditions (3.65) represented another

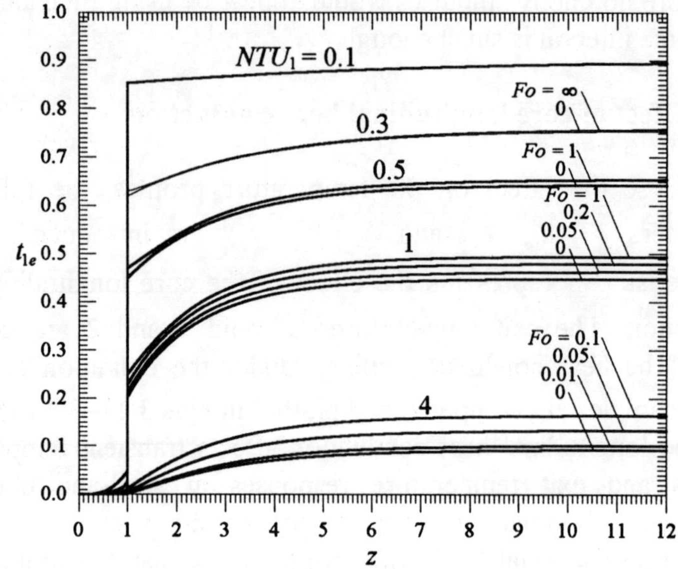

(a) shellside fluid

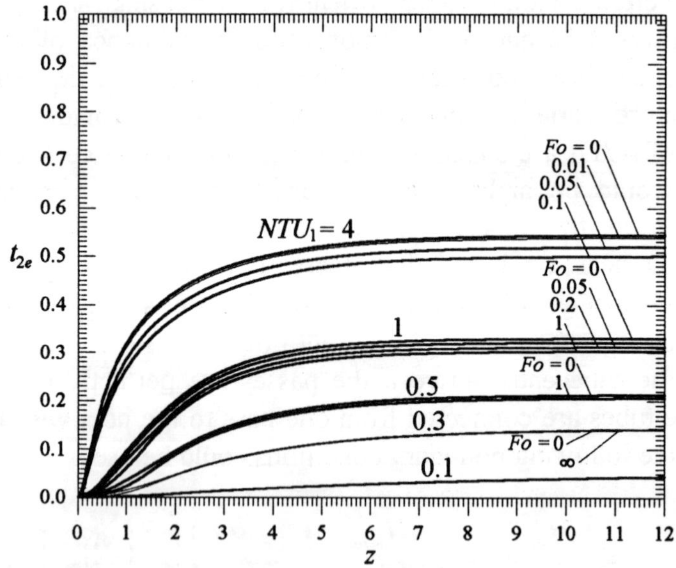

(b) tubeside fluid

Figure 3.11: Exit temperature responses to a shellside step inlet temperature change, countercurrent flow, $R_1 = 0.6$, $R_w = R_\tau = 1$, $U_1 = U_2$, $f_1(z) = 1$, $f_2(z) = 0$

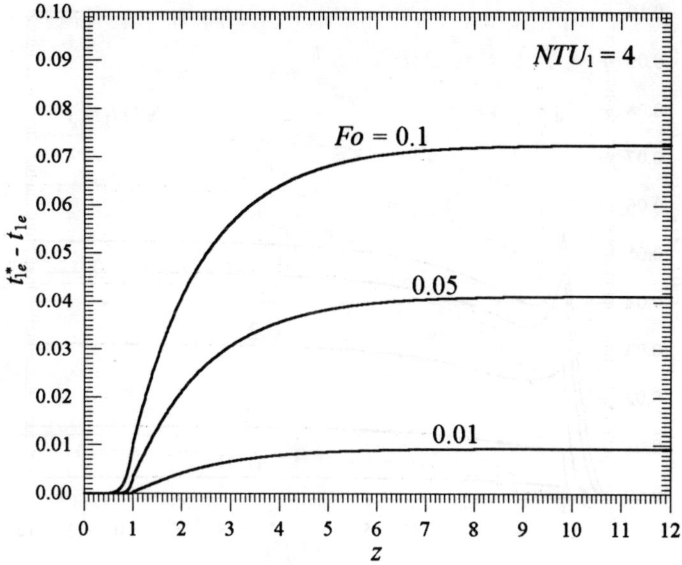

(a) shellside fluid

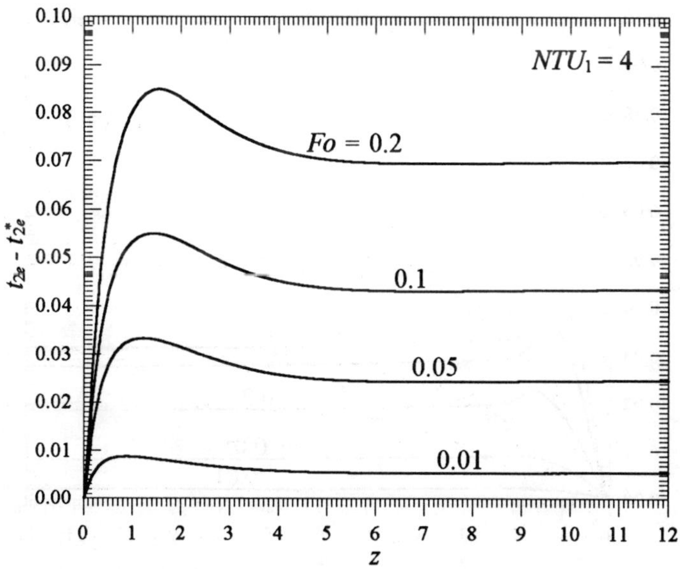

(b) tubeside fluid

Figure 3.12: Effect of longitudinal heat conduction in a countercurrent flow heat exchanger ($NTU_1 = 4$, other parameters are the same as in Fig. 3.11)

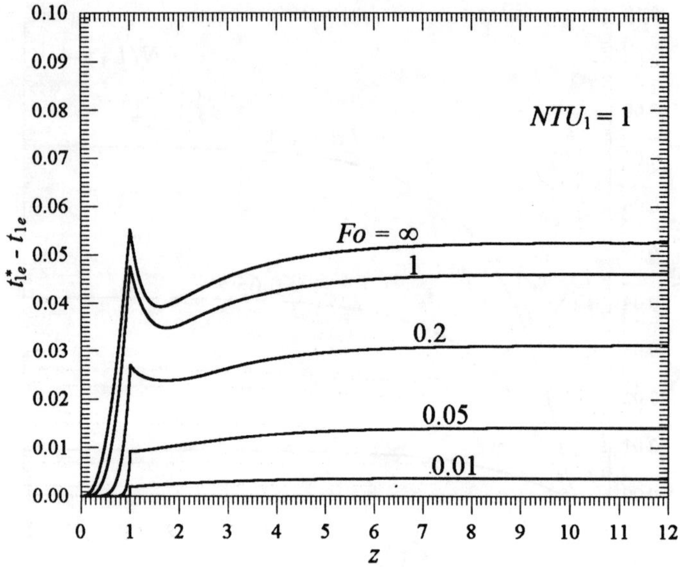

(a) shellside fluid

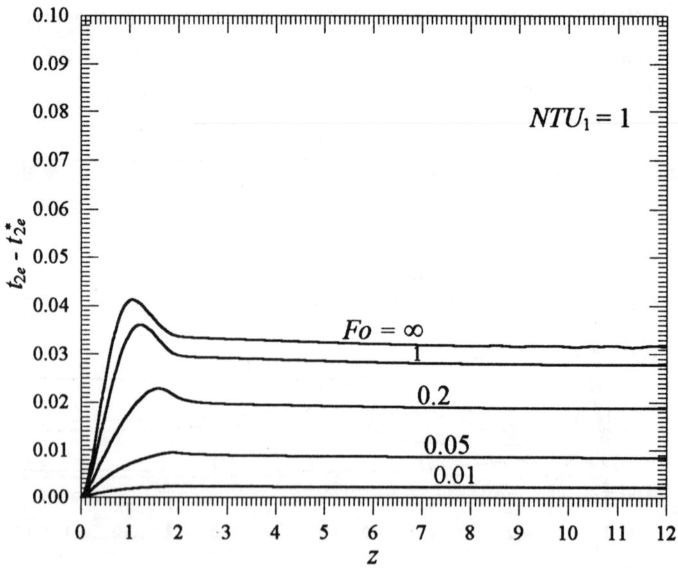

(b) tubeside fluid

Figure 3.13: Effect of longitudinal heat conduction in a countercurrent flow heat exchanger ($NTU_1 = 1$, other parameters are the same as in Fig. 3.11)

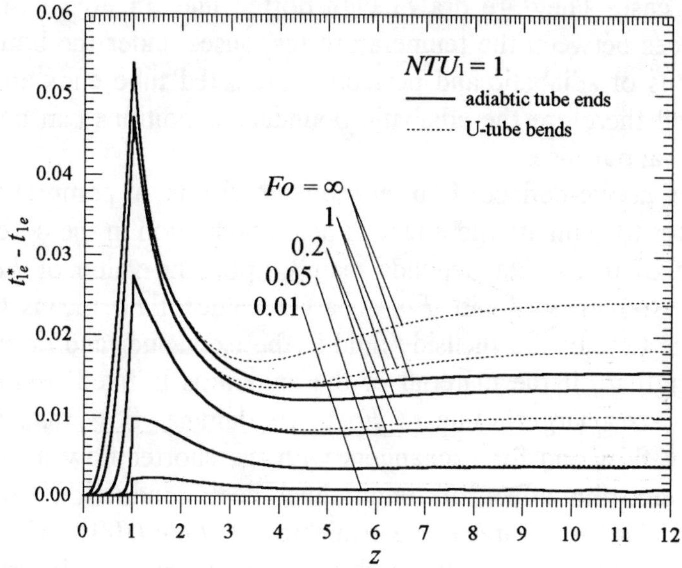

(a) shellside fluid

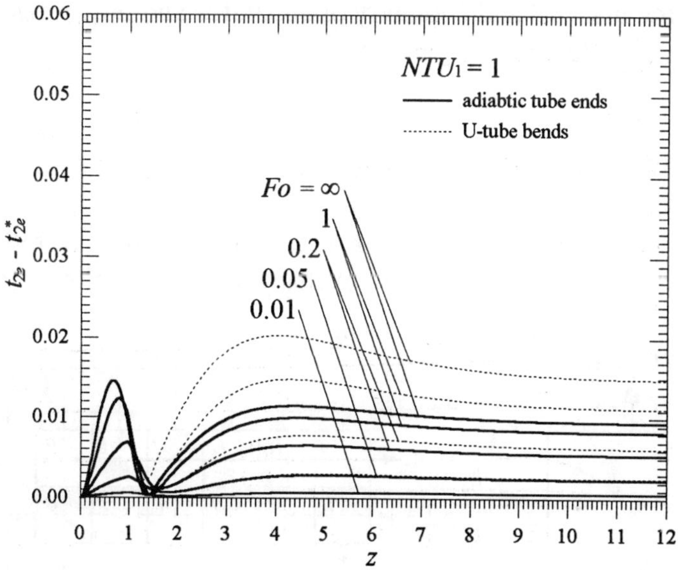

(b) tubeside fluid

Figure 3.14: Effect of longitudinal heat conduction in a 1–2 heat exchanger, flow arrangement II, $\varepsilon_{1i} = \varepsilon_{2i} = \varepsilon_{ci} = \varepsilon_{wi} = 0.5 \ (i = 1, 2)$, $\varepsilon_{a1} = \varepsilon_{a2} = 1$, $R_w = R_\tau = 1$, $NTU_1 = 1$, $f_1(z) = 1$, $f_2(z) = 0$

limiting case. They are drawn with dotted lines in Fig. 3.14. The differences between the temperature responses under the boundary conditions of adiabatic and perfectly connected tube ends are very small and therefore the adiabatic boundary conditions can be used for general purposes.

The above-defined Fourier number *Fo* is a comprehensive parameter to estimate the effect of heat conduction in the core. The strength of this effect depends mainly upon the value of *Fo* and increases remarkably with *Fo*. A larger value of *Fo* means longer residence time of the shellside fluid in the exchanger and/or shorter flow length *L*, if the material of the apparatus is fixed. Evidently, there exists a larger value of *Fo* in the laminar flow than in the turbulent flow and for exchangers with the shorter flow length. In such cases, the effect of longitudinal heat conduction becomes stronger. However, it can be neglected if $Fo < 0.001$. *Fo* varies with the exchanger itself and with operating conditions. In practice, one should first estimate the value of *Fo* from the given conditions such as the material and geometry of apparatus and flow rates to see whether this effect is negligible or not.

3.5 Application of the cell model to transient analysis

Among many different methods used for design and rating of multipass shell-and-tube heat exchangers, the finite-difference

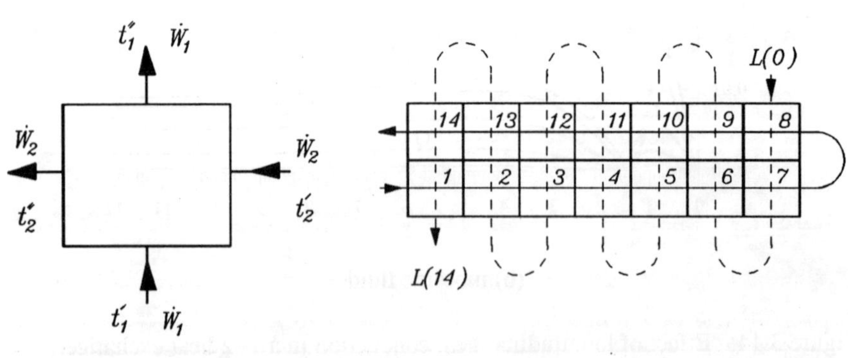

(a) the cell model (b) a multicell model for a 1–2 exchanger

Figure 3.15: The cell model

method may be the most accurate but the most time-consuming. It seems to be impractical to apply it to industrial design and rating. An alternative is the so-called cell model developed by Gaddis and Schlüder (1979). This model is an effective way to simulate multipass shell-and-tube exchangers, especially apparatus with few baffles. It divides the whole apparatus into several elements or cells. The number of these cells is determined by the number of baffles and the number of tubeside passes as well as shellside passes. Each elementary cell is considered as a fictitious exchanger through which both fluids flow with different inlet and outlet temperatures. These elementary units are coupled by the fluids participating in heat transfer process.

The cell model is first used to predict the temperature distribution and the mean temperature difference in multipass shell-and-tube exchangers. Later, Roppo and Ganic (1983) as well as Correa and Marchetti (1987) applied it to dynamic simulation of multipass tubular exchangers. Figure 3.15 shows one cell model and the multicell model for a 1–2 shell-and-tube exchanger.

As pointed out previously, each cell is regarded as an exchanger, and then a set of energy balance equations can be built for it. For the fluids in cell i, the energy equations are given as:

$$C_{2,i}\frac{\mathrm{d}\theta_2(i)}{\mathrm{d}\tau}=\dot{W}_2\,\theta_2(i-1)-\dot{W}_2\,\theta_2(i)+(KA)_i\left[\theta_1(i)-\theta_2(i)\right] \tag{3.65}$$

$$C_1\left[L(i)\right]\frac{\mathrm{d}\theta_1\left[L(i)\right]}{\mathrm{d}\tau}=\dot{W}_1\,\theta_1\left[L(i-1)\right]-\dot{W}_1\,\theta_1\left[L(i)\right]$$
$$+(KA)_i\left\{\theta_2\left[L(i)\right]-\theta_1\left[L(i)\right]\right\} \tag{3.66}$$

where $C_{2,i}$ and $C_1\left[L(i)\right]$ are the heat capacities of tubeside and shellside fluids within cell i. The counter i and $L(i)$ respectively represent the order number of cell by following the trajectory of the tubeside and shellside fluids. There is a relationship between i and $L(i)$ which depends upon cell division approach. An example is shown in Fig. 3.15b. By setting $L(i) = i$, eqns (3.65) and (3.66) express the energy balance between both the fluids on cell *i*. Here the effect of heat capacity of the tube wall is neglected. If necessary, an approximate method is introduced to take metallic components into account, i.e., $C_{2,i}$ in eqn (3.65) is replaced by

$$C_{2,i}^{*} = C_{2,i}\left(1+\omega_i\right) \tag{3.67}$$

where ω_i is a parameter that accounts for the metallic mass of tubes in cell *i*. The dimensionless forms of eqns (3.65) and (3.66) are as follows:

$$\chi_{2,i} R_\tau \frac{\mathrm{d}t_2(i)}{\mathrm{d}z} = t_{2,i-1} - t_{2,i} + \beta_i NTU_2\left(t_{1,i} - t_{2,i}\right) \tag{3.68}$$

$$\chi_{1,L(i)} \frac{\mathrm{d}t_1[L(i)]}{\mathrm{d}z} = t_1[L(i-1)] - t_1[L(i)] + \beta_{L(i)} NTU_1\{t_1[L(i)] - t_2[L(i)]\} \tag{3.69}$$

where $\chi_{2,i} = \dfrac{C_{2,i}}{C_2}$, $\beta_i = \dfrac{(KA)_i}{(KA)}$, $\chi_{1,L(i)} = \dfrac{C_{1,i}}{C_1}$, $\beta_{L(i)} = \dfrac{(KA)_{L(i)}}{(KA)}$. These two equations are valid for any elementary cell. For a 1–*N* shell-and-tube exchanger with $(M-1)$ baffles, an intuitional approach is to divide the whole exchanger into *NM* cells. Hence, there are 2*NM* conjugated first-order ordinary differential equations which describe the dynamic behaviour of the apparatus.

A number of methods can be used to find the solution to the above-mentioned system of 2*NM* equations. A direct procedure is the finite-difference method. Therefore, the time derivative on the left-hand side of eqns (3.68) and (3.69) are approximated by the following finite-difference forms:

$$\frac{\mathrm{d}t_{2,i}}{\mathrm{d}z} \approx \frac{t_{2,i}^{n} - t_{2,i}^{n-1}}{\Delta z} \tag{3.70a}$$

$$\frac{\mathrm{d}t_1[L(i)]}{\mathrm{d}z} \approx \frac{t_{1,L(i)}^{n} - t_{1,L(i)}^{n-1}}{\Delta z}. \tag{3.70b}$$

Furthermore, the arithmetic mean temperatures are applied to the terms on the right-hand side of these equations:

$$t_{2,i} = \frac{1}{2}\left(t_{2,i}^{n} + t_{2,i}^{n-1}\right) \tag{3.71a}$$

$$t_{1,L(i)} = \frac{1}{2}\left(t_{1,L(i)}^{n} + t_{1,L(i)}^{n-1}\right). \tag{3.71b}$$

Substitution of these expressions into the original equations (3.68) and (3.69) gives

$$t_{2,i}^{n} = a_1 t_{1,i}^{n} + a_2 t_{2,i-1}^{n} + a_3 t_{1,i}^{n-1} + a_4 t_{2,i-1}^{n-1} + a_5 t_{2,i}^{n-1}$$
$$(i = 1,\ 2,\ \ldots,\ NM) \tag{3.72}$$

$$t_{1,L(i)}^{n} = b_1 t_{2,L(i)}^{n} + b_2 t_{1,L(i-1)}^{n} + b_3 t_{2,L(i)}^{n-1} + b_4 t_{1,L(i-1)}^{n-1} + b_5 t_{1,L(i)}^{n-1}$$
$$\left(L(i) = 1,\ 2,\ \ldots,\ NM\right) \tag{3.73}$$

where

$$a_1 = \frac{1}{2a_0},\ a_2 = \frac{\beta_i\, NTU_2}{2a_0},\ a_3 = a_1,\ a_4 = a_2,$$

$$a_5 = \frac{2R_\tau \chi_i - \Delta z\left(1 + \beta_i\, NTU_2\right)}{2a_0 \Delta z},\ b_1 = \frac{1}{2b_0},\ b_2 = \frac{\beta_{L(i)}\, NTU_1}{2b_0},$$

$$b_3 = b_1,\ b_4 = b_2,\ b_5 = \frac{2\chi_{L(i)} - \Delta z\left(1 + \beta_{L(i)}\, NTU_1\right)}{2b_0 \Delta z},$$

$$a_0 = \frac{R_\tau \chi_i}{\Delta z} + \frac{1+\beta_i\, NTU_2}{2}, \quad b_0 = \frac{\chi_{L(i)}}{\Delta z} + \frac{1+\beta_{L(i)}\, NTU_1}{2}.$$

For the given initial conditions $t_{2,i}^0 = g_2(i)$ and $t_{1,i}^0 = g_1(i)$ as well as $t_{2,0}^n = f_2(n\Delta z)$ and $t_{1,0}^n = f_1(n\Delta z)$, one is able to find the transient temperature distributions within a multipass exchanger. One should pay attention to the pertinent relationship between the order numbers i and $L(i)$. Since expressions (3.72) and (3.73) are implicit, integrative computation is needed. The detailed procedure and some examples are presented by Correa and Marchetti (1987). A similar method is proposed by Jonsson *et al.* (1992) to modelling and parameter estimation of heat exchangers.

The essential of the previously described cell model is the finite-difference method. It divides an exchanger into a number of lumped elementary cells where each fluid is assumed to be perfectly mixed. This assumption implied that within a cell there is no thermal performance difference among the different individual flow arrangements such as cocurrent flow, countercurrent flow and cross-flow. This model is especially effective and suitable for multipass shell-and-tube heat exchangers with only a few baffles in which the effect of the number of baffles is important. However, with a small number of baffles the assumption of perfect mixing in each cell may be dubious.

3.6 Axial dispersion model

The conventional plug-flow model has widely been used and considered to be the standard model for both steady-state and transient simulations of shell-and-tube heat exchangers. It assumes that the flow is identical everywhere and no flow maldistributions occur — bypass, leakage streams, stagnant flow, and backmixing, for example. Such flow maldistributions will degrade the thermal performance of exchangers, which the plug-flow model fails to describe. The actual flow, especially the shellside flow, is very complicated and various forms of flow maldistribution may exist. The classical description of the shellside flow is made by Tinker (1951). Recently, Pekdemir *et al.* (1993) measured the shellside flow distribution and pressure drops in an E-type shell-and-tube heat exchanger. In the cases of maldistributions, the ideal plug-flow greatly deviates from the real flow pattern. One way to correct this deviation is the dispersion model which is based on a dispersed plug-flow, i.e., the main plug-flow with longitudinal dispersion, named backmixing by Mecklenburg and Hartland (1975). This model has mainly been applied to flow in tubes and packed beds.

The effect of flow maldistributions on thermal performance of the exchanger depends upon the intensity of each maldistribution. Mueller (1987) analysed effects of some types of maldistribution on thermal performance of heat exchangers. The dispersion model is a very effective instrument to estimate the effect of low maldistribution on the performance under the steady-state condition. Roetzel and Spang (1990), Spang (1991), Roetzel and Xuan (1991), and Xuan (1991) as well as Roetzel and Xuan (1992c) analytically and experimentally applied the dispersion model to E-type, G-type and J-type shell-and-tube heat exchangers.

The dispersion model is an improved plug-flow model, i.e., it is based on the main plug-flow with axial diffusion or dispersion which may be both molecular and macroscopic. The synonym of axial dispersion is backmixing, although this terminology should only be used when real axial mixing takes place. In other cases, such as maldistribution caused by a laminar velocity profile in a flow channel, apparent axial dispersion plug-flow occurs but no real backmixing. In the axial dispersion model, an apparent axial heat conduction term is introduced into the energy balance relationship and the effect of flow maldistribution is taken into account by this dispersion term in the energy equation. It is convenient and accessible to use this one-dimensional model with only the axial spatial variable to handle such problems concerning complicated flow patterns. Taylor (1954) may have been the first to develop the dispersion model for mass transfer in turbulent flow through a pipe. In the following context the dispersion model is extended to transient analysis of tubular heat exchangers.

3.6.1 1–1 shell-and-tube heat exchangers

A 1–1 shell-and-tube heat exchanger is schematically shown in Fig. 3.16, in which both cocurrent flow and countercurrent flow arrangements are possible.

Flow maldistribution affects the energy distribution and transfer of a fluid along its flow direction. In this case, the total thermal flow rate is composed of an enthalpy flow and a dispersive or conductive heat flow

$$\dot{m}\tilde{h} = \dot{m}h \pm A_c \mathcal{D} \frac{\partial \theta}{\partial l}. \tag{3.74a}$$

For the single-phase flow, expression (3.74a) may be rewritten as

$$\dot{W}\tilde{\theta} = \dot{W}\theta \pm A_c \mathcal{D} \frac{\partial \theta}{\partial l} \tag{3.74b}$$

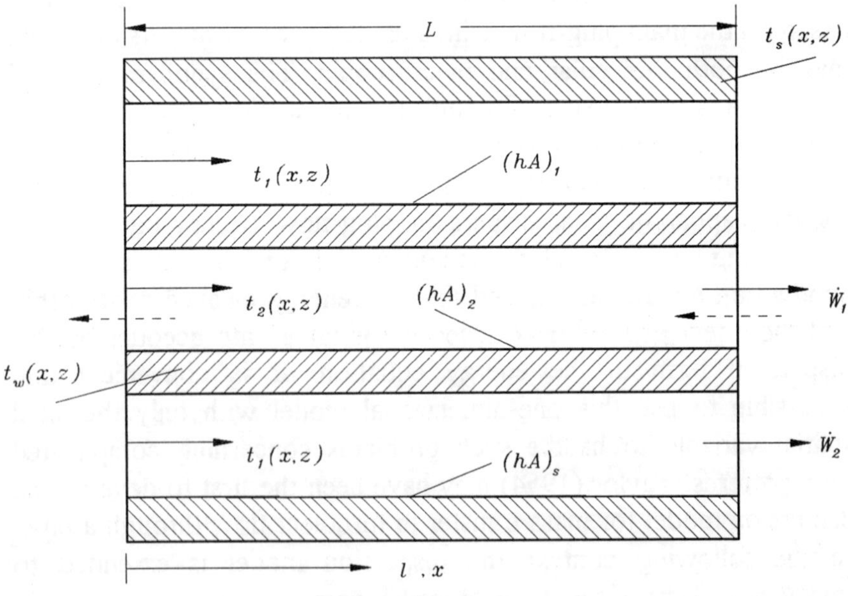

igure 3.16: Schematic representation of cocurrent flow and counterflow heat exchangers

where $\mathcal{D}$ is the dispersion coefficient or apparent thermal conductivity which is a function of properties and flow status of the fluid. If the flow direction is identical to the direction of the co-ordinate system, the negative sign of (±) is assumed, otherwise the positive is used in eqn (3.74). By means of the dispersion model, one is able to derive a set of energy balance equations for both fluids and the tube wall as well as the shell as follows:

$$A_{c1}\mathcal{D}\frac{\partial^2\theta_1}{\partial l^2}-\dot{W}_1\frac{\partial\theta_1}{\partial l}-\frac{C_1}{L}\frac{\partial\theta_1}{\partial\tau}-\frac{(hA)_1}{L}(\theta_1-\theta_w)-\frac{(hA)_s}{L}(\theta_1-\theta_s)=0 \tag{3.75}$$

$$(-1)^{I+1}\dot{W}_2\frac{\partial\theta_2}{\partial l}+\frac{C_2}{L}\frac{\partial\theta_2}{\partial\tau}+\frac{(hA)_2}{L}(\theta_2-\theta_w)=0 \tag{3.76}$$

$$\frac{C_w}{L}\frac{\partial\theta_w}{\partial\tau}-\frac{(hA)_w}{L}(\theta_1-\theta_w)-\frac{(hA)_2}{L}(\theta_2-\theta_w)=0 \tag{3.77}$$

$$\frac{C_s}{L}\frac{\partial\theta_s}{\partial\tau}-\frac{(hA)_s}{L}(\theta_1-\theta_s)=0 \tag{3.78}$$

where $I=0$ and $I=1$ in eqn (3.76) are pertinent to the countercurrent flow and cocurrent flow arrangements, respectively. Equation (3.78) is introduced to take the effect of heat capacity of the shell into account. These equations are rewritten in dimensionless forms:

$$\frac{1}{Pe}\frac{\partial^2 t_1}{\partial x^2}-\frac{\partial t_1}{\partial x}-\frac{\partial t_1}{\partial z}-U_1(t_1-t_w)-U_s(t_1-t_s)=0 \tag{3.79}$$

$$(-1)^{I+1}\frac{\partial t_2}{\partial x}+R_\tau\frac{\partial t_2}{\partial z}+U_2(t_2-t_w)=0 \tag{3.80}$$

$$R_w\frac{\partial t_w}{\partial z}-\alpha_1(t_1-t_w)-\alpha_2(t_2-t_w)=0 \tag{3.81}$$

$$R_s R_w \left(1 + R_{c2}\right)\frac{\partial t_s}{\partial z} - U_1\left(t_1 - t_s\right) = 0 \tag{3.82}$$

where $R_s = C_s / C_w$, $U_s = (hA)_s / \dot{W}_1$, $Pe = \left(\dot{W}_1 L\right) / \left(\mathcal{D} A_{c1}\right)$, and the other parameters are the same as those defined in the previous sections.

The Péclet number *Pe* appearing in eqn (3.79) takes the effect of all types of shellside maldistribution on thermal process into account. Its value depends upon the intensity of axial mixing and maldistribution. The stronger the flow maldistribution, the smaller the Péclet number and the poorer the thermal performance. $Pe \rightarrow 0$ means complete axial mixing; if $Pe \rightarrow \infty$, the dispersion model approaches the conventional plug-flow model. The actual flow pattern in heat exchangers confines values of *Pe* in the whole range $0 < Pe < \infty$. Therefore, one can describe the effect of shellside flow maldistribution by means of *Pe*. It should be emphasized that *Pe* is a comprehensive function of all types of flow maldistribution, not only the axial maldistribution. Generally, it is experimentally determined for tubular exchangers. Here only the shellside maldistribution is taken into account to explain application of the dispersion model. If necessary, one can simultaneously apply this model to the tubeside flow.

For the sake of energy conservation to correspond to the dispersion model, the boundary conditions for the shellside flow should assume the following forms:

$$t_1 - \frac{1}{Pe}\frac{\partial t_1}{\partial x} = f_1(z) \quad \text{at } x = 0 \quad \text{and} \quad \frac{\partial t_1}{\partial x} = 0 \quad \text{at } x = 1. \tag{3.83}$$

These inlet and outlet conditions were first derived by Danckwerts (1953). They are valid if no dispersion or heat conduction takes place in the channel in front of and behind the exchanger, which is a realistic assumption. For the tubeside flow one has

$$t_2(0, z) = f_2(z) \qquad \text{for the cocurrent flow arrangement,} \tag{3.84a}$$

$t_2(1, z) = f_2(z)$ for the countercurrent flow arrangement. (3.84b)

The following general forms of initial temperature profiles are as before:

$$t_1(x, 0) = g_1(x) \tag{3.85a}$$

$$t_2(x, 0) = g_2(x) \tag{3.85b}$$

$$t_w(x, 0) = g_w(x) \tag{3.85c}$$

$$t_s(x, 0) = g_s(x). \tag{3.85d}$$

Transient processes of heat exchangers are induced by inlet temperature and/or flow-rate disturbances. If the flow rate is changed, the corresponding heat transfer coefficient in the transient process differs from that related to the previous state. Using the subscript '*r*' to express the initial state before step change of the flow velocity (from u_r to u) and defining $\sigma_1 = u_1/u_{1r}$ and $\sigma_2 = u_2/u_{2r}$, one can rewrite the thermal flow rates and some dimensionless parameters as

$$\dot{W}_1 = \sigma_1 \dot{W}_{1r},\ \dot{W}_2 = \sigma_2 \dot{W}_{2r},\ \tau_{r1} = \frac{\tau_{r1r}}{\sigma_1},\ \tau_{r2} = \frac{\tau_{r2r}}{\sigma_2},$$

$$R_1 = \frac{\sigma_1}{\sigma_2} R_{1r},\ R_\tau = \frac{\sigma_1}{\sigma_2} R_{\tau r}.$$

If there occur only temperature changes at inlets with constant velocities, $\sigma_1 = \sigma_2 = 1$. If the flow pattern remains the same, i.e., the laminar or turbulent flow before and after the step flow variation, the new heat transfer coefficients h_1 and h_2 can be given as

$$h_1 = \sigma_1^{n_1} h_{1r} \text{ and } h_2 = \sigma_2^{n_2} h_{2r},$$

where the exponent n_1 depends upon the value of the Reynolds number, the geometry, and the assembly of a tube bank. The mean value for in-line banks is $n_1 = 0.63$ and for staggered banks $n_1 = 0.6$ in the range $10^3 < Re_1 < 2 \times 10^5$ (Zukauskas and Ulinskas, 1988). The exponent n_2 depends upon the flow pattern inside the tubes and $n_2 = 0.8$ for the fully developed turbulent flow in a smooth circular tube (Incropera and DeWitt, 1985). Therefore, the new forms of the other parameters follow

$$U_1 = \sigma_1^{n_1-1} U_{1r}, \; U_2 = \sigma_2^{n_2-1} U_{2r}, \; \alpha_1 = \alpha_{1r} \sigma_1^{n_1-1}, \; \alpha_2 = \alpha_{2r} \frac{\sigma_2^{n_2}}{\sigma_1}.$$

Thus, the previously derived differential equations can be used to predict the transient responses to both inlet arbitrary temperature variations and step disturbance of flow rates. According to the given initial conditions, application of the Laplace transform to eqns (3.79)–(3.82) yields a set of ordinary differential equations in the image domain by using the Laplace parameter s with respect to the dimensionless time variable z

$$\frac{\mathrm{d}^2 T_1}{\mathrm{d}x^2} - Pe \frac{\mathrm{d}T_1}{\mathrm{d}x} - b_{11} T_1 - b_{12} T_2 = -G_1(x,s) \tag{3.86}$$

$$\frac{\mathrm{d}T_2}{\mathrm{d}x} - b_{21} T_1 - b_{22} T_2 = G_2(x,s) \tag{3.87}$$

$$T_w = \frac{\alpha_1 T_1 + \alpha_2 T_2}{\alpha_1 + \alpha_2 + R_w s} + G_w(x,s) \tag{3.88}$$

$$T_s = \frac{U_s T_1}{U_s + R_s R_w (1 + R_{c2}) s} + G_s(x,s) \tag{3.89}$$

where

$$b_{11}=\left[s+U_1+U_s-\frac{\alpha_1 U_1}{\alpha_1+\alpha_2+R_w s}-\frac{U_s}{U_s+R_s R_w\left(1+R_{c2}\right)s}\right]Pe,$$

$$b_{12}=-\frac{\alpha_2 U_1 Pe}{\alpha_1+\alpha_2+R_w s},$$

$$b_{21}=\frac{(-1)^{I+1}\alpha_1 U_2}{\alpha_1+\alpha_2+R_w s},$$

$$b_{22}=(-1)^{I+1}\left(\frac{\alpha_2 U_2}{\alpha_1+\alpha_2+R_w s}-R_\tau s-U_2\right),$$

$$G_1(x,s)=Pe\left[g_1(x)+U_1 G_w(x,s)+U_s G_s(x,s)\right],$$

$$G_2(x,s)=(-1)^{I+1}\left[R_\tau g_2(x)+U_2 G_w(x,s)\right],$$

$$G_w(x,s)=\frac{R_w g_w(x)}{\alpha_1+\alpha_2+R_w s},$$

$$G_s(x,s)=\frac{R_s R_w\left(1+R_{c2}\right)g_w(x)}{U_s+R_s R_w\left(1+R_{c2}\right)s}.$$

Introducing the mathematical operator D for "$\mathrm{d}/\mathrm{d}x$", one obtains an algebraic equation

$$\left[\mathrm{D}^3-\left(Pe+b_{22}\right)\mathrm{D}^2+\left(Peb_{22}-b_{11}\right)\mathrm{D}+b_{11}b_{22}-b_{12}b_{21}\right]T_1$$

$$=b_{12}G_2(x,s)-\left(\mathrm{D}-b_{22}\right)G_1(x,s). \tag{3.90}$$

It is easy to derive a general solution of the homogenous differential equation corresponding to the nonhomogeneous form (3.90):

$$T_1^* = A_1\,\mathrm{e}^{\beta_1 x} + A_2\,\mathrm{e}^{\beta_2 x} + A_3\,\mathrm{e}^{\beta_3 x} \tag{3.91}$$

where β_i ($i = 1$, 2 and 3) are three roots of the following equation

$$\beta^3 - (Pe + b_{22})\beta^2 - (Peb_{22} - b_{11})\beta + b_{11}b_{22} - b_{12}b_{21} = 0\,.$$

The nonhomogeneous part of eqn (3.90) should be known in advance. This part depends on the instant when inlet temperature variations and flow disturbances occur. It is a common case that these inlet changes begin after the previous steady state has been reached. Without loss of generality, the solution to eqn (3.90) is derived with such initial temperature distributions. In the light of the procedure of setting up eqns (3.79)–(3.82), one can determine the initial temperature profiles corresponding to the previous steady state

$$g_1(x) = a_1\,\mathrm{e}^{\lambda_1 x} + b_1\,\mathrm{e}^{\lambda_2 x} + \mu_3 \tag{3.92a}$$

$$g_2(x) = a_2\,\mathrm{e}^{\lambda_1 x} + b_2\,\mathrm{e}^{\lambda_2 x} + \mu_3 \tag{3.92b}$$

$$g_w(x) = m_1\,\mathrm{e}^{\lambda_1 x} + m_2\,\mathrm{e}^{\lambda_2 x} + \mu_3 \tag{3.92c}$$

$$g_s(x) = g_1(x) \tag{3.92d}$$

where $a_1 = \dfrac{\mu_1}{\lambda_1}$, $b_1 = \dfrac{\mu_2}{\lambda_2}$, $a_2 = \dfrac{\chi_1\,\mu_1}{\lambda_1(\chi_1 + \lambda_1)}$, $b_2 = \dfrac{\chi_1\,\mu_2}{\lambda_2(\chi_1 + \lambda_2)}$,

$$m_1 = \frac{a_1 R_{1r} U_{1r} + a_2 U_{2r}}{R_{1r} U_{1r} + U_{2r}}\,,\quad m_2 = \frac{b_1 R_{1r} U_{1r} + b_2 U_{2r}}{R_{1r} U_{1r} + U_{2r}}\,,$$

$$\mu_1 = \frac{1}{\dfrac{1}{\lambda_1} - \dfrac{1}{Pe_r} - \dfrac{\chi_1 w_4}{\lambda_1(\chi_1 + \lambda)_1} - \mathrm{e}^{\lambda_1 - \lambda_2}\left[\dfrac{1}{\lambda_1} - \dfrac{1}{Pe_r} - \dfrac{\chi_1 w_5}{\lambda_2(\chi_1 + \lambda_2)}\right]}\,,$$

$$\mu_2 = -\,\mathrm{e}^{\mu_1(\lambda_1-\lambda_2)}, \quad \mu_3 = -\left[\frac{\chi_1 w_4}{\lambda_1(\chi_1+\lambda_1)} - \mathrm{e}^{\lambda_1-\lambda_2}\frac{\chi_1 w_5}{\lambda_2(\chi_1+\lambda_2)}\right]\mu_1,$$

$$\lambda_{1,2} = \frac{Pe_r - \chi_1 \pm \sqrt{(\chi_1 - Pe_r)^2 + 4(\chi_1 Pe_r + \chi_2)}}{2},$$

$$\chi_1 = (-1)^{I+1}\frac{R_{1r}U_{1r}U_{2r}}{R_{1r}U_{1r}+U_{2r}}, \quad \chi_2 = \frac{U_{1r}U_{2r}}{R_{1r}U_{1r}+U_{2r}}Pe_r,$$

$$w_4 = \begin{cases} 1 & \text{for the cocurrent flow} \\ \mathrm{e}^{\lambda_1} & \text{for the countercurrent flow} \end{cases}$$

and

$$w_5 = \begin{cases} 1 & \text{for the cocurrent flow} \\ \mathrm{e}^{\lambda_2} & \text{for the countercurrent flow.} \end{cases}$$

Thus, functions $G_1(x,s)$ and $G_2(x,s)$ in eqns (3.86) and (3.87) are respectively expressed as

$$G_1(x,s) = \gamma_1\ \mathrm{e}^{\lambda_1 x} + \gamma_2\ \mathrm{e}^{\lambda_2 x} + \gamma_3 \tag{3.93}$$

$$G_2(x,s) = \rho_1\ \mathrm{e}^{\lambda_1 x} + \rho_2\ \mathrm{e}^{\lambda_2 x} + \rho_3 \tag{3.94}$$

where

$$\gamma_1 = Pe\left[a_1 + \frac{U_s R_s R_w(1+R_{c2})}{U_s + R_s R_w(1+R_{c2})s}a_1 + \frac{U_1 R_w}{\alpha_1+\alpha_2+R_w s}m_1\right],$$

$$\gamma_2 = Pe\left[b_1 + \frac{U_s R_s R_w(1+R_{c2})}{U_s + R_s R_w(1+R_{c2})s}b_1 + \frac{U_1 R_w}{\alpha_1+\alpha_2+R_w s}m_2\right],$$

$$\gamma_3 = Pe\mu_3\left[1 + \frac{U_s R_s R_w (1 + R_{c2})}{U_s + R_s R_w (1 + R_{c2}) s} + \frac{U_1 R_w}{\alpha_1 + \alpha_2 + R_w s}\right],$$

$$\rho_1 = (-1)^{I+1}\left(R_\tau a_2 + \frac{U_1 R_w}{\alpha_1 + \alpha_2 + R_w s} m_1\right),$$

$$\rho_2 = (-1)^{I+1}\left(R_\tau b_2 + \frac{U_1 R_w}{\alpha_1 + \alpha_2 + R_w s} m_2\right),$$

$$\rho_3 = (-1)^{I+1}\mu_3\left(R_\tau + \frac{U_1 R_w}{\alpha_1 + \alpha_2 + R_w s}\right).$$

Once the initial temperature profiles are derived, one is able to find the solution to the nonhomogeneous differential equations (3.86) and (3.87). Two common cases — transient responses to arbitrary inlet temperature variations and responses to both arbitrary temperature changes and step flow disturbance might be possible. In the first, it is easy to determine the final solution by means of the principle of superposition (Roetzel and Xuan, 1992d). In the second, however, this superposition method fails since the inlet temperatures and flow disturbances simultaneously occur, and the steady and transient temperature profiles cannot be described by the same system of differential equations. To derive the solution to the above-mentioned nonhomogeneous equations subject to all these disturbances, the method of variation of parameters (Boyce and DiPrima, 1977) can be applied to (3.86) and (3.87). According to the given boundary conditions and the previously derived general solution to the related homogeneous equation, the transformed temperature profiles are obtained:

$$T_1 = A_1 e^{\beta_1 x} + A_2 e^{\beta_2 x} + A_3 e^{\beta_3 x} + \frac{D_1}{A_0} e^{\lambda_1 x} + \frac{D_2}{A_0} e^{\lambda_2 x} - \frac{D_3}{A_0} \quad (3.95)$$

$$T_2 = A_1 p_1 \, e^{\beta_1 x} + A_2 \, e^{\beta_2 x} + A_3 p_2 \, e^{\beta_3 x} + p_4 \, e^{\lambda_1 x} + p_5 \, e^{\lambda_2 x} + p_6 \tag{3.96}$$

where

$$A_0 = \beta_2\beta_3(\beta_3 - \beta_2) - \beta_1\beta_3(\beta_3 - \beta_1) + \beta_1\beta_2(\beta_2 - \beta_1),$$

$$A_1 = \frac{1}{\Delta}\left\{E_1(s)\left[p_3 w_3 \beta_2 \, e^{\beta_2} - p_2 w_2 \beta_3 \, e^{\beta_3}\right]\right.$$

$$\left. -\left(1 - \frac{\beta_2}{Pe}\right)\left[p_3 H w_3 - E_2(s)\beta_3 \, e^{\beta_3}\right] + \left(1 - \frac{\beta_3}{Pe}\right)\left[p_2 H w_2 - E_2(s)\beta_2 \, e^{\beta_2}\right]\right\},$$

$$A_2 = \frac{1}{\Delta}\left\{\left(1 - \frac{\beta_1}{Pe}\right)\left[p_3 H w_3 - E_2(s)\beta_3 \, e^{\beta_3}\right]\right.$$

$$\left. - E_1(s)\left[p_3 w_3 \beta_1 \, e^{\beta_1} - p_1 w_1 \beta_3 \, e^{\beta_3}\right] + \left(1 - \frac{\beta_3}{Pe}\right)\left[E_2(s)\beta_1 \, e^{\beta_1} - p_1 H w_1\right]\right\},$$

$$A_3 = \frac{1}{\Delta}\left\{\left(1 - \frac{\beta_1}{Pe}\right)\left[E_2(s)\beta_2 \, e^{\beta_2} - p_2 H w_2\right]\right.$$

$$\left. + E_1(s)\left[p_2 w_2 \beta_1 \, e^{\beta_1} - p_1 w_1 \beta_2 \, e^{\beta_2}\right] - \left(1 - \frac{\beta_2}{Pe}\right)\left[E_2(s)\beta_1 \, e^{\beta_1} - p_1 H w_1\right]\right\},$$

$$\Delta = \left(1 - \frac{\beta_1}{Pe}\right)\left(p_3 w_3 \beta_2 \, e^{\beta_2} - p_2 w_2 \beta_3 \, e^{\beta_3}\right)$$

$$-\left(1 - \frac{\beta_2}{Pe}\right)\left(p_3 w_3 \beta_1 \, e^{\beta_1} - p_1 w_1 \beta_3 \, e^{\beta_3}\right)$$

$$+\left(1 - \frac{\beta_3}{Pe}\right)\left(p_2 w_2 \beta_1 \, e^{\beta_1} - p_1 w_1 \beta_2 \, e^{\beta_2}\right),$$

$$D_1 = \left(\frac{\beta_3 - \beta_2}{\lambda_1 - \beta_1} + \frac{\beta_1 - \beta_3}{\lambda_1 - \beta_2} + \frac{\beta_2 - \beta_1}{\lambda_1 - \beta_3} \right) \left(b_{22}\gamma_1 - \gamma_1 \lambda_1 + b_{12}\rho_1 \right),$$

$$D_2 = \left(\frac{\beta_3 - \beta_2}{\lambda_2 - \beta_1} + \frac{\beta_1 - \beta_3}{\lambda_2 - \beta_2} + \frac{\beta_2 - \beta_1}{\lambda_2 - \beta_3} \right) \left(b_{22}\gamma_2 - \gamma_2 \lambda_2 + b_{12}\rho_2 \right),$$

$$D_3 = \left(\frac{\beta_3 - \beta_2}{\beta_1} + \frac{\beta_1 - \beta_3}{\beta_2} + \frac{\beta_2 - \beta_1}{\beta_3} \right) \left(b_{22}\gamma_3 + b_{12}\rho_3 \right),$$

$$E_1(s) = F_1(s) + \frac{D_1}{A_0}\left(\frac{\lambda_1}{Pe} - 1 \right) + \frac{D_2}{A_0}\left(\frac{\lambda_2}{Pe} - 1 \right) + \frac{D_3}{A_0},$$

$$E_2(s) = F_2(s) - p_4 w_4 - p_5 w_5 - p_6,$$

$$H = -\frac{D_1 \lambda_1}{A_0} \mathrm{e}^{\lambda_1} - \frac{D_2 \lambda_2}{A_0} \mathrm{e}^{\lambda_2},$$

$$p_1 = \frac{\beta_1^2 - Pe\beta_1 - b_{11}}{b_{12}}, \quad p_2 = \frac{\beta_2^2 - Pe\beta_2 - b_{11}}{b_{12}}, \quad p_3 = \frac{\beta_3^2 - Pe\beta_3 - b_{11}}{b_{12}},$$

$$p_4 = \frac{D_1\left(\lambda_1^2 - Pe\lambda_1 - b_{11}\right)}{A_0 b_{12}} + \frac{\gamma_1}{b_{12}}, \quad p_5 = \frac{D_5\left(\lambda_2^2 - Pe\lambda_2 - b_{11}\right)}{A_0 b_{12}} + \frac{\gamma_2}{b_{12}},$$

$$p_6 = \frac{b_{11} D_3}{A_0 b_{12}} + \frac{\gamma_3}{b_{12}}.$$

For the cocurrent flow: $w_1 = w_2 = w_3 = 1$; For the countercurrent flow: $w_1 = \mathrm{e}^{\beta_1}$, $w_2 = \mathrm{e}^{\beta_2}$, $w_3 = \mathrm{e}^{\beta_3}$.

Thus, the transformed temperature profiles in the image domain have been determined according to the nonzero initial conditions (3.92). If the initial conditions take other forms, one can find the solution by using a similar procedure. Evidently, it is almost

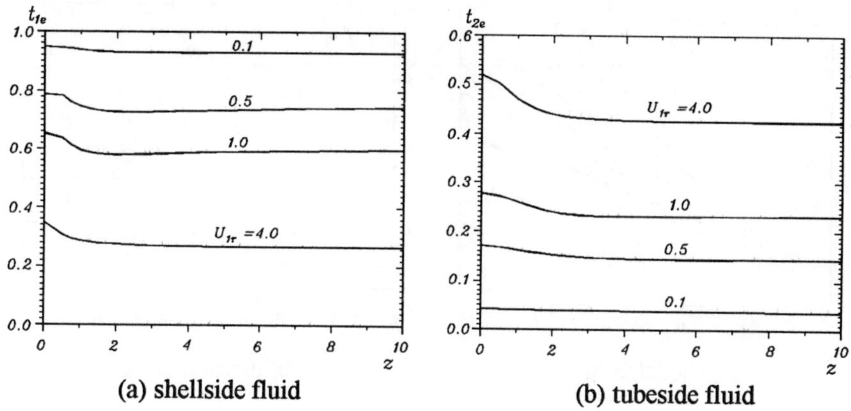

Figure 3.17: Responses of exit temperatures to step flow changes on both sides with countercurrent flow arrangement ($R_{1r} = 0.8$, $R_w = 0.5$, $R_s = 0.2$, $U_s/U_1 = 0.15$, $Pe = 4$, $f_1(z) = 1$, $f_2(z) = 0$, $\sigma_1 = 0.8$ and $\sigma_2 = 1.1$)

impossible to perform inverse Laplace transform analytically. The numerical inverse transform is used to obtain the transient behaviour of exchangers in the real time-domain. As before, the calculation experience recommends two adaptable algorithms of the inversion — the Gaver–Stehfest algorithm (Stehfest, 1970) and the Fourier-series method (Ichikawa and Kiswhima, 1972). According to the forms of inlet disturbance, one can choose either of them.

Following the previously outlined procedure, one can determine the transient performance of heat exchangers subject to arbitrary temperature variations and step flow disturbances, with consideration of the effect of flow maldistribution, heat capacities of both fluids, tube bank and the shell. These disturbances may occur at inlets separately or simultaneously on both sides. Under the conditions that $U_{1r} = U_{2r}$, $R_r = 1$, $Pe_r = Pe$, $n_1 = 0.6$ and $n_2 = 0.8$, some examples are calculated. Figure 3.17 illustrates the exit responses to step change of velocities on both sides with counterflow arrangement and Figure 3.18 shows the responses to linear temperature variations and step flow disturbances which take place simultaneously on both sides of a cocurrent flow exchanger.

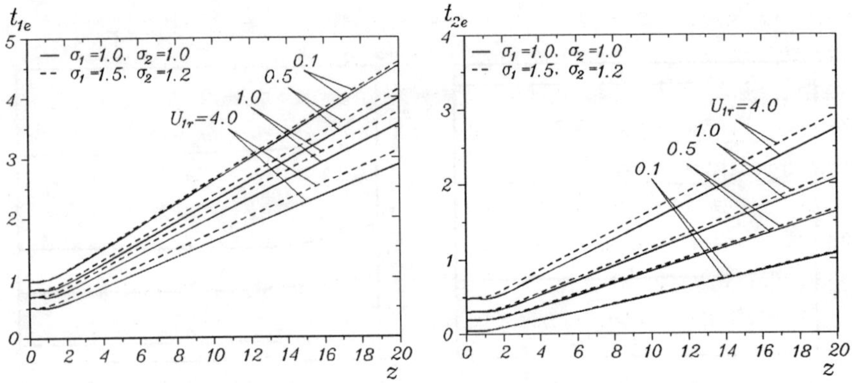

(a) shellside fluid (b) tubeside fluid

Figure 3.18: Responses of exit temperature to linear temperature variations and step flow changes on both side with cocurrent flow arrangement ($R_{1r}=1$, $R_w=0.3$, $U_s/U_1=0.2$, $Pe=6$, $f_1(z)=1+0.2z$, $f_2(z)=0.05z$)

In these figures, the initial states corresponds to the previous steady state which exactly satisfies the energy balance, that is, $t_{2e}=R_{1r}\left(1-t_{1e}\right)$ at z=0. Figure 3.17 shows that a transient process quickly approaches a new steady state under the given conditions. The energy balance between both fluids is also satisfied at this new steady state, which numerically demonstrates the feasibility that one can use the presented method to predict the transient responses to arbitrary temperature variations and step flow disturbances. As an example, Fig. 3.18 shows the superposition of effects of temperature variations and step flow disturbances on the dynamic responses of an apparatus. Under the conditions used in the computation the influence of temperature changes is evidently dominant and the effect of flow disturbances increases with the values of U_{1r} (or U_{2r}).

3.6.2 Multipass shell-and-tube heat exchangers

The scheme of multipass shell-and-tube heat exchangers is shown in Fig. 3.6, in which different possible flow arrangements are presented and the origin-point of the co-ordinate system is always located at the entrance of shellside fluid. In these 1-*N* shell-and-tube heat exchangers, generally, flow maldistribution can take place on both shell and tube sides. The complicated shellside geometrical structure and manufacturing clearances induce non-uniform distributions of fluid such as leakage, bypassing and backmixing. On the other hand, the tubeside fluid may not be evenly distributed among parallel channels in a pass because of different pressure drops and possible manufacturing flaws among channels, which is more severe for laminar flow. For fixed-tubesheet or floating-head types of exchangers, the tube-side fluid is stirred in the headers and this effect on steady and transient processes cannot be described by the conventional plug-flow model. In general, poor flow distribution causes degradation in performance of exchangers, especially for heat exchangers with greater values of number of transfer units *NTU*.

Taking the shellside and tubeside flow maldistributions into account, Roetzel and Xuan (1992d) as well as Xuan and Roetzel (1993b) applied the dispersion model to simulation of multipass shell-and-tube heat exchangers subject to arbitrary temperature variations and step flow disturbances. They provided a versatile solution which includes the effects of heat capacities of both fluids, capacities of shell and tube wall, different flow arrangements, arbitrary number and different size of tube passes, and nonzero initial conditions. With the dispersion model, auxiliary variables $\tilde{\theta}_1$ and $\tilde{\theta}_{2i}$ $(i = 1, 2, \ldots, N)$ are introduced as

$$\dot{W}_1\tilde{\theta}_1 = \dot{W}\theta_1 - A_{c1}\mathcal{D}_1\frac{\partial\theta_1}{\partial l}$$

$$\dot{W}_{2i}\tilde{\theta}_{2i} = \dot{W}\theta_{2i} \pm A_{c2}\mathcal{D}_2\frac{\partial\theta_{2i}}{\partial l}\,. \tag{3.97}$$

In these two expressions the first term on the right-hand side corresponds to the convective part for the plug-flow and the second term the dispersive part for the effect of flow maldistribution. Thus, the governing equations are written:

$$A_{c1}\mathcal{D}_1\frac{\partial^2\theta_1}{\partial l^2}-\dot{W}_1\frac{\partial\theta_1}{\partial l}-\frac{C_1}{L}\frac{\partial\theta_1}{\partial\tau}-\sum_{i=1}^{N}\frac{(hA)_1}{L}(\theta_1-\theta_{wi})$$

$$-\frac{(hA)_s}{L}(\theta_1-\theta_s)=0 \tag{3.98}$$

$$A_{c2i}\mathcal{D}_{2i}\frac{\partial^2\theta_{2i}}{\partial l^2}+(-)^{i+I}\dot{W}_{2i}\frac{\partial\theta_{2i}}{\partial l}-\frac{C_{2i}}{L}\frac{\partial\theta_{2i}}{\partial\tau}-\frac{(hA)_{2i}}{L}(\theta_{2i}-\theta_{wi})=0. \tag{3.99}$$

where $I=0$ and $I=1$ represent the flow arrangements I and II, respectively. Equation (3.99) is for any tube pass $(i=1,2,\ldots,N)$. With respect to the influence of heat capacities of shell and tube walls, the corresponding energy equations are as follows:

$$\frac{C_{wi}}{L}\frac{\partial\theta_{wi}}{\partial\tau}-\frac{(hA)_{wi}}{L}(\theta_1-\theta_{wi})-\frac{(hA)_{2i}}{L}(\theta_{2i}-\theta_{wi})=0 \tag{3.100}$$

$$\frac{C_s}{L}\frac{\partial\theta_s}{\partial\tau}-\frac{(hA)_s}{L}(\theta_1-\theta_s)=0 \tag{3.101}$$

The axial (or longitudinal) dispersion coefficient (or apparent heat conductivity) $\mathcal{D}$ consists of two parts: the heat conductivity k_f of fluid and the macro-eddy diffusion coefficient $\mathcal{D}_f$ caused by flow maldistribution, i.e., $\mathcal{D}=\mathcal{D}_f+k_f$. If the flow is pure plug-flow and no eddy diffusion occurs, $\mathcal{D}_f=0$. With regard to baffles and end plates, one can approximately distribute the heat capacities of these components among all tube passes. In this case, C_{wi} is replaced by the apparent heat capacity C^*_{wi} obtained from eqn (3.53). As pointed out previously, the thermal flow rate $\dot{W}_{2i}$ does not change from pass to pass, but the heat capacity C_{2i} may be different among tube passes according to the given geometric dimension of a heat exchanger. These equations take the following dimensionless forms:

$$\frac{1}{Pe_1}\frac{\partial^2 t_1}{\partial x^2}-\frac{\partial t_1}{\partial x}-\frac{\partial t_1}{\partial z}-\sum_{i=1}^{N}U_{1i}\left(t_1-t_{wi}\right)-U_s\left(t_1-t_s\right)=0 \tag{3.102}$$

$$\frac{1}{Pe_{2i}}\frac{\partial^2 t_{2i}}{\partial x^2}+\left(-1\right)^{i+1}\frac{\partial t_{2i}}{\partial x}-\varepsilon_{ci}R_\tau\frac{\partial t_{2i}}{\partial z}-U_{2i}\left(t_{2i}-t_{wi}\right)=0$$

$$(i=1,\ 2,\ \dots,\ N) \tag{3.103}$$

$$R_{wi}\frac{\partial t_{wi}}{\partial z}-\alpha_{1i}\left(t_1-t_{wi}\right)-\alpha_{2i}\left(t_{2i}-t_{wi}\right)=0 \quad (i=1,\ 2,\ \dots,\ N) \tag{3.104}$$

$$R_s R_w\left(1+R_{c2}\right)\frac{\partial t_s}{\partial z}-U_1\left(t_1-t_s\right)=0. \tag{3.105}$$

The Péclet numbers $Pe=\left(\dot{W}L\right)/\left(A_c\mathcal{D}\right)$ quantitatively describe the effect of flow maldistributions on transient behaviour. For the plug-flow $\mathcal{D}=0$ (heat conduction and dispersion in a fluid is negligible) and $Pe\to\infty$, eqns (3.102) and (3.103) are reduced to the same form derived by the ideal plug-flow model in Section 3.3.1. The actual flow pattern in multi-pass heat exchangers interposes between axially unmixed plug-flow and perfect axial mixing and the value of Pe lies in the range of $0<Pe<\infty$. The general forms of initial temperature are expressed by

$$t_1(x,0)=g_1(x) \tag{3.106}$$

$$t_{2i}(x,0)=g_{2i}(x) \qquad (i=1,\ 2,\ \dots,\ N) \tag{3.107}$$

$$t_{wi}(x,0)=g_{wi}(x) \qquad (i=1,\ 2,\ \dots,\ N) \tag{3.108}$$

$$t_s(x,0)=g_s(x). \tag{3.109}$$

All these functions are the temperature distributions at the instant of a new transient process. Equating the flux due to convection just outside the entrance (or exit) of heat exchangers to that due to convection and dispersion just inside the entrance (or exit) of the exchangers (Danckwerts, 1953), one obtains the suitable boundary conditions pertinent to the dispersion model. The shellside boundary conditions have the similar form to eqn (3.83)

$$t_1 - \frac{1}{Pe_1}\frac{\partial t_1}{\partial x} = f_1(z) \quad \text{at } x = 0 \quad \text{and} \quad \frac{\partial t_1}{\partial x} = 0 \quad \text{at } x = 1 \tag{3.110}$$

The tubeside boundary conditions vary with the flow arrangement and can be expressed as

$$\begin{aligned} & t_{21} + \frac{(-1)^I}{Pe_{21}}\frac{\partial t_{21}}{\partial x} = f_2(z) \quad \text{at } x = 1 - I \\ & \frac{\partial t_{2N}}{\partial x} = 0 \qquad\qquad\qquad \text{at } x = \frac{1}{2}\left[1 + (-1)^{N+I}\right] \end{aligned} \tag{3.111a}$$

The arbitrary shellside and tubeside temperature variations $f_1(z)$ and $f_2(z)$ may take place simultaneously or separately. Evidently, $(2N+2)$ boundary conditions are necessary to solve the whole system of eqns (3.102)–(3.105). The other $(2N-2)$ conditions are interfacial conditions between adjacent tube passes as location $x = 0$ and $x = 1$ for the temperature $t_{2i}(x,z)$ and the corresponding derivative,

$$t_{2i} = t_{2,i-1}, \quad \frac{1}{Pe_{2i}}\frac{\partial t_{2i}}{\partial x} = -\frac{1}{Pe_{2,i-1}}\frac{\partial t_{2,i-1}}{\partial x} \quad \text{at } x = \frac{1}{2}\left[1 - (-1)^{i+I}\right]$$
$$(i = 2, 3, \ldots, N) \tag{3.111b}$$

If step disturbances of flow rates occur, a method similar to that in Section 3.6.1 is used. The transformed forms of eqns (3.102)–(3.105) in the image-domain are as follows:

$$\frac{d^2 T_1}{dx^2} - Pe_1 \frac{dT_1}{dx}$$

$$= Pe_1 \left[s + U_1 + U_s - \sum_{i=1}^{N} \frac{\varepsilon_{1i}\alpha_{1i}U_1}{\alpha_{1i} + \alpha_{2i} + R_{wi}s} - \frac{U_s^2}{R_s R_w (1 + R_{c2})s + U_s} \right] T_1$$

$$- Pe_1 \sum_{i=1}^{N} \frac{\varepsilon_{2i}\alpha_{2i}U_1}{\alpha_{1i} + \alpha_{2i} + R_{wi}s} T_{2i} + h_1(x, s) \tag{3.112}$$

$$\frac{d^2 T_{2i}}{dx^2} + (-1)^{i+I} Pe_{2i} \frac{dT_{2i}}{dx} = -\frac{Pe_{2i}\varepsilon_{2i}\alpha_{1i}U_2}{\alpha_{1i} + \alpha_{2i} + R_{wi}s} T_1$$

$$- Pe_{2i} \left(\frac{\varepsilon_{2i}\alpha_{2i}U_2}{\alpha_{1i} + \alpha_{2i} + R_{wi}s} - \varepsilon_{ci}R_\tau s - U_2\varepsilon_{2i} \right) T_{2i} + h_{2i}(x, s)$$

$$(i = 1,\ 2,\ \ldots,\ N) \tag{3.113}$$

$$T_{wi} = \frac{\alpha_{1i}T_1 + \alpha_{2i}T_{2i}}{\alpha_{1i} + \alpha_{2i} + R_{wi}s} + h_{wi}(x, s) \qquad (i = 1,\ 2,\ \ldots,\ N) \tag{3.114}$$

$$T_s = \frac{U_s T_1}{U_s + R_s R_w (1 + R_{ci})s} + h_s(x, s) \tag{3.115}$$

where

$$h_1(x, s) = -Pe_1 \left[g_1(x) + \sum_{i=1}^{N} \frac{U_1 R_{wi}}{\alpha_{1i} + \alpha_{2i} + R_{wi}s} g_{wi}(x) \right.$$

$$\left. + \frac{R_s R_w (1 + R_{c2})}{U_s + R_s R_w (1 + R_{c2})s} g_s(s) \right],$$

$$h_{2i}(x,s) = -Pe_{2i}\left[\varepsilon_{ci}R_\tau g_{2i}(x) + \frac{U_{2i}R_{wi}}{\alpha_{1i}+\alpha_{2i}+R_{wi}s}g_{wi}(x)\right],$$

$$h_{wi}(x,s) = \frac{R_{wi}}{\alpha_{1i}+\alpha_{2i}+R_{wi}s}g_{wi}(x),$$

$$h_s(x,s) = \frac{R_sR_w(1+R_{c2})}{U_s+R_sR_w(1+R_{c2})s}g_s(s).$$

By introducing the following variables

$$Y_1 = \frac{\mathrm{d}T_1}{\mathrm{d}x} \quad \text{and} \quad Y_{2i} = \frac{\mathrm{d}T_{2i}}{\mathrm{d}x} \quad (i = 1,\ 2,\ \ldots,\ N)$$

eqns (3.112)–(3.113) are replaced by $(2N+2)$ first-order differential equations which are expressed in matrix notation

$$\frac{\mathrm{d}\mathbf{T}}{\mathrm{d}x} = \mathbf{AT} + \mathbf{H}(x) \tag{3.116}$$

where vectors

$$\mathbf{T} = (T_{21},\ T_{22},\ \ldots,\ T_{2N},\ T_1,\ Y_{21},\ Y_{22},\ \ldots,\ Y_{2N},\ Y)^{\mathrm{T}}$$

and

$$\mathbf{H}(x) = [0,\ 0,\ \ldots,\ 0,\ h_{21}(x,\ s),\ h_{22}(x,\ s),\ \ldots,\ h_{2N}(x,\ s),\ h_1(x,\ s)]^{\mathrm{T}}$$

and **A** is a $(2N+2)\times(2N+2)$ matrix whose elements are given by eqns (3.111) and (3.112). Using the similar procedure described in Section 3.3.2, one obtains a general form of the solution to eqn (3.116)

$$\mathbf{T} = \mathrm{e}^{\mathbf{A}x}\,\mathbf{D}^{*} + \mathrm{e}^{\mathbf{A}x}\int \mathrm{e}^{-\mathbf{A}x'}\mathbf{H}(x')\,\mathrm{d}x' \tag{3.117a}$$

or

$$\mathbf{T} = \mathbf{U}\,\mathrm{e}^{\mathbf{L}x}\,\mathbf{D} + \int \mathbf{U}\,\mathrm{e}^{\mathbf{L}(x-x')}\,\mathbf{U}^{-1}\mathbf{H}(x')\,\mathrm{d}x' \tag{3.117b}$$

where the diagonal matrix $\mathrm{e}^{\mathbf{L}x} = \mathrm{diag}\{\mathrm{e}^{\lambda_1 x},\ \mathrm{e}^{\lambda_2 x},\ \ldots,\ \mathrm{e}^{\lambda_{2N+2} x}\}$, $\mathbf{U}$ is a $(2N+2)\times(2N+2)$ matrix whose columns are the eigenvectors $\{\mathbf{u}_j\}$ $(j = 1,\ 2,\ \ldots,\ 2N+2)$ which correspond to the eigenvalues $\{\lambda_j\}$ $(j = 1,\ 2,\ \ldots,\ 2N+2)$ of **A**, and **D** (or $\mathbf{D}^{*}$) is a coefficient vector which is determined according to the boundary and interfacial conditions. The necessary condition for the validity of the solution (3.117) is that all eigenvalues are distinct. In general, the eigenvalues of **A** in eqn (3.116) describing transient behaviour of multipass shell-and-tube heat exchangers are different from each other. If there exist some eigenvalues of multiplicity m $(m > 1)$, one must build a new solution to eqn (3.116) and it is a somewhat awkward task. A simple approximate method can avoid this difficulty. Introducing an infinitesimal parameter δ (for example, $\delta = 10^{-4}$), one alters the value of ε_i by adding or subtracting δ under the constraint $\sum_{i=1}^{N} \varepsilon_i = 1$, so that one may have $(2N+2)$ distinct eigenvalues again. Such a slight change causes no remarkable changes of calculated temperature profiles and the overall accuracy is not affected.

Insertion of the boundary and interfacial conditions into eqn (3.117) yields

$$\mathbf{WD} = \mathbf{F} + \mathbf{S} \tag{3.118}$$

where **W** is a $(2N+2)\times(2N+2)$ matrix whose elements are prescribed by the aforementioned determinant conditions. It varies with the number N of tube passes as well as the flow arrangement.

Vector $\mathbf{F} = \left[0,\ 0,\ \dots,\ F_2(s),\ F_1(s)\right]^{\mathrm{T}}$ if the inlet conditions are placed in the last two rows of matrix equation (3.118), and vector $\mathbf{S} = \left(s_1,\ s_2,\ \dots,\ s_{2N+2}\right)^{\mathrm{T}}$ is obtained from the second term on the right-hand side of eqn (3.117), i.e., $s_i = 0$ or s_i takes the integration value from $x = 0$ to $x = 1$. The position of s_i in vector $\mathbf{S}$ corresponds to arrangement order of matrix $\mathbf{W}$. Thus, the coefficient vector $\mathbf{D}$ follows

$$\mathbf{D} = \mathbf{W}^{-1}(\mathbf{F} + \mathbf{S}). \tag{3.119}$$

So far the transformed temperature profiles and temperature gradients of both fluids in the Laplace image domain have been found. Their explicit expressions are

$$T_i(x,s) = \sum_{j=1}^{2N+2} d_j u_{ij}\, \mathrm{e}^{\lambda_j x} + \sum_{k\neq 1}^{2N+2} \sum_{j=1}^{2N+2} u_{ij} v_{jk} \int \mathrm{e}^{\lambda_j (x-x')} h_k(x')\,\mathrm{d}x' \quad \left(i = 1,\ 2,\ \dots,\ 2N+2\right) \tag{3.120}$$

where d_j, u_{ij}, v_{jk} and $h_k(x)$ are elements of $\mathbf{D}$, $\mathbf{U}$, $\mathbf{U}^{-1}$ and $\mathbf{H}(x)$, respectively. From this expression, one can easily determine the transformed responses at exits. The related profiles of the tube wall and shell are already given by eqns (3.114) and (3.115). Functions $h_k(x)$ in (3.120) consist of initial temperature distributions $g_1(x)$, $g_s(x)$, $g_{2i}(x)$ and $g_{wi}(x)$. These distributions may appear in different forms according to the instant state of transient process. Three typical forms of the initial temperature profiles are described as follows:

(1) Uniform initial conditions

This type of initial conditions is the simplest case. Zero conditions $g_1(x) = g_s(x) = 0$ and $g_{2i}(x) = g_{wi}(x) = 0$ make eqn (3.116) homogeneous and its general solution is

$$\mathbf{T} = \mathbf{U}\,\mathrm{e}^{\mathbf{L}x}\,\mathbf{D}. \tag{3.121}$$

(2) Nonzero steady-state distributions for inlet temperature variations

When only temperature variations take place at inlets after the preceding process has reached a steady state, one can find the solution vector **T** by means of the principle of superposition. In this case, thermal flow rates $\dot{W}_1$ and $\dot{W}_{2i}$ as well as heat transfer coefficients h_1 and h_{2i} are constant and the same system of governing differential equations can be used to describe both the previous steady-state process (by eliminating time derivatives of temperature) and transient process.

Assuming

$$t_m(x,z) = g_m(x) + \phi_m(x,z) \quad \text{or} \quad T_m(x,s) = g_m(x) + \Phi_m(x,s)$$

$$(m = 1,\ s,\ 2i \text{ or } wi) \qquad (3.122)$$

one inserts these expressions into eqns (3.102)–(3.105) and the pertinent determinant conditions, bearing in mind the fact that $g_m(x)$ is the steady-state solution of the preceding process as $z \to \infty$. Therefore, transient part $\phi_m(x,z)$ is described by

$$\frac{1}{Pe_1}\frac{\partial^2 \phi_1}{\partial x^2} - \frac{\partial \phi_1}{\partial x} - \frac{\partial \phi_1}{\partial z} - \sum_{i=1}^{N} U_{1i}(\phi_1 - \phi_{wi}) - U_s(\phi_1 - \phi_s) = 0 \qquad (3.123)$$

$$\frac{1}{Pe_{2i}}\frac{\partial^2 \phi_{2i}}{\partial x^2} + (-1)^{i+1}\frac{\partial \phi_{2i}}{\partial x} - \varepsilon_{ci} R_\tau \frac{\partial \phi_{2i}}{\partial z} - U_{2i}(\phi_{2i} - \phi_{wi}) = 0$$

$$(i = 1,\ 2,\ \ldots,\ N) \qquad (3.124)$$

$$R_{wi}\frac{\partial \phi_{wi}}{\partial z} - \alpha_{1i}(\phi_1 - \phi_{wi}) - \alpha_{2i}(\phi_{2i} - \phi_{wi}) = 0 \qquad (i = 1,\ 2,\ \ldots,\ N)$$

$$(3.125)$$

$$R_s R_w (1 + R_{c2})\frac{\partial \phi_s}{\partial z} - U_1(\phi_1 - \phi_s) = 0. \qquad (3.126)$$

The zero initial conditions for $\phi_m(x,z)$ are

$$\phi_1(x,0) = \phi_s(x,0) = 0 \text{ and } \phi_{2i}(x,0) = \phi_{wi}(x,0) = 0 \qquad (i = 1, 2, \ldots, N) \qquad (3.127)$$

and the boundary conditions yield

$$\phi_1 - \frac{1}{Pe_1}\frac{\partial \phi_1}{\partial x} = f_1(z) - \left(g_1 - \frac{1}{Pe_1}\frac{\mathrm{d}g_1}{\mathrm{d}x}\right) \quad \text{at } x = 0 \qquad (3.128a)$$

$$\frac{\partial \phi_1}{\partial x} = -\frac{\mathrm{d}g_1}{\mathrm{d}x} \quad \text{at } x = 1 \qquad (3.128b)$$

$$\phi_{21} + \frac{(-1)^I}{Pe_{21}}\frac{\partial \phi_{21}}{\partial x} = f_2(z) - \left[g_{21} + \frac{(-1)^I}{Pe_{21}}\frac{\mathrm{d}g_{21}}{\mathrm{d}x}\right] \quad \text{at } x = 1 - I \qquad (3.129a)$$

$$\frac{\partial \phi_{2N}}{\partial x} = -\frac{\mathrm{d}g_{2N}}{\mathrm{d}x} \quad \text{at } x = \frac{1}{2}\left[1 + (-1)^{N+I}\right]. \qquad (3.129b)$$

In general, the steady-state solutions $g_1(x)$ and $g_{2i}(x)$ are found at the inlet conditions:

$$g_1 - \frac{1}{Pe_1}\frac{\mathrm{d}g_1}{\mathrm{d}x} = 1 \quad \text{at } x = 0$$

$$\frac{\partial \phi_1}{\partial x} = -\frac{\mathrm{d}g_1}{\mathrm{d}x} \quad \text{at } x = 1$$

$$g_{21} + \frac{(-1)^I}{Pe_{21}}\frac{\mathrm{d}g_{21}}{\mathrm{d}x} = 0 \quad \text{at } x = 1 - I$$

$$\frac{\mathrm{d}g_{2N}}{\mathrm{d}x} = 0 \quad \text{at } x = \frac{1}{2}\left[1 + (-1)^{N+I}\right].$$

Thus, the boundary conditions for $\phi_1(x,z)$, $\phi_{21}(x,z)$ and $\phi_{2N}(x,z)$ are simplified as

$$\phi_1 - \frac{1}{Pe_1}\frac{\partial \phi_1}{\partial x} = f_1(z) - 1 \qquad \text{at } x = 0 \tag{3.128c}$$

$$\frac{\partial \phi_1}{\partial x} = 0 \qquad \text{at } x = 1 \tag{3.128d}$$

$$\phi_{21} + \frac{(-1)^I}{Pe_{21}}\frac{\partial \phi_{21}}{\partial x} = f_2(z) \qquad \text{at } x = 1 - I \tag{3.129c}$$

$$\frac{\partial \phi_{2N}}{\partial x} = 0 \qquad \text{at } x = \frac{1}{2}\left[1 + (-1)^{N+I}\right]. \tag{3.129d}$$

By means of the Laplace transform, one can obtain without question a homogeneous system of ordinary differential equations according to the uniform initial conditions (3.127). Following the previously described procedure, one is able to find a transformed solution $\Phi_m(x,s)$ whose form is similar to eqn (3.121), and then $T_m(x,s)$ from eqn (3.122). In this way, the tedious task of calculating the integration in eqn (3.120) is avoided.

(3) Nonzero steady-state distributions for inlet temperature and flow rate variations

Because of flow disturbances, thermal flow rate $\dot{W}_1$ and $\dot{W}_{2i}$ as well as heat transfer coefficients h_1 and h_{2i} are subjected to corresponding changes. The preceding treatment for pure temperature variations is not applicable. Here one should resort to expression (3.120). In this case, general forms of the initial distribution of both fluids can be expressed as

$$g_1(x) = \sum_{j=1}^{2N+2} c_j b_{N+1,j}\, e^{\beta_j x} \tag{3.130a}$$

$$g_{2i}(x) = \sum_{j=1}^{2N+2} c_j b_{ij} \, e^{\beta_j x} \qquad (i = 1,\ 2,\ \dots,\ N) \qquad (3.130b)$$

where $\{\beta_j\}$ and $\{\mathbf{B}_j\}$ are $(2N+2)$ eigenvalues and the corresponding eigenvectors. They are functions of the operating parameters of the previous process. $\{c_j\}$ are $(2N+2)$ coefficients determined from the subjected boundary and interfacial conditions related to the previous process. The steady-state temperature profiles in shell and tube wall are:

$$g_s(x) = g_1(x) \qquad (3.130c)$$

$$g_{wi}(x) = \frac{U_{1ir} g_1(x) + U_{2ir} g_{2i}(x)}{U_{1ir} + U_{2ir} R_{2r}} \qquad (i = 1,\ 2,\ \dots,\ N). \qquad (3.130d)$$

Henceforth, the nonhomogeneous parts in matrix equation (3.116) are already known. Having carried out some derivations, one finds the explicit transformed transient responses resulting simultaneously from both arbitrary temperature variations and step flow rate disturbances under the nonzero initial conditions which coincide with the steady-state temperature distributions of the previous process

$$T_i(x,s) = \sum_{j=1}^{2N+2} d_j u_{ij} \, e^{\lambda_j x} + \sum_{m=1}^{2N+2} \sum_{k=1}^{2N+2} u_{ik} v_{km} \sum_{j=1}^{2N+2} \omega_{mj} c_j \frac{e^{\beta_j x}}{\beta_j - \lambda_k}$$

$$(i = 1,\ 2,\ \dots,\ 2N+2) \qquad (3.131)$$

where

$$\omega_{mj} = -Pe_{2m} \left[\varepsilon_{cm} R_\tau b_{mj} + \frac{U_{2m} R_{wm} \left(U_{1m\,r}\, b_{N+1,j} + U_{2m\,r} R_{2r} b_{mj} \right)}{\left(\alpha_{1m} + \alpha_{2m} + R_{wm} s \right) \left(U_{1m\,r} + U_{2m\,r} R_{2r} \right)} \right]$$

$$\left(1 \le m \le N,\ 1 \le j \le 2N+2\right)$$

$$\omega_{N+1,j} = -Pe_1\left\{\left[1+\frac{U_s R_s R_w(1+R_{c2})}{U_s + R_s R_w(1+R_{c2})s}\right]b_{N+1,j}\right.$$

$$\left.+\sum_{i=1}^{N}\frac{U_{1i}R_{wi}(U_{1ir}b_{N+1,j}+U_{2ir}R_{2r}b_{ij})}{(\alpha_{1i}+\alpha_{2i}+R_{wi}s)(U_{1ir}+U_{2ir}R_{2r})}\right\} \qquad (1 \le j \le 2N+2)$$

$\omega_{mj} = 0$ for the other m and $1 \le j \le 2N+2$.

The previously derived solutions related to the different cases are all in the image-domain. The numerical inverse transform are to be applied to obtaining real-time transient responses of heat exchangers. The inverse transform results as $z \to \infty$ correspond to new steady-state temperature distributions. Under the following given parameters $R_\tau = 1$, $U_1 = U_2$, $R_s = 0.2$ and $U_s = 0.1U_1$, some examples are computed and illustrated in Figs 3.19–3.21. Figure 3.19 describes the shellside temperature $t_1(x, z)$ subject to a step temperature change in a countercurrent flow exchanger. The curved surface in the diagram evidently shows how the temperature wave propagates from the inlet to the exit (from the high to the low). The temperature steeply increases in the range $z < 3$ and the performance approaches a steady state for $z > 10$. The slope of profile $t_1(x, z)$ depends mainly upon R_1, NTU_1, and Pe.

Figure 3.20 introduces an example in which two sine temperature waves $f_1(z) = \sin z$ and $f_2(z) = \sin 1.5z$ are generated at the shell and tubeside inlets, respectively, of an exchanger with two tube passes. The remarkable attenuation and superposition of two temperature waves take place for greater values of NTU_1, especially for $NTU_1 > 1$. In such cases the interference of two waves with different periods may greatly cut down the amplitude of the synthesized wave. The curves show that the damping function of both fluids and tube wall depends not only upon heat capacities but also the value of NTU_1. The damping increases with greater NTU_1 and the amplitude at the exit becomes lower. The effect of shell on

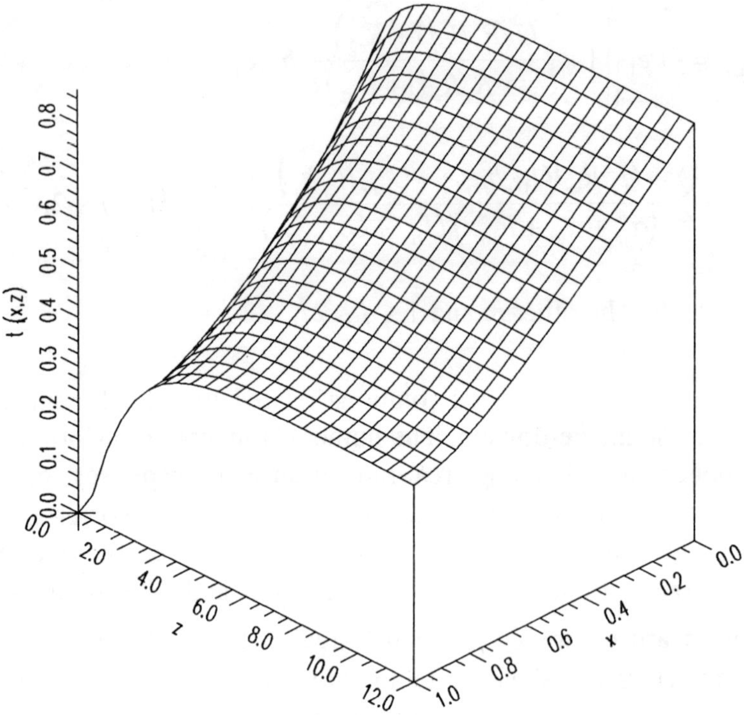

Figure 3.19: Shellside temperature distribution subject to a step temperature variation in a countercurrent flow heat exchanger ($R_1 = 1$, $R_w = 1$, $NTU_1 = 2.5$, $Pe_1 = 4$, $Pe_2 = 10$, $\sigma_1 = \sigma_2 = 1$, $f_1(z) = 1$ and $f_2(z) = 0$)

transient process has the similar feature. In brief, one should distinguish the effect of such solid components as shell and tubes as well as baffles and tubesheets on transient behaviour according to the heat capacities and heat transfer coefficients.

An example of step flow disturbances ($\sigma_1 = 0.6$ and $\sigma_2 = 1.2$) is plotted in Fig. 3.21. In this example the nonzero initial conditions are identical to the steady-state temperature distributions of the previous process. The value of t_{2e} or t_{1e} at $z = 0$ corresponds to the thermal effectiveness of the previous process. These curves show that a decrease of shellside flow rate causes reduction of exit temperature temperatures and that this reduction becomes more

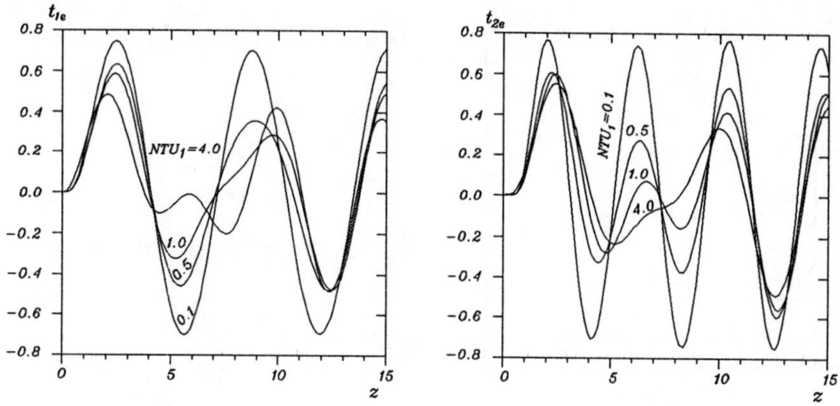

(a) shellside fluid (b) tubeside fluid

Figure 3.20: Superposition of exit responses to periodic inlet temperatures $f_1(z) = \sin z$ and $f_2(z) = \sin 1.5z$ in a 1–2 exchanger with flow arrangement I ($\varepsilon_{11} = \varepsilon_{12} = 0.5$, $\varepsilon_{21} = \varepsilon_{22} = 0.5$, $\varepsilon_{c1} = \varepsilon_{c2} = 0.5$, $\varepsilon_{w1} = \varepsilon_{w2} = 0.5$, $R_1 = 1$, $R_w = 0.6$, $Pe_1 = 4$, $Pe_{21} = Pe_{22} = 10$, $\sigma_1 = \sigma_2 = 1$)

remarkable for $NTU_1 > 0.1$, as expected. The transient process almost approaches a new steady state for $z > 8$. Fig. 3.21(a) and (b) show the energy balance $\left[R_1(1 - t_{1e}) = t_{2e}\right]$ between the shell and tubeside fluids for $z > 10$.

As pointed out, flow maldistribution incurs degeneration of thermal performance of heat exchangers and the measure of flow maldistribution is the Péclet number (here Pe_1 and Pe_{2i}). Figure 3.22 illustrates tubeside exit temperature at $z = 1.5$ *vs* NTU_1 corresponding to different values of Pe_1 and shows how maldistribution exerts its influence on transient behaviour. The quantitative description in Fig. 3.22 reveals that the effect of maldistribution is considerably strong if $Pe_1 < 25$ and its effect increases with increasing NTU_1. With regard to possible maldistribution, designers provide exchangers with higher values of NTU_1 and measures against occurrence of maldistribution to reach the specified effectiveness. On the other hand, the effect of

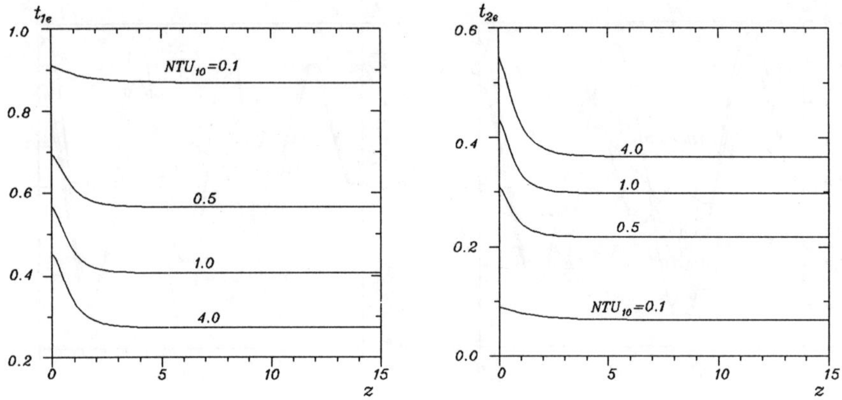

(a) shellside fluid (b) tubeside fluid

Figure 3.21: Exit responses to step disturbances of flow rates in a 1–4 exchanger with flow arrangement II, $\varepsilon_{1i} = \varepsilon_{2i} = \varepsilon_{ci} = \varepsilon_{wi} = 0.25$ $(i = 1,\ 2,\ 3,\ 4)$, $\sigma_1 = 0.6$, $\sigma_2 = 1.2$, $R_{1r} = 1$, $R_w = 0.4$, $Pe_1 = 6$, $Pe_{21} = Pe_{22} = Pe_{23} = Pe_{24} = 8$, $f_1(z) = 1$ and $f_2(z) = 0$

maldistribution is negligible if $Pe_1 > 55$. In fact, the dispersion model is almost identical to the conventional plug-flow model for $Pe_1 > 55$. In other words, application of the dispersion model is of importance for smaller Pe_1, especially for $Pe_1 < 25$. It is predicted that the effect of Pe_{2i} is identical to that of Pe_1. For tubeside laminar flow, the effect of maldistribution becomes more evident, and then application of the dispersion model to the tubeside fluid is more important.

3.6.3 The improved cell model

An alternative to take the effect of the shellside flow maldistribution on transient behaviour of shell-and-tube heat exchangers into account is the so-called improved cell model proposed by Kukral and Stephan (1994) and Kukral (1994), in which the whole exchanger is divided into a number of cells. Among the three submodels for the tubeside flow , the solid body and the shellside flow, the first two are similar to most of the existing publications.

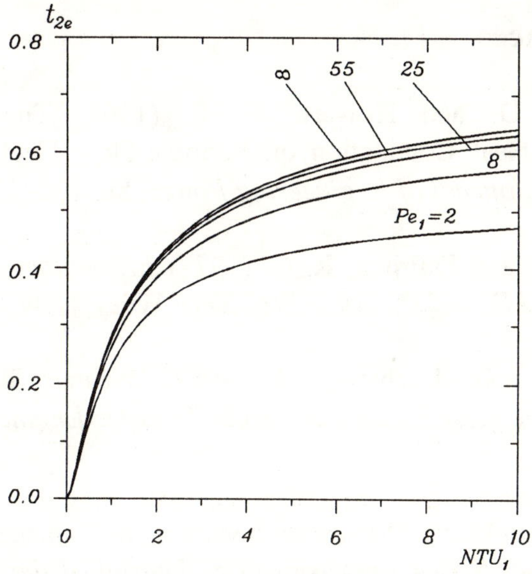

Figure 3.22: Effect of Pe_1 on transient behaviour of a countercurrent flow exchanger, $R_1 = 0.8$, $R_w = 0.3$, $Pe_1 = 2$, $Pe_2 = 60$, $\sigma_1 = \sigma_2 = 1$, $f_1(z) = 1$ and $f_2(z) = 0$

The main model feature is that attention is paid to the shellside flow. In the light of the flow model of Tinker (1951), the shellside flow is regarded as a network of segregated flow streams and the energy equation is built on every cell for each of these streams designated by Tinker, although it is difficult to know the flow rate of each stream. In addition, the shellside fluid is allowed to be a mixture composed of a few components and changes of compositions at the inlets can similarly be treated to the inlet temperature variations.

The mathematical description of transient equations are formally similar to that based on the conventional cell model. As a whole, all balance equations among the tubeside and shellside fluids, the fluid component and the solid body constitute a system of differential-algebraic equations that can be integrated with respect to time by standard solvers. For further detailed information, readers may refer to the above-mentioned literature.

3.7 References

Bahnke, G. D. and Howard, C. P. (1964) The Effect of Longitudinal Heat Conduction on Periodic-Flow Heat Exchanger Performance, *Journal of Engineering Power*, **86**, 105–120.

Boyce, W. E. and DiPrima, R. C. (1977) *Elementary Differential Equations and Boundary Value Problems,* Wiley, New York.

Correa, D. J. and Marchetti, J. L. (1987) Dynamic Simulation of Shell-and-Tube Heat Exchangers, *Heat Transfer Engineering*, **8**(1), 50–59.

Crump, K. S. (1976) Numerical Inversion of Laplace Transform Using a Fourier Series Approximation, *Journal of the Association for Computing Machinery,* **23**, 89–96.

Danckwerts, P. V. (1953) Continuous Flow Systems — Distribution of Residence Times, *Chemical Engineering Science*, **2**, 1–13.

Gaver, D. P. (1966) Observation Stochastic Process, and Approximate Transform Inversion, *Operations Research,* **14**(3), 444–459.

Gaddis, E. S. and Schlüder, E. U. (1979) Temperature Distribution and Heat Exchange in Multipass Shell and Tube Exchangers with Baffles, *Heat Transfer Engineering,* **1**(1), 42–52.

Hausen, H. (1976) *Wärmeübertragung im Gegenstrom, Gleichstrom und Kreuzstrom,* Springer-Verlag, Berlin.

Heggs, P. J. and Render, C. L. (1983) Transient Response of Heat Exchangers with One Infinite Capacitance Fluid, *Heat Transfer Engineering*, **4**(2), 19–27.

Ichikawa, S. and Kishima, A. (1972) Applications of Fourier Series

Technique to Inverse Laplace Transform, *Kyoto University Memoires,* **34**, Part I, 53–67.

Incropera, F. P. and DeWitt, D. P. (1985) *Fundamentals of Heat and Mass Transfer*, Wiley, New York.

Jacquot, R. G., Steadman, J. W. and Rhodine, C. N. (1983) The Gaver–Stehfest Algorithm for Approximate Inversion of Laplace Transforms, *IEEE Circuits and Systems Magazine*, **5**, 4–8.

Jonsson, G., Palsson, O. P. and Sejling, K. (1992) Modeling and Parameter Estimation of Heat Exchangers — A Statistical Approach, *Journal of Dynamic Systems, Measurement, and Control*, **114**, 673–679.

Kawata, S., Kanoh, H. and Masubuchi, M. (1989) A Correlation between Steady-State and Dynamic Response of a Counterflow Heat Exchanger, *Journal of Dynamic Systems, Measurement and Control,* **11**, 115–118.

Kays, W. M. and London, A. L. (1984) *Compact Heat Exchangers*, McGraw-Hill Book Company, New York.

Kukral, R. (1994) *Modelle zur Beschreibung der Zustandänderungen in Rohrbündel-Wärmeübertragern bei zeitlich veränderlichen Betriebsbedingungen,* Fortschritt-Berichte VDI, Reihe 19, Wärmetechnik/Kältetechnik, Nr. 76, VDI-Verlag GmbH, Düsseldorf.

Kukral, R. and Stephan, K. (1994) The Effect of Internal Leakage on Steady-State and Transient Behaviour of Shell-and-Tube Heat Exchangers, in *Proceedings of the 10th International Heat Transfer Conference,* **4**, 9-HE-13.

Li, C. H. (1983) A Numerical Finite Difference Method for Performance Evaluation of a Periodic-Flow Heat Exchanger, *Journal of Heat Transfer*, **105**, 611–617.

Mecklenburg, J. C. and Hartland, S. (1975) *The Theory of Backmixing*, Wiley, London.

Minkowycz, W. J, Sparrow, E. M., Schneider, G. E. and Pletcher, R. H. (1988), *Handbook of Numerical Heat Transfer*, John Wiley & Sons, INC., New York.

Mondt, J.R. (1964) Vehicular Gas Turbine Periodic-Flow Heat Exchanger Solid and Fluid Temperature Distributions, *Journal of Engineering Power*, **86**, 121–126.

Mueller, A. C. (1987) Effects of Some Types of Maldistribution on the Performance of Heat Exchangers, *Heat Transfer Engineering*, **8**(2), 75-86.

Pekdemir, T., Davies, T. W., Haseler, L. E., and Diaper, A. D. (1993) Flow Distribution on the Shellside of a cylindrical Shell and Tube Heat Exchanger, *International Journal of Heat and Fluid Flow*, **14**(1), 76–85.

Roetzel, W. (1988) Thermische Berechnung von dreigängigen Rohrbündelwärmeüber-tragern mit zwei Gegenstromdurchgängen gleicher Größe, *Wärme- und Stoffübertragung*, **22**, 3–11.

Roetzel, W. and Spang, B. (1990) Effective Mean Temperature Difference in Segmentally Baffled Shell and Tube Heat Exchangers, in *Proceedings of the 9th Int. Heat Transfer Conference*, **5**, 79–84.

Roetzel, W. and Xuan Y. (1991) Dispersion Model for Divided-Flow Heat Exchangers, in *Design and Operation of Heat Exchangers* (ed. by Roetzel, W., Heggs, P. J. and Butterworth, D.), pp. 98–110, Springer-Verlag, Berlin.

Roetzel, W. and Xuan, Y. (1992a) Transient Response of Parallel and Counterflow Heat Exchangers, *Journal of Heat Transfer*, **114**, 510–512.

Roetzel, W. and Xuan, Y. (1992b) Transient Behaviour of Multipass Shell-and-Tube Heat Exchangers, *International Journal of Heat and Mass Transfer*, **35**, 703–710.

Roetzel, Y. and Xuan, Y. (1992c) Dispersion Model for Split-Flow Heat Exchangers, in *Recent Advances in Heat Transfer* (ed. by Sundén, B. and Zukauskas, A.), pp. 561–575, Elsevier, Amsterdam.

Roetzel, W. and Xuan, Y. (1992d) Analysis of Transient Behaviour of Multipass Shell and Tube Heat Exchangers with the Dispersion Model, *International Journal of Heat and Mass Transfer*, **35**, 2953–2961.

Roetzel, W. and Xuan, Y. (1993a) The Effect of Core Longitudinal Heat Conduction on the Transient Behaviour of Multipass Shell-and-Tube Heat Exchangers, *Heat Transfer Engineering*, **14**(1), 52–61.

Romie, F. E. (1984) Transient Response of the Counterflow Heat Exchanger, *Journal of Heat Transfer*, **106**, 620–626.

Romie, F. E. (1985) Transient Response of the Parallel-Flow Heat Exchanger, *Journal of Heat Transfer*, **107**, 727–730.

Roppo, M. N. and Ganic, E. N. (1983) Time-Dependent Heat Exchanger Modelling, *Heat Transfer Engineering*, **4**(2), 42–46.

Skiepko, T. (1988) The Effect of Matrix Longitudinal Heat Conduction on the Temperature Fields in the Rotary Heat Exchangers, *International Journal of Heat and Mass Transfer*, **31**, 2227–2238.

Spang, B. (1991) *Über das thermische Verhalten von Rohrbündelwärmeübertragern mit Segmentumlenkblechen*, Fortschritt-Berichte VDI, Reihe 19, Wärmetechnik/Kältetechnik, Nr. 48, VDI-Verlag GmbH, Düsseldorf.

Stehfest, H. (1970) Numerical Inversion of Laplace Transforms, *Communications of the ACM*, **13**, 47–49.

Tan, H. K. S. (1994) Continuous Cocurrent Processes in the Nonsteady State, *AIChE Journal*, **40**, 369–372.

Tinker, T. (1951) Shell Side Characteristics of Segmentally Baffled Shell-and-Tube Heat Exchangers, Part I, II and III, in *Proceedings of General Discussion on Heat Transfer*, Institution of Mechanical Engineers, London, 89–116.

Taylor, S. G. (1954) The Dispersion of Matter in Turbulent Flow through A Pipe, *Proceeding of the Royal Society*, **A223**, 447–468.

Xuan, Y. (1991) *Thermische Modellierung mehrgängiger Rohrbündelwärmeübertrager mit Umlenkblechen und geteiltem Mantelstrom*, Fortschritt-Berichte VDI, Reihe 19, Wärmetechnik/ Kältetechnik, Nr. 52, VDI-Verlag GmbH, Düsseldorf.

Xuan, Y. and Roetzel, W. (1993a) Dynamics of Shell-and-Tube Heat Exchangers to Arbitrary Temperature and Step Flow Variations, *AIChE Journal*, **39**, 413–421.

Xuan, Y, and Roetzel, W. (1993b) Stationary and Dynamic Simulation of Multipass Shell and Tube Heat Exchangers with the Dispersion Model for Both Fluids, *International Journal of Heat and Mass Transfer*, **36**, 4221–4231.

Xuan, Y., Spang, B. and Roetzel, W. (1991) Thermal Analysis of Shell-and-Tube Exchangers with Divided-Flow Pattern, *International Journal of Heat and Mass Transfer*, **34**, 853–861.

Zukauskas, A. and Ulinskas, R. (1988) *Heat Transfer in Tube Banks in Crossflow*, Hemisphere Publishing, New York.

Chapter 4

Transient analysis of cross-flow heat exchangers

Cross-flow heat exchangers refer to thermal units in which fluids involved in heat transfer process flow over the individual surfaces in a cross-flow direction. In most cases, the cross-flow arrangement is used in compact heat exchangers, for example, the cross-flow exchangers attached to aircraft gas turbines, vehicles, process plants and air-conditioning systems. According to the demands for thermal loads, a variety of types of cross-flow units are put into operation. Besides other factors, the flow pattern, i.e., whether a fluid is transversely mixed or unmixed, affects the thermal performance of cross-flow heat exchanges.

Compared with other types of heat exchangers, the transient response of cross-flow units has seldom been handled because of its complexity. Two spatial variables and one time variable are involved for the simulation of the transient processes of this type of apparatus. An early work on the transient behaviour of a cross-flow exchanger was published by Dusinberre (1959), in which a general finite-difference solution for a gas-to-gas exchanger with neither fluid mixed was developed. Under certain assumptions and constraints, Myers *et al.* (1967), Yamashita *et al.* (1978), Romie (1983), Gvozdenac (1986, 1991) described transient responses of cross-flow exchangers to inlet temperature step changes. Taking heat capacities of both fluids as well as the heat capacity and heat conduction resistance of the core wall into account, Xuan (1994) proposed a three-dimensional dynamic simulation of cross-flow heat exchangers subject to arbitrary inlet temperature variations. Roetzel and Luo (1993) applied the dispersion model to analysing the effect of longitudinal dispersion in cross-flow. Luo (1998) developed his work to predict the transient responses to arbitrary inlet temperature variations. By means of the discrete and continuous approaches, Xuan (1996) developed a dynamic simulation algorithm for multipass cross-flow heat exchangers with an arbitrary number of rows per pass and accounted for the effect of flow maldistribution with the dispersed plug-flow model.

The common idealizations used in dynamic simulation of cross-flow heat exchangers are as follows:

(1) thermal properties of fluids and heat transfer coefficients are constant in the whole exchanger;

(2) the thermal flow rates remain constant;

(3) axial and transverse mixing and heat conduction in the fluid and conduction in the wall is negligible;

(4) the exchanger shell or shroud is adiabatic on the inside surface;

(5) no heat generation and viscous dissipation exist within the fluid.

Among these assumptions, some can be further relaxed according to the particularity of the transient problem and the used mathematical method. For example, assumptions (1) and (2) may not be necessary if a finite-difference scheme or other numerical methods are applied. For liquid-to-gas or gas-to-gas cross-flow heat exchangers, the heat capacity of the gas may be neglected to simplify the problem. The mathematical complexity varies with the assumption of whether a fluid is assumed to be transversely mixed or unmixed. In addition, the inlet disturbance forms (i.e., step or arbitrary variations) affect both the transient behaviour and the simulation approach.

4.1 Dynamic behaviour of single pass cross-flow heat exchangers

There are three possible flow patterns (Kays and London, 1984) in a single pass cross-flow heat exchanger:

(1) Both fluids unmixed. This type of cross-flow is named pure cross-flow. Each fluid is assumed to be broken up into a large number of separate flow channels through an exchanger without cross-mixing. The cross-flow in a one-pass finned-tube bundle with a high number (e.g., 10 or more) of tube rows may be regarded as the pure cross-flow.

(2) One fluid transversally mixed, the other unmixed. This case corresponds to that in which one fluid is separated into many parallel streams without transverse mixing, while the other is perfectly cross-mixed. This type of cross-flow is realized in the one-pass (finned-tube) cross-flow exchanger with one tube row where the tubeside streams can be assumed to be transversally mixed within each tube of the row. Each tube flow stream experiences the same heat transfer process, provided the mass flow rates are uniform.

(3) Both fluids transversally mixed. This is an uncommon flow pattern in which the steady-state temperature of each fluid is a function of only one spatial coordinate along its flow direction. Such a flow pattern has the lowest thermal effectiveness of the three one-pass arrangements discussed above. It is approximated in a multipass E-type shell-and-tube heat exchanger with a high even number of tubeside passes. Here the terms *mixed* and *unmixed* refer only to transverse mixing of a fluid. Axial mixing is excluded.

4.1.1 The governing equations

The scheme of an one-pass cross-flow unit is described in Fig. 4.1, in which fluid 1 flows in the l_1 (or x) direction, and fluid 2 in the l_2 (or y) direction. Consider an infinitesimal element shown in Fig. 4.1. According to the aforementioned assumptions, the energy balance equations in the given coordinate system are as follows:

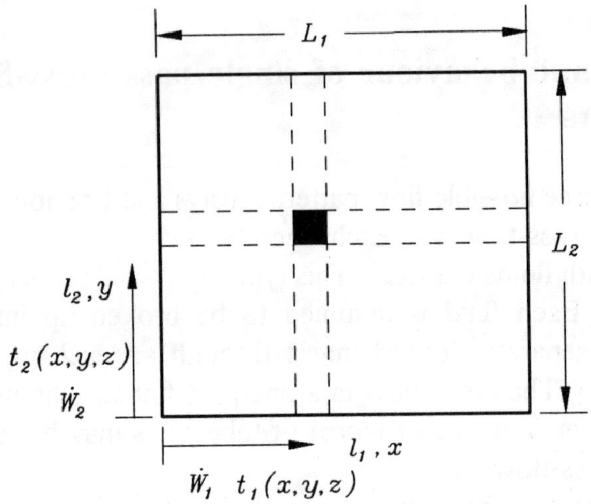

Figure 4.1: Schematic representation of a single pass cross-flow heat exchanger

$$\dot{W}_1 \frac{\partial \theta_1}{\partial l_1} + \frac{C_1}{L_1} \frac{\partial \theta_1}{\partial \tau} + \frac{(hA)_1}{L_1} (\theta_1 - \theta_w) = 0 \tag{4.1}$$

$$\dot{W}_2 \frac{\partial \theta_2}{\partial l_2} + \frac{C_2}{L_2} \frac{\partial \theta_2}{\partial \tau} + \frac{(hA)_2}{L_2} (\theta_2 - \theta_w) = 0 \tag{4.2}$$

$$C_w \frac{\partial \theta_w}{\partial \tau} + (hA)_1 (\theta_w - \theta_1) + (hA)_2 (\theta_w - \theta_2) = 0 . \tag{4.3}$$

These equations are derived according to the conventional plug-flow model. If the heat capacity of one fluid is much smaller than that of the other fluid (for example, in a liquid-to-gas cross-flow heat exchanger) or heat capacities of both fluids are very small (for example, in a gas-to-gas cross-flow exchanger), the energy balance equations for the fluids can further be simplified by omitting the terms containing the time derivatives. Introducing the dimensionless variables and parameters used for the cocurrent and countercurrent flow as well as $x = l_1/L_1$ and $y = l_2/L_2$, one converts eqns (4.1)–(4.3) into the dimensionless forms:

$$\frac{\partial t_1}{\partial z}+\frac{\partial t_1}{\partial x}+U_1(t_1-t_w)=0 \tag{4.4}$$

$$R_\tau\frac{\partial t_2}{\partial z}+\frac{\partial t_2}{\partial y}+U_2(t_2-t_w)=0 \tag{4.5}$$

$$R_w\frac{\partial t_w}{\partial z}+\alpha_1(t_w-t_1)+\alpha_2(t_w-t_2)=0. \tag{4.6}$$

Equations (4.1) and (4.2) as well as (4.4) and (4.5) are valid only for the case of neither fluid mixed. In this case transient temperatures of the fluids are two-dimensional functions $t_1(x,y,z)$ and $t_2(x,y,z)$. For the case of one fluid, e.g., fluid 1, transversely mixed, the transient temperature of the mixed fluid is a one-dimensional function, $t_1=t_1(x,z)$, and the local wall temperature in eqn (4.4) has to be replaced by the mean wall temperature $\bar{t}_w(x,z)=\int_0^1 t_w(x,y,z)\,\mathrm{d}y$. Similarly, if fluid 2 is transversely mixed, the local wall temperature in eqn (4.5) should be replaced by the mean wall temperature $\bar{t}_w(y,z)=\int_0^1 t_w(x,y,z)\,\mathrm{d}x$.

The following three groups of the general initial and boundary conditions are given. For the case of neither fluid mixed

$$t_1(x,y,0)=g_1(x,y),\ t_2(x,y,0)=g_2(x,y),\ t_w(x,y,0)=g_w(x,y) \tag{4.7a}$$

$$t_1(0,y,z)=f_1(y,z),\ t_2(x,0,z)=f_2(x,z). \tag{4.7b}$$

For the case of fluid 1 mixed

$$t_1(x,0)=g_1(x),\ t_2(x,y)=g_2(x,y),\ t_w(x,y,0)=g_w(x,y) \tag{4.8a}$$

$$t_1(0,z)=f_1(z),\ t_2(x,0,z)=f_2(x,z). \tag{4.8b}$$

For the case of fluid 2 mixed:

$$t_1(x,y,0)=g_1(x,y),\; t_2(y,0)=g_2(y),\; t_w(x,y,0)=g_w(x,y) \quad (4.9a)$$

$$t_1(0,y,z)=f_1(y,z),\; t_2(0,z)=f_2(z) \quad (4.9b)$$

For the case of both fluids mixed, both latter cases apply.

According to the flow pattern in the cross-flow exchanger under consideration, one of the above three groups can be selected. Depending on the type of disturbance, these conditions can further be simplified. If only a step temperature change occurs at the entrance of fluid 1, for example, one may have $f_1(z)=1$ and $f_2(z)=0$. In accordance with the particular problem, eqns (4.4)–(4.6) combined with the given initial and boundary conditions are solved. For this purpose, two methods, the Laplace transform and the finite-difference method, are frequently used, but attention has mainly been paid to the transient response to step temperature changes and uniform initial conditions, especially in cases in which the Laplace transform is applied (Romie, 1983; Gvozdenac, 1986, 1991; Chen and Chen, 1991).

4.1.2 Cross-flow, both fluids mixed

This flow pattern is uncommon. However, it is the simplest example to explain simulation principle of dynamic behaviour of cross-flow heat exchangers. In this case, the temperature distributions of both fluids are one-dimensional functions, i.e., $t_1(x,z)$ and $t_2(y,z)$. Application of the Laplace transform to eqns (4.4)–(4.6) yields

$$\frac{\mathrm{d}T_1}{\mathrm{d}x}+a_1T_1=a_2T_2 \quad (4.10)$$

$$\frac{\mathrm{d}T_2}{\mathrm{d}y}+b_2T_2=b_1T_1 \quad (4.11)$$

$$T_w = \frac{\alpha_1 T_1 + \alpha_2 T_2}{\alpha_1 + \alpha_2 + R_w s} \tag{4.12}$$

where

$$a_1 = s + U_1 - \frac{\alpha_1 U_1}{\alpha_1 + \alpha_2 + R_w s}, \quad a_2 = \frac{\alpha_1 U_1}{\alpha_1 + \alpha_2 + R_w s},$$

$$b_1 = \frac{\alpha_1 U_2}{\alpha_1 + \alpha_2 + R_w s}, \qquad b_2 = R_\tau s + U_2 - \frac{\alpha_2 U_2}{\alpha_1 + \alpha_2 + R_w s}.$$

The uniform initial conditions are introduced to obtain eqns (4.10)–(4.12). The ordinary differential equations (4.10) and (4.11) can be integrated. The general solutions are given as

$$T_1(x,s) = c_1\,\mathrm{e}^{-a_1 x} + \frac{a_2}{a_1}\int_0^1 T_2(y,s)\,\mathrm{d}y \tag{4.13}$$

$$T_2(y,s) = c_2\,\mathrm{e}^{-b_2 y} + \frac{b_1}{b_2}\int_0^1 T_1(x,s)\,\mathrm{d}x. \tag{4.14}$$

According to these solutions, one has

$$\overline{T}_1 = \int_0^1 T_1(x,s)\,\mathrm{d}x = d_1 c_1 + d_2 c_2 \tag{4.15}$$

$$\overline{T}_2 = \int_0^1 T_2(y,s)\,\mathrm{d}y = e_1 c_1 + e_2 c_2 \tag{4.16}$$

where, $d_1 = \dfrac{b_2\left(1-\mathrm{e}^{-a_1}\right)}{a_1\left(b_2 - b_1\right)}$, $d_2 = \dfrac{a_2\left(1-\mathrm{e}^{-b_2}\right)}{a_1\left(b_2 - b_1\right)}$, $e_1 = \dfrac{b_1\left(1-\mathrm{e}^{-a_1}\right)}{a_1 b_2 - a_2 b_1}$, $e_2 = \dfrac{a_1\left(1-\mathrm{e}^{-b_2}\right)}{a_1 b_2 - a_2 b_1}$. According to the boundary conditions

$T_1(0,s) = F_1(s)$ and $T_2(0,s) = F_2(s)$. In this case, the unknown coefficients c_1 and c_2 can be determined as

$$c_1 = \frac{(1+b_1d_2)F_1(s) - a_2e_2F_2(s)}{(1+a_2e_1)(1+b_1d_2) - a_2b_1d_1e_2},$$

$$c_2 = -\frac{b_1d_1F_1(s) - (1+a_2e_1)F_2(s)}{(1+a_2e_1)(1+b_1d_2) - a_2b_1d_1e_2}.$$

Thus, the transformed temperatures of cross-flow exchangers with both fluids mixed in the image-domain have been derived. By using the numerical inverse transform, one finds transient responses to arbitrary inlet temperature disturbances. Similarly, one is able to determine the dynamic behaviour of cross-flow exchangers with only one fluid mixed.

4.1.3 Cross-flow, with neither fluid mixed

The common flow pattern appearing in cross-flow exchangers is the cross-flow with neither fluid mixed, the so-called pure cross-flow. The temperatures of fluids are two-dimensional functions. The dual Laplace transform will convenience the solution process. By using the Laplace transform variable *s* with respect to the variable *z*, eqns (4.4)–(4.6) become

$$\frac{\partial T_1}{\partial x} + AT_1 - BT_2 = H_1(x,y,s) \tag{4.17}$$

$$\frac{\partial T_2}{\partial y} - CT_1 + DT_2 = H_2(x,y,s) \tag{4.18}$$

$$T_w = \frac{\alpha_1 T_1 + \alpha_2 T_2}{\alpha_1 + \alpha_2 + R_w s} + \frac{R_w}{\alpha_1 + \alpha_2 + R_w s} g_w(x,y) \tag{4.19}$$

where

$$A = s + U_1 - \frac{\alpha_1 U_1}{\alpha_1 + \alpha_2 + R_w s},$$

$$B = \frac{\alpha_2 U_1}{\alpha_1 + \alpha_2 + R_w s}, \; C = \frac{\alpha_1 U_2}{\alpha_1 + \alpha_2 + R_w s},$$

$$D = R_\tau s + U_2 - \frac{\alpha_2 U_2}{\alpha_1 + \alpha_2 + R_w s}$$

$$H_1(x, y, s) = g_1(x, y) + \frac{R_w U_1}{\alpha_1 + \alpha_2 + R_w s} g_w(x, y),$$

$$H_2(x, y, s) = R_\tau g_2(x, y) + \frac{R_w U_2}{\alpha_1 + \alpha_2 + R_w s} g_w(x, y).$$

By means of the Laplace transform variable q with respect to y, these equations become:

$$\frac{\mathrm{d}\widehat{T}_1}{\mathrm{d}x} + \left(A - \frac{BC}{q+D}\right)\widehat{T}_1 = \widehat{H}_1(x, p, s) + \frac{B}{q+D}\left[F_2(x, s) + \widehat{H}_2(x, q, s)\right] \tag{4.20}$$

$$\widehat{T}_2 = \frac{C}{q+D}\widehat{T}_1 + \frac{F_2(x, s) + \widehat{H}_2(x, q, s)}{q+D} \tag{4.21}$$

$$\widehat{T}_w = \frac{\alpha_1 \widehat{T}_1 + \alpha_2 \widehat{T}_2}{\alpha_1 + \alpha_2 + R_w s} + \frac{R_w}{\alpha_1 + \alpha_2 + R_w s} G_w(x, q) \tag{4.22}$$

where $\widehat{H}$ and $\widehat{T}$ are the forms corresponding to the dual Laplace transform. Once $\widehat{T}_1$ is found, the other transformed temperatures can also be determined. According to the given boundary condition $f_1(y, z)$, the solution to eqn (4.20) yields

$$\widehat{T}_1 = \widehat{F}_1(q,s)\ \mathrm{e}^{-ax}$$

$$+\,\mathrm{e}^{-ax}\int_0^x \mathrm{e}^{ax'}\left\{\widehat{H}_1(x',q,s)+\frac{B}{p+D}\left[F_2(x',s)+\widehat{H}_2(x',q,s)\right]\right\}\mathrm{d}\,x' \tag{4.23}$$

where $a = A - \dfrac{BC}{q+D}$. The next task is to perform the inverse transform of the solutions (4.21)–(4.23). For this purpose, the numerical inverse transform can be extended to the dual Laplace transform (Xuan, 1994). For the dual Laplace transform pair:

$$\widehat{G}(q,s) = \int_0^\infty\int_0^\infty g(y,z)\ \mathrm{e}^{-qy}\ \mathrm{e}^{-sz}\ \mathrm{d}\,y\,\mathrm{d}\,z$$

and

$$g(y,z) = \frac{1}{2\pi i}\int_{c_1-i\infty}^{c_1+i\infty}\frac{1}{2\pi i}\int_{c_2-i\infty}^{c_2+i\infty}\widehat{G}(q,s)\ \mathrm{e}^{-qy}\mathrm{e}^{-sz}\,\mathrm{d}\,y\,\mathrm{d}\,z,$$

the formula of the numerical inverse algorithm can be expressed as

$$g(y,z) = \sum_{j=1}^{N}\sum_{k=1}^{M}\chi_{jk}\left[a_k(y),\ b_j(z)\right] \tag{4.24}$$

where, χ_{jk}, $a_k(y)$ and $b_j(z)$ may take different forms according to the inverse algorithm selected. M and N depend upon the desired accuracy of the numerical inverse transform. In fact, the numerical inverse transform of the dual Laplace transform is successively performed versus the Laplace variables p and s and the inverse order does not affect the final result. Thus, the inverse algorithms which are suitable for the Laplace transform can directly be applied to the dual Laplace transform. By setting $\widehat{G}(q,s) = \widehat{T}_1(x,q,s)$ and

$\hat{G}(q,s)=\hat{T}_2(x,q,s)$ as well as $\hat{G}(q,s)\hat{T}_w(x,q,s)$ in (4.24), respectively, one is able to find the transient temperature distributions $t_1(x,y,z)$, $t_2(x,y,z)$ and $t_w(x,y,z)$ of cross-flow heat exchangers with neither fluid mixed. To estimate dynamic responses to inlet disturbances, the following mean exit temperatures of both fluids can be calculated

$$t_{1e}=\int_0^1 t_1(1,y,z)\,\mathrm{d}y \tag{4.25a}$$

$$t_{2e}=\int_0^1 t_2(x,1,z)\,\mathrm{d}x. \tag{4.25b}$$

The analytical solutions in time domain were obtained by Spiga and Spiga (1992) and Gvozdenac (1991), in which the dual or threefold integration appears and should be calculated numerically.

For the uniform initial temperature distribution the form of the solution is simpler (Spiga and Spiga, 1988), but the numerical integration is still needed for arbitrary inlet temperature variations. Instead of numerical integration Luo (1998) used a Fast Fourier Transform algorithm to perform the numerical inverse Laplace transform. The solutions in the Laplace domain are given as follows:

$$T_1(x,y,s)=F_1(s)\left[\mathrm{e}^{-\left(A-\frac{BC}{D}\right)x}-\mathrm{e}^{-(Ax+Dy)}\sum_{n=1}^{\infty}\frac{(BCx/D)^n}{n!}\sum_{k=0}^{n-1}\frac{(Dy)^k}{k!}\right] \tag{4.26}$$

$$T_2(x,y,s)=F_1(s)\frac{C}{D}\left[\mathrm{e}^{-\left(A-\frac{BC}{D}\right)x}-\mathrm{e}^{-(Ax+Dy)}\sum_{n=0}^{\infty}\frac{(BCx/D)^n}{n!}\sum_{k=0}^{n}\frac{(Dy)^k}{k!}\right] \tag{4.27}$$

$$T_w(x,y,s)=\frac{\alpha_1T_1(x,y,s)+\alpha_2T_2(x,y,s)}{\alpha_1+\alpha_2+R_ws} \tag{4.28}$$

$$T_{1e}(s) = F_1(s)\left[\left(1-\frac{BC}{D^2}\right)\mathrm{e}^{\frac{BC}{D}-A} + \frac{\mathrm{e}^{-(A+D)}}{D}\sum_{n=1}^{\infty}\frac{(BC/D)^n}{n!}\sum_{k=0}^{n}(n-k)\frac{D^k}{k!}\right] \tag{4.29}$$

$$T_{2e}(s) = \frac{F_1(s)C}{AD}$$

$$\times\left[\frac{1-\mathrm{e}^{\frac{BC}{D}-A}}{1-\dfrac{BC}{AD}} - \sum_{n=0}^{\infty}\left(\frac{BC}{AD}\right)^n\left(\mathrm{e}^{-D}\sum_{k=0}^{n}\frac{D^k}{k!}\right)\left(1-\mathrm{e}^{-A}\sum_{k=0}^{n}\frac{A^k}{k!}\right)\right] \tag{4.30}$$

where A, B, C and D are the same as those used in eqns (4.17) and (4.18). The above solutions represent the temperature responses to the inlet temperature variation in fluid 1. Exchanging the co-ordinates and fluid indices as well as all the corresponding parameters, one obtains the temperature responses to the inlet temperature variation of the other fluid. The superposition of these two solutions yields the dynamic responses to the inlet temperature disturbances occurring in both fluids.

In the aforementioned models it is assumed that the flow distributions in both fluids are uniform. This assumption often fails because of the flow maldistributions in one or both fluids. In some types of cross-flow heat exchangers, e.g., tube bundle exchangers, the flow maldistribution happens mainly in the shellside, whereas the tubeside flow can be considered to be uniform. As has been mentioned in Section 3.6, the flow maldistributions in cross-flow heat exchangers can be described by a general dispersion model (Roetzel, 1993) which is solved numerically. For the case that axial dispersion takes place only in fluid 1, Luo (1998) gave transient solutions for uniform initial fluid and wall temperatures.

If the inlet temperature variation occurs only in fluid 1, the following solutions are valid in the Laplace domain:

$$T_1(x,y,s) = F_1(s)\,\mathrm{e}^{ax}\left\{2a\frac{a\sinh[b_0(1-x)] + b_0\cosh[b_0(1-x)]}{(a^2+b_0^2)\sinh(b_0) + 2ab_0\cosh(b_0)}\right.$$

$$+\sum_{j=1}^{\infty}\frac{2\beta_j\left[(\alpha_1+\alpha_2+R_w s)(R_\tau s+U_2+q_j)-\alpha_2 U_2\right]^2}{\alpha_1 U_1 \alpha_2 U_2 q_j\left(\beta_j^2+a^2+2a\right)}$$

$$\left.\times\left[(a\sin\beta_j x+\beta_j\cos\beta_j x)\,\mathrm{e}^{q_j y}\right]\right\} \tag{4.31}$$

$$T_2(x,y,s) = F_1(s)\,\mathrm{e}^{ax}\left\{\frac{\alpha_1 U_2}{(\alpha_1+\alpha_2+R_w s)(R_\tau s+U_2)-\alpha_2 U_2}\right.$$

$$\times\left[2a\frac{a\sinh[b_0(1-x)] + b_0\cosh[b_0(1-x)]}{(a^2+b_0^2)\sinh(b_0) + 2ab_0\cosh(b_0)}\right]$$

$$+\sum_{j=1}^{\infty}\frac{2\beta_j\left[(\alpha_1+\alpha_2+R_w s)(R_\tau s+U_2+q_j)-\alpha_2 U_2\right]}{\alpha_1 \alpha_2 q_j\left(\beta_j^2+a^2+2a\right)}$$

$$\left.\times\left[(a\sin\beta_j x+\beta_j\cos\beta_j x)\,\mathrm{e}^{q_j y}\right]\right\} \tag{4.32}$$

$$T_w(x,y,s) = F_1(s)\,\mathrm{e}^{ax}\left\{\frac{\alpha_1(R_\tau s+U_2)}{(\alpha_1+\alpha_2+R_w s)(R_\tau s+U_2)-\alpha_2 U_2}\right.$$

$$\times\left[2a\frac{a\sinh[b_0(1-x)] + b_0\cosh[b_0(1-x)]}{(a^2+b_0^2)\sinh(b_0) + 2ab_0\cosh(b_0)}\right]$$

$$+\sum_{j=1}^{\infty}\frac{2\beta_j\left(R_\tau s+U_2+q_j\right)\left[\left(\alpha_1+\alpha_2+R_w s\right)\left(R_\tau s+U_2+p_j\right)-\alpha_2 U_2\right]}{U_1\alpha_2 U_2 q_j\left(\beta_j^2+a^2+2a\right)}$$

$$\left.\times\left(a\sin\beta_j x+\beta_j\cos\beta_j x\right)e^{q_j y}\right\}. \tag{4.33}$$

$$T_{1e}(s)=F_1(s)\,e^a\left\{\frac{2ab_0}{\left(a^2+b_0^2\right)\sinh\left(b_0\right)+2ab_0\cosh\left(b_0\right)}\right.$$

$$\left.-2\sum_{j=1}^{\infty}\frac{(-1)^j\beta_j^2\left[\left(\alpha_1+\alpha_2+R_w s\right)\left(R_\tau s+U_2+q_j\right)-\alpha_2 U_2\right]^2}{\alpha_1 U_1\alpha_2 U_2 q_j^2\left(\beta_j^2+a^2+2a\right)}\left(e^{q_j}-1\right)\right\} \tag{4.34}$$

$$T_{2e}(s)=F_1(s)\left\{\frac{2a\alpha_1 U_2}{\left(b_0^2-a^2\right)\left[\left(\alpha_1+\alpha_2+R_w s\right)\left(R_\tau s+U_2\right)-\alpha_2 U_2\right]}\right.$$

$$\times\left[1-\frac{2ab_0\,e^a}{\left(b_0^2+a^2\right)\sinh\left(b_0\right)+2ab_0\cosh\left(b_0\right)}\right]$$

$$\left.-4a\,e^a\sum_{j=1}^{\infty}\frac{(-1)^j\beta_j^2\left[\left(\alpha_1+\alpha_2+R_w s\right)\left(R_\tau s+U_2+q_j\right)-\alpha_2 U_2\right]}{U_1\alpha_2 q_j\left(\beta_j^2+a^2\right)\left(\beta_j^2+a^2+2a\right)}e^{q_j}\right\} \tag{4.35}$$

where

$$a=\frac{Pe}{2}$$

$$b_0 = \sqrt{a^2 + 2a\left[s + U_1 - \frac{\alpha_1 U_1 (R_\tau s + U_2)}{(\alpha_1 + \alpha_2 + R_w s)(R_\tau s + U_2) - \alpha_2 U_2}\right]}$$

$$q_j = \frac{\alpha_2 U_2 (\beta_j^2 + a^2 + 2as + 2aU_1)}{(\alpha_1 + \alpha_2 + R_w s)(\beta_j^2 + a^2 + 2as + 2aU_1) - 2a\alpha_1 U_1} - (R_\tau s + U_2)$$

and β_j is the jth positive root of the eigenfunction

$$\tan\beta_j = \frac{2a\beta_j}{\beta_j^2 - a^2} \qquad (j = 1,\ 2,\ \ldots,\ \infty). \tag{4.36}$$

If the inlet temperature variation occurs in the fluid without axial dispersion (i.e., in fluid 2), one has the following temperature responses:

$$T_1(x, y, s) = \sum_{j=1}^{\infty} \frac{8a^2 U_1 \alpha_2 F_2(s)}{(\beta_j^2 + a^2)(\beta_j^2 + a^2 + 2a)} \times \left[\frac{\beta_j (a \sin\beta_j x + \beta_j \cos\beta_j x)\,\mathrm{e}^{ax + q_j y}}{(\alpha_1 + \alpha_2 + R_w s)(\beta_j^2 + a^2 + 2as + 2aU_1) - 2a\alpha_1 U_1}\right] \tag{4.37}$$

$$T_2(x, y, s) = \sum_{j=1}^{\infty} \frac{4aF_2(s)\beta_j (a \sin\beta_j x + \beta_j \cos\beta_j x)}{(\beta_j^2 + a^2)(\beta_j^2 + a^2 + 2a)}\,\mathrm{e}^{ax + q_j y} \tag{4.38}$$

$$T_w(x, y, s) = \sum_{j=1}^{\infty} \frac{4a\alpha_2 (\beta_j^2 + a^2 + 2as + 2aU_1) F_2(s)}{(\beta_j^2 + a^2)(\beta_j^2 + a^2 + 2a)} \times \left[\frac{\beta_j (a \sin\beta_j x + \beta_j \cos\beta_j x)\,\mathrm{e}^{ax + q_j y}}{(\alpha_1 + \alpha_2 + R_w s)(\beta_j^2 + a^2 + 2as + 2aU_1) - 2a\alpha_1 U_1}\right] \tag{4.39}$$

$$T_{1e}(s) = \sum_{j=1}^{\infty} \frac{8a^2\, e^a\, U_1 \alpha_2 F_2(s)}{q_j \left(\beta_j^2 + a^2\right)\left(\beta_j^2 + a^2 + 2a\right)}$$

$$\times \left[\frac{(-1)^j \beta_j^2 \left(1 - e^{q_j}\right)}{\left(\alpha_1 + \alpha_2 + R_w s\right)\left(\beta_j^2 + a^2 + 2as + 2aU_1\right) - 2a\alpha_1 U_1} \right] \tag{4.40}$$

$$T_{2e}(s) = \sum_{j=1}^{\infty} \frac{8a^2 F_2(s)(-1)^{j+1} \beta_j^2}{\left(\beta_j^2 + a^2\right)^2 \left(\beta_j^2 + a^2 + 2a\right)} e^{a+q_j} \tag{4.41}$$

The effect of axial dispersion on the dynamic outlet temperature responses to a sudden change in inlet temperature of fluid 1 and fluid 2 are shown in Figs 4.2(a) and 4.2(b), respectively. The solution for the arbitrary inlet temperature variations in both fluids is the sum of the aforementioned solutions according to the principle of the superposition.

4.1.4 Inlet flow rate disturbances

In general, both inlet temperature variations and flow rate disturbances induce the transient process of heat exchangers. For simulation of dynamic behaviour of heat exchangers due to flow disturbances, one must pay attention to the consequent alteration of heat transfer coefficients. The typical flow disturbance is the step change. In Chapter 3 a method is proposed to deal with transient analysis of shell-and-tube heat exchangers subject to step disturbances of flow rates. This method can also be used for the prediction of transient responses to a step flow disturbance at inlets of cross-flow heat exchangers. If other disturbance forms of flow rates occur, the problem becomes more complicated and the assumption of constant heat transfer coefficients fails. In this case, eqns (4.1)–(4.3) are still valid. But one should keep in mind that $\dot{W}_i$ and h_i vary with time because the flow rates are functions of time. By redefining some of the dimensionless parameters as follows

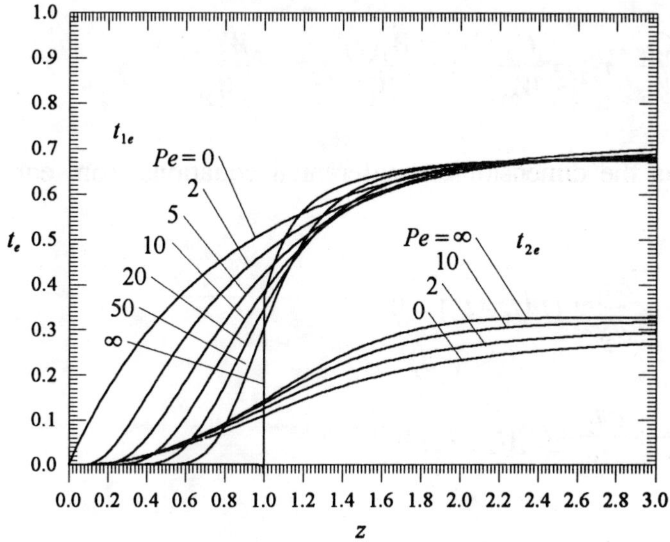

(a) $f_1(z)=1,\ f_2(z)=0$

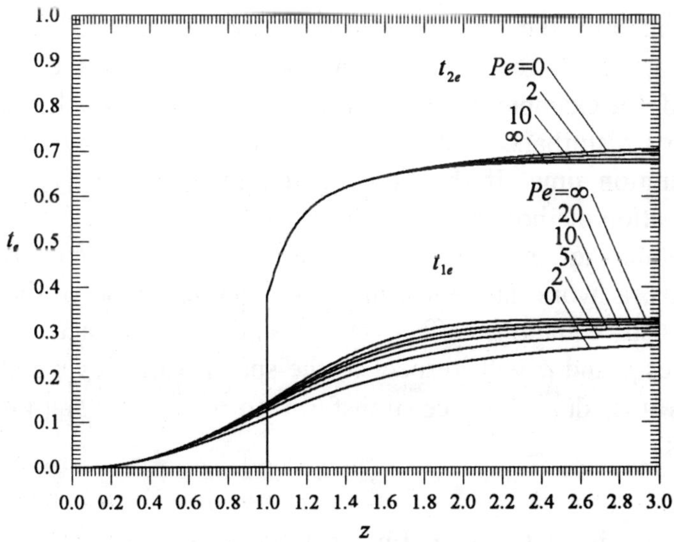

(b) $f_1(z)=0,\ f_2(z)=1$

Figure 4.2: Effect of axial dispersion in one fluid in a cross-flow heat exchanger, $R_1 = R_\tau = 1$, $R_w = 0.1$, $U_1 = U_2 = 1$

$$\tau_{r1} = \frac{C_1}{\dot{W}_{10}},\ \tau_{r2} = \frac{C_2}{\dot{W}_{20}},\ \sigma_1 = \frac{\dot{W}_1(\tau)}{\dot{W}_{10}},\ \sigma_2 = \frac{\dot{W}_2(\tau)}{\dot{W}_{20}},$$

one has the dimensionless differential equations from eqns (4.1)–(4.3):

$$\frac{1}{\sigma_1}\frac{\partial t_1}{\partial z} + \frac{\partial t_1}{\partial x} + U_1(t_1 - t_w) = 0 \tag{4.42}$$

$$\frac{R_\tau}{\sigma_2}\frac{\partial t_2}{\partial z} + \frac{\partial t_2}{\partial y} + U_2(t_2 - t_w) = 0 \tag{4.43}$$

$$R_w\frac{\partial t_w}{\partial z} + \frac{U_1\sigma_1}{1+R_{c2}}(t_w - t_1) + \frac{U_2\sigma_2}{R_\tau(1+R_{c1})}(t_w - t_2) = 0\,. \tag{4.44}$$

Evidently, the Laplace transform can be applied to them. A direct approach for solving eqns (4.42)–(4.44) is the numerical method, for example, the finite-difference method, which can handle arbitrary disturbances of the flow rates. But it may cost much computation time. If the flow fluctuation is relatively small, the perturbation method can be used to derive an approximate solution. By combining the finite-difference and the Laplace transform methods, a hybrid numerical method is derived to solve the above-mentioned equations effectively. Using the Laplace transform variables p and q with respect to the spatial variables x and y, one applies the dual Laplace transform to eqns (4.42)–(4.44) and obtains

$$\frac{1}{\sigma_1}\frac{\mathrm{d}\widehat{T}_1}{\mathrm{d}z} + p\widehat{T}_1 - \widehat{F}_1(q,z) + U_1(\widehat{T}_1 - \widehat{T}_w) = 0 \tag{4.45}$$

$$\frac{R_\tau}{\sigma_2}\frac{\mathrm{d}\widehat{T}_2}{\mathrm{d}z} + q\widehat{T}_2 - \widehat{F}_2(p,z) + U_2(\widehat{T}_2 - \widehat{T}_w) = 0 \tag{4.46}$$

$$R_w \frac{\mathrm{d}\widehat{T}_w}{\mathrm{d}z} + \left[\frac{U_1\sigma_1}{1+R_{c2}} + \frac{U_2\sigma_2}{R_\tau\left(1+R_{c1}\right)}\right]\widehat{T}_w - \frac{U_1\sigma_1}{1+R_{c2}}\widehat{T}_1 - \frac{U_2\sigma_2}{R_\tau\left(1+R_{c1}\right)}\widehat{T}_2 = 0 \tag{4.47}$$

where the boundary conditions (4.7) are simultaneously transformed and inserted in eqns (4.45) and (4.46). According to the given initial conditions, these ordinary differential equations can be solved by the finite-difference method in the image-domain of the dual Laplace transform. Either the explicit or the implicit algorithm can be used. After the finite-difference solutions to eqns (4.45)–(4.47) have been found, the numerical inverse algorithm (4.24) should be applied to them in order to determine the transient temperature distributions in the real-time domain. In this way one can effectively obtain transient temperature profiles at any locations without stepwise computation along the spatial coordinate. It is especially useful if only the exit temperatures of both fluids need to be estimated to monitor dynamic behaviour of heat exchangers subject to arbitrary temperature and flow rate disturbances.

4.2 Three-dimensional dynamic simulation of cross-flow heat exchangers

In most of the existing publications on dynamic simulation of cross-flow exchangers, no heat conduction resistance perpendicular to the surface of the core wall is taken into account. In fact, such conduction resistance of the wall often has a noticeable impact on the thermal performance of heat exchangers, for example, when convective resistances on both sides are relatively small or fouling becomes severe as the deposit-layer on the wall thickens. Using the conventional plug-flow model and accounting for the finite heat capacities of both fluids and the wall as well as the conduction resistance of the wall, Xuan (1994) proposed a dynamic simulation method of cross-flow exchangers with neither fluid mixed.

The three-dimensional scheme of a cross-flow heat exchanger is shown in Fig. 4.3. Each fluid may be transversely mixed or unmixed. The temperature distributions of both fluids and the core wall are transient three-dimensional functions which are respectively expressed as $t_1(x,y,z)$, $t_2(x,y,z)$ and $t_w(x,y,\zeta,z)$. In practice, these temperatures may be simplified to be two- or one-dimensional according to the case of application.

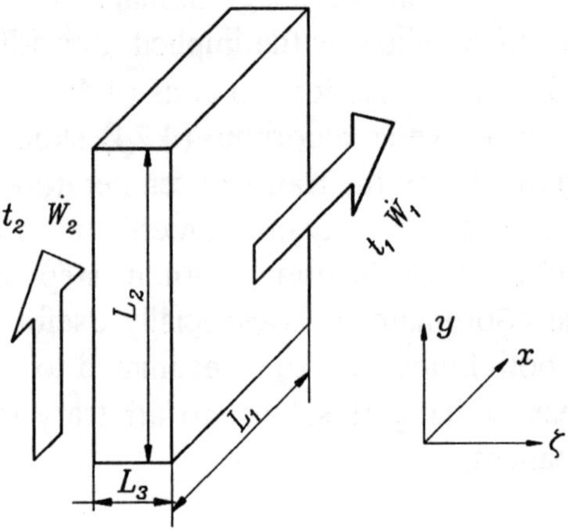

Figure 4.3: Schematic representation of a cross-flow heat exchanger

4.2.1 The accurate approach

By means of a similar procedure to that described in Section 4.1, the energy balance equations for both the fluids can easily be derived as follows

$$\frac{\partial t_1}{\partial z}+\frac{\partial t_1}{\partial x}+U_1(t_1-t_{w1})=0 \tag{4.48}$$

$$R_\tau\frac{\partial t_2}{\partial z}+\frac{\partial t_2}{\partial y}+U_2(t_2-t_{w2})=0 \tag{4.49}$$

with the related initial and boundary conditions

$$t_1(x,y,0)=g_1(x,y),\quad t_2(x,y,0)=g_2(x,y) \tag{4.50a}$$

$$t_1(0,y,z)=f_1(y,z),\quad t_2(x,0,z)=f_2(x,z). \tag{4.50b}$$

The temperatures t_{w1} and t_{w2} in eqns (4.48) and (4.49) are different wall temperatures on both the sides because of the nonzero conduction resistance. As for the transverse heat conduction in the core wall, the governing equation and the determinant conditions yield:

$$\frac{\partial t_w}{\partial z}-Fo_w\frac{\partial^2 t_w}{\partial \zeta^2}=0 \tag{4.51a}$$

$$t_w(x,y,\zeta,0)=g_w(x,y,\zeta) \tag{4.51b}$$

$$-\frac{\partial t_w}{\partial \zeta}+Bi_1(t_w-t_1)=0 \qquad \text{at } \zeta=0 \tag{4.51c}$$

$$\frac{\partial t_w}{\partial \zeta}+Bi_2(t_w-t_2)=0 \qquad \text{at } \zeta=1 \tag{4.51d}$$

where three dimensionless numbers are defined as $Fo_w=a_w\tau_{r1}/L_3^2$, $Bi_1=h_1L_3/k_w$ and $Bi_2=h_2L_3/k_w$. To account for the effect of fouling resistance on the dynamic behaviour of exchangers, the mean constant fouling resistances $R_{f1,2}$ on both sides are introduced, although fouling itself is a slow transient process. Therefore, the apparent conductivity of the wall of original thickness L_3 is defined as

$$\lambda_w^*=\frac{\lambda_w L_3}{L_3+\lambda_w(R_{f1}+R_{f2})}. \tag{4.52}$$

The Fourier number Fo_w and the Biot number $Bi_{1,2}$ introduced characterize the transient heat transfer process in cross-flow heat exchangers. The former determines temperature response of the core wall, and the latter the influence of transverse conduction resistance of the core wall on the transient behaviour. The Biot number Bi is composed of the convective resistance on the side of fluid 1 or fluid 2 and the transverse conduction resistance. $Bi = 0$ means no temperature gradient in the wall and $t_{w1} = t_{w2} = t_w$, which is identical to the conventional assumption of no conduction resistance of the wall. The following relationships exist between the parameters Fo_w and $Bi_{1,2}$:

$$Fo_w Bi_1 = \frac{U_1}{R_w\left(1+R_{c2}\right)}, \quad Bi_2 = Bi_1 R_2 \frac{U_2}{U_1}. \tag{4.53}$$

Introducing the Laplace transform variable s with respect to the dimensionless time z, one derives the transformed wall temperature

$$T_w(x,y,\zeta,s) = \left(a_1\,\mathrm{e}^{m\zeta} + b_1\,\mathrm{e}^{-m\zeta}\right)T_1 + \left(a_2\,\mathrm{e}^{m\zeta} + b_2\,\mathrm{e}^{-m\zeta}\right)T_2 + G_w(x,y,\zeta) \tag{4.54}$$

where

$$m^2 = \frac{s}{Fo_w},$$

$$a_0 = \left(Bi_1 - m\right)\left(Bi_2 - m\right)\mathrm{e}^{-m} - \left(Bi_1 + m\right)\left(Bi_2 + m\right)\mathrm{e}^{m},$$

$$a_1 = \frac{Bi_1\left(Bi_2 - m\right)\mathrm{e}^{-m}}{a_0}, \quad a_2 = -\frac{Bi_2\left(Bi_1 + m\right)}{a_0},$$

$$b_1 = -\frac{Bi_1\left(Bi_2 + m\right)\mathrm{e}^{m}}{a_0}, \quad b_2 = \frac{Bi_2\left(Bi_1 - m\right)}{a_0},$$

$$G_w(x,y,\zeta,s) = H_w(x,y,s)\left(1+\frac{m-Bi_1}{m+Bi_1}\right)\mathrm{e}^{m\zeta} + \frac{Q'_w(x,y,0,s)}{m+Bi_1}\mathrm{e}^{-m\zeta}$$

$$-Q_w(x,y,\zeta,s),$$

$$H_w(x,y,s) = \frac{1}{a_0}\{(Bi_2-m)Q'_w(x,y,0,s)\mathrm{e}^{-m}$$

$$-(Bi_1+m)[Q'_w(x,y,1,s)+Bi_2Q_w(x,y,1,s)]\},$$

$$Q_w(x,y,\zeta,s) = \frac{1}{mFo_w}\int_0^{\zeta}\mathrm{sh}[m(\zeta-\zeta')]g_w(x,y,\zeta')\mathrm{d}\zeta',$$

$$Q'_w(x,y,\zeta,s) = \frac{\partial Q_w(x,y,\zeta,s)}{\partial\zeta}.$$

Therefore, the transformed temperatures T_{w1} and T_{w2} are obtained as

$$T_{w1} = A_1T_1 + A_2T_2 + G_w(x,y,0,s) \tag{4.55a}$$

$$T_{w2} = B_1T_1 + B_2T_2 + G_w(x,y,1,s) \tag{4.55b}$$

where,

$$A_1 = a_1 + b_1,\ A_2 = a_2 + b_2,\ B_1 = a_1\mathrm{e}^{m} + b_1\mathrm{e}^{-m},\ B_2 = a_2\mathrm{e}^{m} + b_2\mathrm{e}^{-m}.$$

Insertion of (4.55) into the transformed forms of eqns (4.48) and (4.49) yields

$$\frac{\partial T_1}{\partial x} + AT_1 - BT_2 = H_1(x,y,s) \tag{4.56a}$$

$$\frac{\partial T_2}{\partial y} - CT_1 + DT_2 = H_2(x,y,s) \tag{4.56b}$$

where

$$A = s + U_1 - A_1U_1,\ B = A_2U_1,\ C = B_1U_2,\ D = R_\tau s + U_2 - B_2U_2,$$

$$H_1(x,y,s) = g_1(x,y) + U_1G_w(x,y,0,s),$$

$$H_2(x,y,s) = g_2(x,y) + U_2G_w(x,y,1,s).$$

Both these equations are formally the same as eqns (4.17) and (4.18). By means of the Laplace transform variable p with respect to y, the dual transform of them yields

$$\frac{\mathrm{d}\widehat{T}_1}{\mathrm{d}x} + \left(A - \frac{BC}{p+D}\right)\widehat{T}_1 = \widehat{H}_1(x,p,s) + \frac{B}{p+D}\left[F_2(x,s) + \widehat{H}_2(x,p,s)\right] \tag{4.57}$$

$$\widehat{T}_2 = \frac{C}{p+D}\widehat{T}_1 + \frac{F_2(x,s) + \widehat{H}_2(x,p,s)}{p+D}. \tag{4.58}$$

The solution to eqn (4.57) is

$$\widehat{T}_1 = \widehat{F}_1(p,s)e^{-ax}$$

$$+ e^{-ax}\int_0^x e^{ax'}\left\{\widehat{H}_1(x',p,s) + \frac{B}{p+D}\left[F_2(x',s) + \widehat{H}_2(x',p,s)\right]\right\}\mathrm{d}x'. \tag{4.59}$$

So far the transformed temperatures in the image-domain have been found, in which the effect of transverse conduction resistance is involved. Application of the inverse algorithm (4.24) results in transient temperature distributions. With the same definition as (4.25), one can analyse transient responses to arbitrary temperature variations and step flow changes as well as the effect of conduction resistance in cross-flow heat exchangers.

4.2.2 The approximate approach

Since the heat conduction equation (4.51) for the wall is involved, the above-described method costs more computation time than the conventional lumped heat capacity model of the wall (no heat conduction resistance) does. With regard to the fact that the lumped parameter model of the wall has been applied to transient simulation of cross-flow heat exchangers and other flow arrangements for a long time, an approximate method of treating transverse conduction of the wall is proposed. In this case, the governing equations remain the same as eqns (4.5)–(4.6) from the lumped heat capacity model, repeated here:

$$\frac{\partial t_1}{\partial z}+\frac{\partial t_1}{\partial x}+U_1(t_1-t_w)=0 \tag{4.4}$$

$$R_\tau\frac{\partial t_2}{\partial z}+\frac{\partial t_2}{\partial y}+U_2(t_2-t_w)=0 \tag{4.5}$$

$$R_w\frac{\partial t_w}{\partial z}+\alpha_1(t_w-t_1)+\alpha_2(t_w-t_2)=0 \tag{4.6}$$

To treat the conduction resistance of the core wall approximately, this type of resistance is artificially divided into two parts, either of which is added to the convective resistance on the side of fluid 1 or fluid 2, respectively. Two apparent coefficients h_1^* and h_2^* yield:

$$\frac{1}{h_1^* A_1}=\frac{1}{(hA)_1}+\varphi\frac{L_3}{\lambda_w A_w} \tag{4.60a}$$

$$\frac{1}{h_2^* A_2} = \frac{1}{(hA)_2} + (1-\varphi)\frac{L_3}{\lambda_w A_w} . \tag{4.60b}$$

The sum of both the apparent resistances is equal to the true overall resistance. The distribution coefficient φ $\left(0 \le \varphi \le 1\right)$ controls the distribution of the conduction resistance. Thus,

$$U_1^* = \frac{U_1}{1 + \varphi\, Bi_1 \dfrac{A_1}{A_w}} \tag{4.61a}$$

$$U_2^* = \frac{U_2}{1 + (1-\varphi)\, Bi_2 \dfrac{A_2}{A_w}} . \tag{4.61b}$$

By substituting U_1^* and U_2^* for U_1 and U_2 in the governing eqns (4.4)–(4.6), the conduction resistance of the wall is approximately taken into account. In this way, the original three-dimensional problem is reduced to a two-dimensional one and the solution approach becomes simpler. Thus, by means of the existing solution to the transient simulation of cross-flow exchangers from the lumped parameter model, one can approximately but quickly estimate the influence of transverse conduction resistance. This approximation can also be applied to tubular exchangers with different flow arrangements (Xuan and Roetzel, 1993).

4.2.3 Examples and discussions

As pointed out, the combination of the transformed solution in the image-domain and the numerical inverse algorithm (4.24) leads to transient temperature distributions in the real-time domain. Two examples are illustrated in Figs 4.4 and 4.5. Both refer to the case of an inlet temperature step change of fluid 1 and uniform initial conditions in cross-flow exchangers with neither fluid mixed. The surface in Fig. 4.4 shows the exit temperature variation of fluid 1

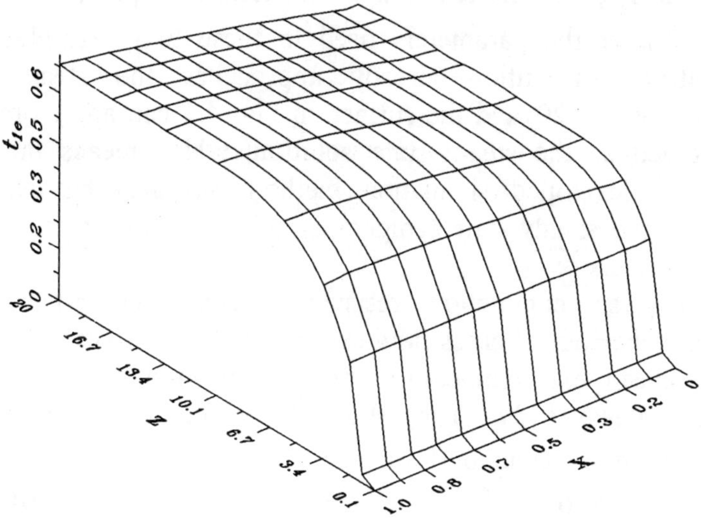

Figure 4.4: Exit temperature of fluid 1
($U_1 = U_2 = 1.5$, $R_1 = R_\tau = 1$, $R_w = 1.5$, $Bi_1 = 0.1$)

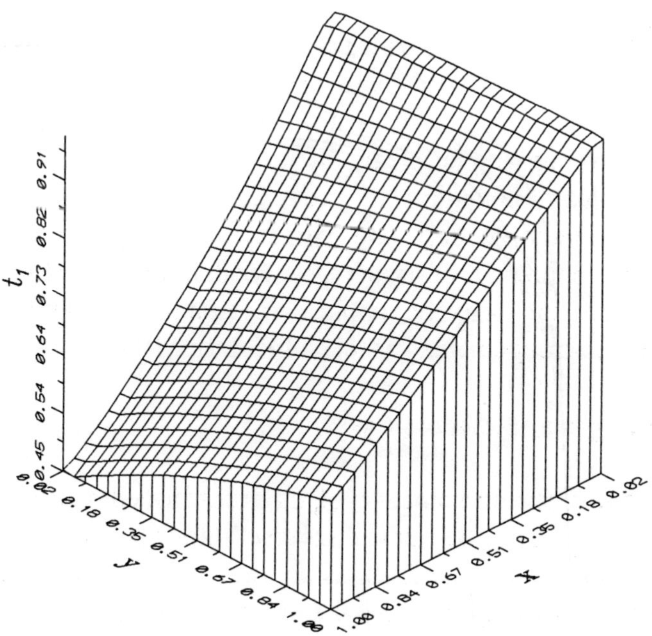

Figure 4.5: Temperature distribution of fluid 1 at $z = 25$
($U_1 = U_2 = 2$, $R_1 = R_\tau = 1$, $R_w = 2$, $Bi_1 = 2$)

and that in Fig. 4.5 the temperature distribution $t_1(x, y, z = 25)$ of fluid 1. Under the parameters used in these two examples, the temperature distribution already approaches the steady-state response for $z \geq 20$ and the surface in Fig. 4.5 can approximately be regarded as the steady-state solution, which means that the previously developed simulation method can also be used to determine the steady-state temperature distributions of cross-flow heat exchangers.

Transverse conduction resistance remarkably affects the transient behaviour of cross-flow exchangers. The curves in Fig. 4.6 are the mean exit temperature profiles of fluid 1 pertinent to different values of Bi_1. They show the effect of the conduction resistance on dynamic performance. $Bi_1 = 0$ corresponds to the ideal case of zero conduction resistance. With increasing Bi_1, this

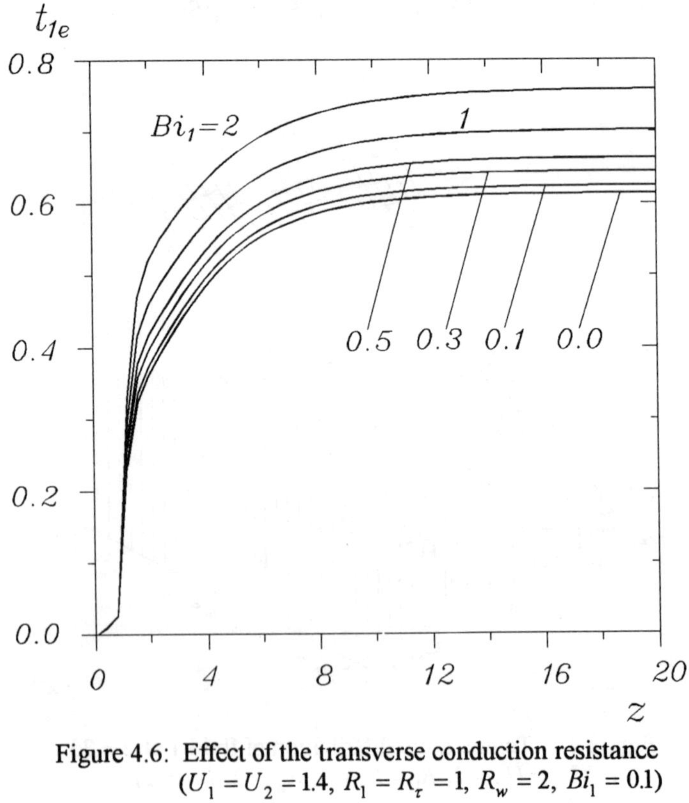

Figure 4.6: Effect of the transverse conduction resistance ($U_1 = U_2 = 1.4$, $R_1 = R_\tau = 1$, $R_w = 2$, $Bi_1 = 0.1$)

effect becomes noticeable and the deviation from the lumped heat capacity model enlarges. If $Bi_1 > 2$, neglecting the conduction resistance leads to an error of more than 25% in the computed dimensionless outlet temperature. Only if $Bi_1 < 0.1$ can this resistance be neglected. It is recommended that the Biot number Bi_1 is calculated and then it is decided whether or not the conduction resistance of the wall as well as the fouling resistance has to be taken into account. In addition, parameters U_1 and U_2 as well as R_1 manipulate the effect of the conduction resistance on the transient behaviour. Further computation shows that the effect of the conduction resistance remarkably increases with decreasing U_1 or U_2, especially for the dimensionless time span $z < 5$.

Figure 4.7 describes the comparison between the accurate and approximate methods of treating the conduction resistance. In this

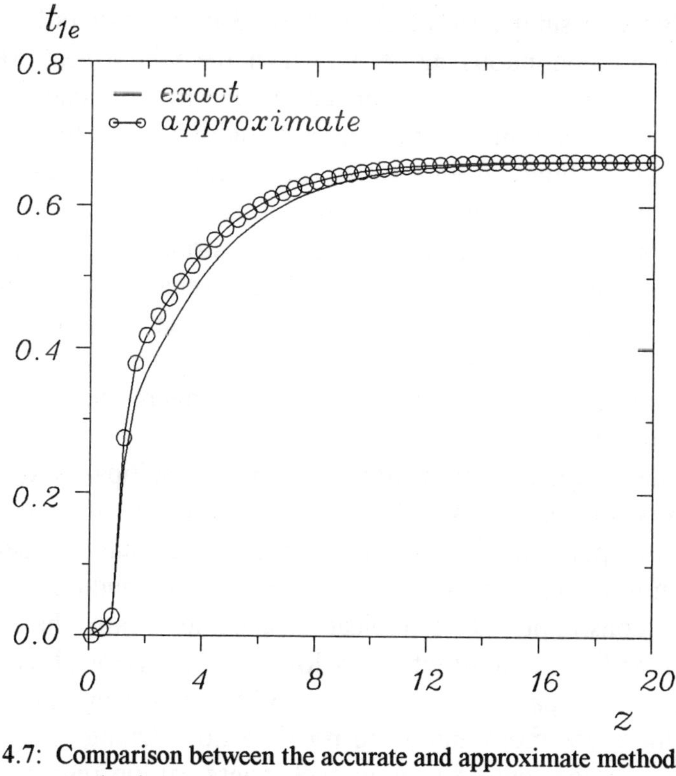

Figure 4.7: Comparison between the accurate and approximate methods of the conduction resistance ($U_1 = U_2 = 1.5$, $R_1 = R_\tau = 1$, $R_w = 2$, $Bi_1 = 0.5$, $\varphi = 0.6$)

example, $A_1 = A_2 = A_w$ is assumed for the sake of simplicity. Distinctly, the results obtained from both the approaches fundamentally coincide with each other. As for the distribution coefficient φ involved in the approximate method, the first intuitive choice may be $\varphi = 0.5$. In fact, a great number of calculation results reveals that any other value of φ between $0 \le \varphi \le 1$ is also suitable. Hence, a short cut method for handling the conduction resistance of the core wall is found. In such a way, one can take the conduction resistance into account without much computation time which may be of importance for real-time simulation and control of exchangers.

4.3 Transient analysis of multipass cross-flow heat exchangers

So far dynamic simulation of cross-flow heat exchangers has been based on the continuous model in which the whole heat transfer surface consisting of a finite number of tube rows in each pass is considered as a continuous surface, although transient temperature distribution of fluids from row to row may give designers and operators much more insight into performance of cross-flow heat exchangers. Furthermore, the effect of maldistribution of both fluids on transient behaviour of cross-flow heat exchangers is generally neglected by most researchers. Considering axial and lateral flow mixing, Roetzel (1996) proposed a set of governing equations describing dynamics of cross-flow heat exchangers and gave some solution examples.

Besides single pass cross-flow exchangers, multipass cross-flow exchangers are widely used. For multipass cross-flow units, there are various possible flow arrangements which may be selected according to the process demands and installation conditions as well as other constraints. Two typical arrangements are the overall cocurrent and countercurrent cross-flow, in which two-, three-, four and even more passes are involved. With increasing number of passes, there are more and more possible flow arrangements. The overall countercurrent cross-flow exchangers can be regarded as a compromise between the desiderata of efficiency and ease of

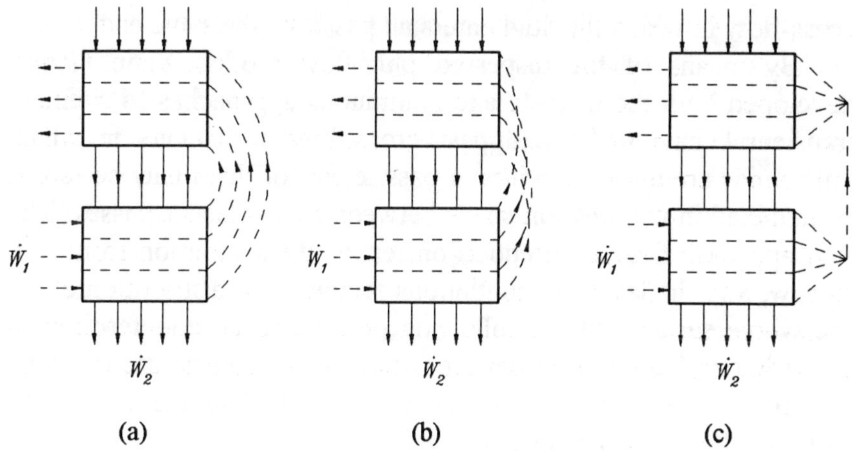

Figure 4.8: Schematic representation of two-pass countercurrent cross-flow heat exchangers with two counterdirected passes in different coupling orders

construction. The greater the number of passes, the closer is the approach to countercurrent flow (Schlünder, 1986). For a given multipass cross-flow heat exchanger, a fluid may be in different coupling orders between two adjacent passes. Some different forms of fluid coupling order between adjacent passes were listed by Gvozdenac (1986). Here the identical order coupling refers to that in which a fluid leaving a pass enters the next pass from the same location as that in the previous pass; the inverted order coupling means that a fluid leaving a pass enters the next pass from the opposite location to that in the previous pass. For the codirected cross-flow the fluid enters different passes at the same end; for the counterdirected cross-flow the fluid alternatively enters different passes at both ends.

Figure 4.8 shows three examples of possible flow arrangements in two-pass countercurrent exchangers, in which fluid 2 is always in the inverted coupling order. Figure 4.8(a) is for the inverted coupling order of fluid 1; Figure 4.8(b) refers to the identical coupling order of fluid 1; Figure 4.8(c) stands for the case that fluid 1 is mixed between passes. In addition, all these examples have a common feature of the counterdirected cross-flow because the flow direction of fluid 1 alternates from pass to pass. One can also illustrate other coupling orders of a fluid in the codirected cross-flow in which the fluid enters all passes at the same end.

By means of the dispersed plug-flow model, Xuan (1996) developed both the discrete and continuous approaches to simulate transient behaviour of multipass cross-flow exchangers in which both fluids are unmixed within a pass, either of them may be mixed or unmixed in conjunction space between two adjacent passes. The first approach focuses attention on temperature variation from row to row, and the latter on continuous temperature distribution along the whole surface. In the following derivation the counterdirected cocurrent and countercurrent cross-flow arrangements are involved and different fluid coupling orders are allowed. The method is also valid for the other arrangements.

4.3.1 The discrete approach

With the discrete approach, the conduction resistance of the wall is neglected. In the following derivation, pass one refers to the first pass through which fluid 1 flows into the cross-flow heat exchanger. The co-ordinate l follows the flow direction of tubeside fluid 1 in the first tube pass and the other variable ω along the flow path of outside fluid 2. From the view of the discrete approach the governing equations should be derived for each row in each pass. The dispersed plug-flow model is applied to fluid 1 to consider possible flow maldistribution within one tube row, The energy equation in row i of pass j is obtained as

$$A_{c1i,j}\mathcal{D}_{1i,j}\frac{\partial^2\theta_{1i,j}}{\partial l_1^2}+sign\dot{W}_{1i,j}\frac{\partial\theta_{1i,j}}{\partial l_1}-\frac{C_{1i,j}}{L_1}\frac{\partial\theta_{1i,j}}{\partial\tau}$$

$$-\frac{2\pi r_1 L_1 N_T h_{1i,j}}{L_1}\left(\theta_{1i,j}-\theta_{wi,j}\right)=0$$

$$\left(i=1,\ 2,\ \ldots,\ N_R,\ j=1,\ 2,\ \ldots,\ N\right) \qquad (4.62)$$

where *sign* depends upon flow direction of the tubeside fluid: $sign=-1$ if the tubeside fluid flows in the direction of l_1 or x, and $sign = 1$ otherwise. If it is assumed that the flow maldistribution of fluid 2 outside the tubes is not severe, the governing equation of fluid 2 is derived according to the conventional plug-flow model

$$\dot{W}_{2i,j}\frac{\partial\theta_{2i,j}}{r_2\partial\omega}+\frac{C_{2i,j}}{\pi}\frac{\partial\theta_{2i,j}}{\partial\tau}+\frac{2r_2L_1N_Th_{2i,j}}{L_1}\left(\theta_{2i,j}-\theta_{wi,j}\right)=0$$

$$\left(i=1,\ 2,\ \dots,\ N_R,\ j=1,\ 2,\ \dots,\ N\right). \quad (4.63)$$

The transient equation of the core wall is obtained from the lumped parameter model

$$\frac{C_{wi,j}}{N_T}\frac{\partial\theta_{wi,j}}{\partial\tau}-2\pi r_1L_1h_{1i,j}\left(\theta_{1i,j}-\theta_{wi,j}\right)$$

$$\int_0^{\pi}2r_2L_1h_{2i,j}\left(\theta_{2i,j}-\theta_{wi,j}\right)d\omega=0$$

$$\left(i=1,\ 2,\ \dots,\ N_R,\ j=1,\ 2,\ \dots,\ N\right). \quad (4.64)$$

Defining the following dimensionless parameters and variables

$$\tau_{r1}=\frac{C_{1i,j}}{\dot{W}_{1i,j}},\ z=\frac{\tau}{\tau_{r1}},\ x=\frac{l_1}{L_1},\ \varpi=\frac{\omega}{\pi},\ t=\frac{\theta-\theta_{r2}}{\theta-\theta_{r2}},\ R_{2i,j}=\frac{\dot{W}_{2i,j}}{\dot{W}_{1i,j}},$$

$$Pe_{1i,j}=\frac{\dot{W}_{1i,j}L_1}{\mathcal{D}_{1i,j}A_{c1i,j}},\ R_{ci,j}=\frac{C_{2i,j}}{C_{1i,j}},\ R_{wi,j}=\frac{C_{wi,j}}{C_{1i,j}},$$

$$U_{1i,j}=\frac{2\pi r_1L_1N_Th_{1i,j}}{\dot{W}_{1i,j}},\ U_{2i,j}=\frac{2\pi r_2L_1N_Th_{2i,j}}{\dot{W}_{2i,j}},$$

one can invert the above equations into the following dimensionless forms:

$$\frac{1}{Pe_{1i,j}}\frac{\partial^2 t_{1i,j}}{\partial x^2}+sign\frac{\partial t_{1i,j}}{\partial x}-\frac{\partial t_{1i,j}}{\partial z}-U_{1i,j}\left(t_{1i,j}-t_{wi,j}\right)=0 \tag{4.65}$$

$$\frac{\partial t_{2i,j}}{\partial \varpi}+\frac{R_{ci,j}}{R_{2i,j}}\frac{\partial t_{2i,j}}{\partial z}+U_{2i,j}\left(t_{2i,j}-t_{wi,j}\right)=0 \tag{4.66}$$

$$R_{wi,j}\frac{\partial t_{wi,j}}{\partial z}-U_{1i,j}\left(t_{2i,j}-t_{wi,j}\right)-\int_0^1 U_{2i,j}R_{2i,j}\left(t_{2i,j}-t_{wi,j}\right)\mathrm{d}\,\varpi=0\,. \tag{4.67}$$

Evidently, the temperature distributions of both fluids and the wall are presented as $t_{1i,j}(x,z)$, $t_{2i,j}(x,\varpi,z)$ and $t_{wi,j}(x,z)$. The general initial conditions are prescribed as follows

$$t_{1i,j}(x,\,0)=g_{1i,j}(x),\;t_{2i,j}(x,\varpi,\,0)=g_{2i,j}(x,\varpi),\;t_{wi,j}(x,\,0)=g_{wi,j}(x)\,. \tag{4.68}$$

For multipass cross-flow exchangers, no general forms of the boundary conditions can be given. They depends upon the number of passes and rows per pass and flow arrangements such as cocurrent or countercurrent cross-flow (for a cross-flow heat exchanger with more than one pass) as well as the flow patterns and coupling orders of either fluid between adjacent rows within a pass or passes. In order to coincide with the dispersed plug-flow model, the inlet and outlet boundary conditions of fluid 1 are described as

$$t_{1i,J}-\frac{1}{Pe_{1i,J}}\frac{\partial t_{1i,J}}{\partial x}=f_{1i}(z)\qquad \text{at } x=0 \tag{4.69a}$$

$$\frac{\partial t_{1i,J}}{\partial x}=0\qquad \text{at } x=0 \text{ (even } N\text{)} \quad \text{or} \quad \text{at } x=1 \text{ (odd } N\text{)} \tag{4.69b}$$

where subscript $J=1$ for the cocurrent cross-flow and $J=N$ for countercurrent cross-flow. Because no maldistribution of fluid 2 is assumed and the conventional plug-flow model is used, the related inlet boundary condition is

$$t_{2i,J}(x,0,z) = t_{2i,J}^{I} = f_{2i}(x,z) \tag{4.70}$$

where superscript I designates the state of fluid 2 at the entrance of row i. The following superscript F represents the state of fluid 2 at the outlet of row i. Besides these boundary conditions, the other interfacial conditions between adjacent tube rows as well as passes are needed. The interfacial conditions for fluid 1 are listed in Table 4.1 in which $t_{1I(j,j+1)}$ refers to the intermediate temperature of fluid 1 between row I of pass j and row $\bar{I}$ of pass $j+1$. The values of I and $\bar{I}$ depend upon the coupling order of fluid 1 between two adjacent passes.

If the tubeside fluid is mixed in the header between adjacent passes, the interfacial conditions in Table 4.1 are independent of row number, for example, $t_{1I(j,j+1)} = t_{1(j,j+1)}$. Typically, this may correspond to the case that two adjacent passes have different numbers of rows. The interfacial conditions related to fluid 2 may take different forms from row to row within a given pass or from pass to pass. For two adjacent rows within a pass

$$t_{2(i+)1,j}^{I}(x,0,z) = t_{2i,j}^{F}(x,1,z) \quad (i = 1,\ 2,\ \ldots,\ N_R, j = 1,\ 2,\ \ldots,\ N). \tag{4.71a}$$

For the first row of a pass and the last row of the previous pass, one has

$$t_{21,j+1}^{I}(x,0,z) = t_{2N_R,j}^{F}(x,1,z) \qquad (j = 1,\ 2,\ \ldots,\ N-1)$$

for the inverted coupling order (4.71b)

$$t_{21,j+1}^{I}(x,0,z) = t_{2N_R,j}^{F}(1-x,1,z) \qquad (j = 1,\ 2,\ \ldots,\ N-1)$$

for the identical coupling order (4.71c)

Table 4.1: The interfacial conditions for $t_{1i,j}(x,z)$

$x=0,\ z>0$	$x=1,\ z>0$
$t_{1I,j}=t_{1\bar{I},j+1}=t_{1I(j,j+1)}$ $\dfrac{1}{Pe_{1I,j}}\dfrac{\partial t_{1I,j}}{\partial x}=-\dfrac{1}{Pe_{1\bar{I},j+1}}\dfrac{\partial t_{1I,j+1}}{\partial x}$	$t_{1I,j}=t_{1\bar{I},j+1}=t_{1I(j,j+1)}$ $\dfrac{1}{Pe_{1I,j}}\dfrac{\partial t_{1I,j}}{\partial x}=-\dfrac{1}{Pe_{1\bar{I},j+1}}\dfrac{\partial t_{1I,j+1}}{\partial x}$
even N	
$j=2,\ 4,\ \ldots,\ N-2$	$j=1,\ 3,\ \ldots,\ N-1$
odd N	
$j=2,\ 4,\ \ldots,\ N-1$	$j=1,\ 3,\ \ldots,\ N-2$

$I=i$ and $\bar{I}=i$ for the identical coupling order or $I=i$ and $\bar{I}=N_R+1-i$ for the inverted coupling order. The order number i of row $i=1,\ 2,\ \ldots,\ N_R$)

If fluid 2 is transversely mixed within the space between two adjacent rows in a pass or between the first row of a pass and the last row of the previous pass, the interfacial conditions (4.71) are independent of the spatial variable x. So far one is able to determine dynamic behaviour of cross-flow exchangers from the aforementioned differential equations and related determinant conditions. But one should start the derivation row by row and pass by pass. By applying the Laplace transform, one inverts eqn (4.66) as

$$T_{2i,j}=\frac{U_{2i,j}}{\beta_{i,j}}\left(1-\mathrm{e}^{\beta_{i,j}\varpi}\right)T_{wi,j}(x,s)$$

$$+\mathrm{e}^{-\beta_{i,j}\varpi}\left[T_{2i,j}(x,\ 0,s)+\int_0^{\varpi}R_{1i,j}R_{ci,j}\mathrm{e}^{\beta_{i,j}y}\,g_{2i,j}(x,y)\,\mathrm{d}\,y\right] \quad (4.72)$$

where $\beta_{i,j} = R_{1i,j}R_{ci,j}s + U_{2i,j}$. Thus, the transformed outlet temperature of fluid 2 in row i is

$$T_{2i,j}^{F} = \frac{U_{2i,j}}{\beta_{i,j}}\left(1-\mathrm{e}^{\beta_{i,j}}\right)T_{wi,j}(x,s)$$

$$+\mathrm{e}^{-\beta_{i,j}}\left[T_{2i,j}(x,0,s)+\int_0^1 R_{1i,j}R_{ci,j}\mathrm{e}^{\beta_{i,j}y}\,g_{2i,j}(x,y)\,\mathrm{d}\,y\right]. \quad (4.73)$$

Application of the Laplace transform to eqn (4.67) yields

$$T_{wi,j}(x,s) = a_{1i,j}T_{1i,j}(x,s) + a_{2i,j}T_{2i,j}^{I}(x,0,s) + H_{wi,j}(x,s) \quad (4.74)$$

where

$$a_{1i,j} = \frac{U_{1i,j}}{U_{1i,j}+U_{2i,j}R_{2i,j}-A_1U_{2i,j}R_{2i,j}+R_{wi,j}s},$$

$$a_{2i,j} = \frac{A_1U_{1i,j}R_{2i,j}}{U_{1i,j}+U_{2i,j}R_{2i,j}-A_2U_{2i,j}R_{2i,j}+R_{wi,j}s},$$

$$H_{wi,j}(x,s) = \frac{U_{2i,j}R_{2i,j}G_{2i,j}(x)+R_{wi,j}g_{wi,j}(x)}{U_{1i,j}+U_{2i,j}R_{2i,j}-A_2U_{2i,j}R_{2i,j}+R_{wi,j}s},$$

$$A_1 = \frac{1}{\beta_{i,j}} - \frac{\mathrm{e}^{-\beta_{i,j}}}{\beta_{,ij}}, \quad A_2 = \frac{U_{2i,j}}{\beta_{i,j}} + \frac{U_{2i,j}}{\beta_{i,j}^2}\left(\mathrm{e}^{-\beta_{i,j}}-1\right),$$

$$G_{2i,j}(x) = \int_0^1 \mathrm{e}^{-\beta_{i,j}\varpi}\left[\int_0^{\varpi} R_{1i,j}R_{ci,j}\mathrm{e}^{\beta_{i,j}y}\,g_{2i,j}(x,y)\,\mathrm{d}\,y\right]\mathrm{d}\,\varpi.$$

Insertion of expressions (4.72) and (4.74) into the transformed form of eqn (4.65) leads to

$$\frac{1}{Pe_{1i,j}}\frac{\mathrm{d}^2 T_{1i,j}}{\mathrm{d}x^2}+sign\frac{\mathrm{d}T_{1i,j}}{\mathrm{d}x}+\sigma T_{1i,j}$$

$$=-U_{1i,j}a_{2i,j}T^{I}_{2i,j}-g_{1i,j}(x)-U_{1i,j}H_{wi,j}(x,s) \tag{4.75}$$

where $\sigma=U_{1i,j}a_{1i,j}-U_{1i,j}-s$. By means of the principle of parameter variation, one is able to obtain the general solution to the nonhomogeneous equation (4.75)

$$T_{1i,j}(x,s)=c_{1i,j}\,\mathrm{e}^{m_1 x}+c_{2i,j}\,\mathrm{e}^{m_2 x}+\mathrm{e}^{m_1 x}\int_0^x\frac{\mathrm{e}^{-m_1 t}}{m_2-m_1}P(t,s)\,\mathrm{d}t$$

$$-\,\mathrm{e}^{m_2 x}\int_0^x\frac{\mathrm{e}^{-m_2 t}}{m_2-m_1}P(t,s)\,\mathrm{d}t \tag{4.76}$$

where $P(t,s)=Pe_{1i,j}\left[U_{1i,j}a_{2i,j}T^{I}_{2i,j}(t,\,0,s)+g_{1i,j}(t)+U_{1i,j}H_{wi,j}(t,s)\right]$. m_1 and m_2 are two distinct roots of the following eigenfunction

$$m^2+signPe_{1i,j}m+\sigma Pe_{1i,j}=0\,. \tag{4.77}$$

The unknown coefficients $c_{1i,j}$ and $c_{2i,j}$ are determined by means of the given boundary and interfacial conditions related to the flow arrangement and the coupling order. Since the dispersed plug-flow model is applied, the outlet boundary condition of fluid 1 is needed. Generally, it is tedious to find these coefficients which are coupled by the recursive relationships among rows and passes. If fluid 2 is transversely mixed between adjacent rows, expressions (4.71) is independent of the variable *x*. The matrix form of the linear algebraic equations about these coefficients may be written as follows:

$$\mathbf{AC}=\mathbf{B} \tag{4.78}$$

where **C** is the coefficient vector; matrix **A** and vector **B** are obtained from expressions (4.72) and (4.76). Once these coefficients have been found, the transformed temperatures $T_{1i,j}$, $T_{2i,j}$ and $T_{wi,j}$ are derived. If the conventional plug-flow model is also applied to fluid 1, $T_{1i,j}$ can directly obtained as

$$T_{1i,j}(x,s) = \mathrm{e}^{-mx}\left[c_{1i,j} + sign\int_0^x \mathrm{e}^{mx} P(t,s)\,\mathrm{d}t\right] \tag{4.79}$$

where

$$m = sign\sigma\,,$$

$$P(t,s) = sign\left[U_{1i,j}a_{2i,j}T^I_{2i,j}(t,\,0,\,s) + g_{1i,j}(t) + U_{1i,j}H_{wi,j}(t,s)\right].$$

In this case no outlet boundary conditions are needed. By means of the recursive relationship, hence, one can calculate $T_{1i,j}$ and $T_{2i,j}$ as well as $T_{wi,j}$ row by row in a pass and from pass to pass to obtain the solutions corresponding to the specified row in the specified pass. Therefore, the given inlet boundary conditions of both fluids yield

$$c_{1i,1} = F_{1i}(s)\,, \qquad T^I_{21,1}(x,\,0,\,s) = F_2(x,s)\,. \tag{4.80}$$

The other coefficients $c_{1i,j}$ and $T_{2i,j}(x,\,0,\,s)$ $(1 \le i \le N_R,\ 1 < j \le N)$ are stepwise found from the interfacial conditions and the recursive relationships. If the identical coupling order is assumed in a cross-flow with the even pass number N, for example, one has

$$c_{1i,j} = T_{1i,j-1}(1,s) \qquad (j = 2,\ 4,\ \ldots,\ N-2) \tag{4.81a}$$

$$c_{1i,j} = T_{1i,j-1}(0,s) \qquad (j = 3,\ 5,\ \ldots,\ N-1) \tag{4.81b}$$

If fluid 2 is mixed in the space between two adjacent rows, $T^I_{2i,j}$ is independent of x:

$$T^{I}_{2i,j} = \int_0^1 T_{2i,j-1}(x, 1, s)\,\mathrm{d}x\,. \tag{4.82}$$

Thus, the transformed solution in the image-domain are found by the discrete approach. The following task is to perform their inverse transform into the real-time domain. The numerical inverse algorithm is used. For the Laplace transform pair

$$F(\xi, s) = \int_0^\infty f(\xi, \eta)\,\mathrm{e}^{-s\eta}\,\mathrm{d}\eta \tag{4.83a}$$

$$f(\xi, \eta) = \frac{1}{2\pi i}\int_{c_1 - i\infty}^{c_1 + i\infty} F(\xi, s)\,\mathrm{e}^{s\eta}\,\mathrm{d}s \tag{4.83b}$$

the formula of the numerical inversion can be expressed as

$$f(\xi, \eta) = \sum_{k=1}^{M} \phi_k F[\xi, \varphi_k(\eta)] \tag{4.84}$$

where ξ is considered as a parameter; ϕ_k and φ_k may take different forms according to the numerical inversion algorithm. M is selected according to the desired accuracy. Two of the typical algorithms are introduced in Chapter 3. Application of formula (4.84) to the above-derived transformed solutions results in transient responses to arbitrary inlet temperature variations and step flow disturbances in the time domain.

4.3.2 The continuous approach

Evidently, the discrete approach of handling dynamic simulation of cross-flow exchangers is considerably complicated. In most cases, such a problem has so far been treated by the continuous approach without paying attention to a single row in any pass. According to this approach it is assumed that all rows in a pass construct a continuous heat transfer surface. In the light of the similar procedure to that described in Sections 4.1. or 4.2, the continuous approach results in the following energy equations of both fluids in any pass of a multipass cross-flow heat exchanger:

$$\frac{1}{Pe_{1j}}\frac{\partial^2 t_{1j}}{\partial x^2}+sign\frac{\partial t_{1j}}{\partial x}-\frac{\partial t_{1j}}{\partial z}-U_{1j}\left(t_{1j}-t_{wj1}\right)=0 \tag{4.85}$$

$$\frac{\partial t_{2j}}{\partial y}+\frac{R_{cj}}{R_{2j}}\frac{\partial t_{2j}}{\partial z}+U_{2j}\left(t_{2j}-t_{wj2}\right)=0\,. \tag{4.86}$$

The dimensionless spatial variable y is along the flow direction of fluid 2 and is defined as $y=l_2/L_2$. Similarly to the discrete approach, here the dispersed plug-flow model is applied to fluid 1 to account for all possible types of tubeside flow maldistribution. With regard to the transverse conduction resistance of the core wall, the conduction equation of the wall yields:

$$\frac{\partial t_{wj}}{\partial z}-Fo_{wj}\frac{\partial^2 t_{wj}}{\partial \zeta^2}=0 \tag{4.87a}$$

$$t_{wj}(x,y,\zeta,0)=g_{wj}(x,y,\zeta) \tag{4.87b}$$

$$-\frac{\partial t_{wj}}{\partial \zeta}+Bi_{j1}\left(t_{wj}-t_{1j}\right)-0 \qquad \text{at } \zeta=0 \tag{4.87c}$$

$$\frac{\partial t_{wj}}{\partial \zeta}+Bi_{j2}\left(t_{wj}-t_{2j}\right)=0 \qquad \text{at } \zeta=1 \tag{4.87d}$$

where $\zeta=l_3/L_3$, $Fo_{wj}=a_{wj}\tau_{r1j}/L_3^2$, $Bi_{1j}=h_{1j}L_3/\lambda_{wj}$ and $Bi_{2j}=h_{2j}L_3/\lambda_{wj}$. If fouling of the heat transfer surface is severe, expression (4.52) can be used to substitute an apparent conductivity for the heat conductivity of the core wall:

$$\lambda_w^* = \frac{\lambda_w L_3}{L_3 + \lambda_w (R_{f1} + R_{f2})} \tag{4.52}$$

There exist the following relationships between parameters Fo_{wj}, Bi_{1j} and Bi_{2j}:

$$Fo_{wj} Bi_{1j} = \frac{U_{1j}}{R_{wj}}, \qquad Bi_{2j} = Bi_{1j} R_2 \frac{U_{2j}}{U_{1j}}. \tag{4.88}$$

Under the given determinant conditions, application of the Laplace transform to eqn (4.87) yields the following transformed solution

$$T_{wj}(x, y, \zeta, s) = c_{1j} e^{\gamma_{1j}\zeta} + c_{2j} e^{\gamma_{2j}\zeta} + G_{wj}(x, y, \zeta, s) \tag{4.89}$$

where

$$G_{wj}(x, y, \zeta, s) = \frac{1}{(\gamma_{2j} - \gamma_{1j}) Fo_{wj}}$$

$$\times \left[\mathrm{e}^{\gamma_{1j}\zeta} \int_0^\zeta \mathrm{e}^{-\gamma_{1j} t} g_{wj}(x, y, t) \mathrm{d}t - \mathrm{e}^{\gamma_{2j}\zeta} \int_0^\zeta \mathrm{e}^{-\gamma_{2j} t} g_{wj}(x, y, t) \mathrm{d}t \right],$$

$$c_{1j} = a_{0j} + a_{1j} T_{1j} + a_{2j} T_{2j}, \qquad c_{2j} = b_{0j} + b_{1j} T_{1j} + b_{2j} T_{2j},$$

$$a_{0j} = \frac{(Bi_{j1} - \gamma_{2j}) \left[Bi_{j2} G_{wj}(x, y, 1, s) + G'_{wj}(x, y, 1, s) \right]}{\Delta_{wj}},$$

$$a_{1j} = \frac{Bi_{j1} (Bi_{j2} + \gamma_{2j}) e^{\gamma_{2j}}}{\Delta_{wj}}, \qquad a_{2j} = -\frac{Bi_{j2} (Bi_{j1} - \gamma_{2j})}{\Delta_{wj}},$$

$$b_{0j} = -\frac{\left(Bi_{j1} - \gamma_{1j}\right)\left[Bi_{j2} G_{wj}(x, y, 1, s) + G'_{wj}(x, y, 1, s)\right]}{\Delta_{wj}},$$

$$b_{1j} = -\frac{Bi_{j1}\left(Bi_{j2} + \gamma_{1j}\right)e^{\gamma_{1j}}}{\Delta_{wj}}, \quad b_{2j} = \frac{Bi_{j2}\left(Bi_{j1} - \gamma_{1j}\right)}{\Delta_{wj}},$$

$$\Delta_{wj} = \left(Bi_{j1} - \gamma_{1j}\right)\left(Bi_{j2} + \gamma_{2j}\right)e^{\gamma_{2j}} - \left(Bi_{j1} - \gamma_{2j}\right)\left(Bi_{j2} + \gamma_{1j}\right)e^{\gamma_{1j}},$$

$$G'_{wj}(x, y, \zeta, s) = \frac{\partial G_{wj}(x, y, \zeta, s)}{\partial \zeta}, \; \gamma_{i,2j} = \pm\sqrt{\frac{s}{Fo_{wj}}}.$$

Therefore, T_{wj1} and T_{wj2} are obtained as follows

$$T_{wj1} = T_{wj}(x, y, 0, s) = \left(a_{1j} + b_{1j}\right)T_{1j} + \left(a_{2j} + b_{2j}\right)T_{2j} + H_{wj}(x, y, 0, s) \tag{4.90a}$$

$$T_{wj2} = T_{wj}(x, y, 1, s) = \left(a_{1j}e^{\gamma_{1j}} + b_{1j}e^{\gamma_{2j}}\right)T_{1j} + \left(a_{2j}e^{\gamma_{1j}} + b_{2j}e^{\gamma_{2j}}\right)T_{2j} + H_{wj}(x, y, 1, s) \tag{4.90b}$$

where $H_{wj}(x, y, \zeta, s) = a_{0j}\, e^{\gamma_{1j}\zeta} + b_{0j}\, e^{\gamma_{2j}\zeta} + G_{wj}(x, y, \zeta, s)$. Insertion of expressions (4.90) into the transformed forms of eqns (4.85) and (4.86) yields

$$\frac{1}{Pe_{1j}}\frac{\partial^2 T_{1j}}{\partial x^2} + sign\frac{\partial T_{1j}}{\partial x} + \sigma_1 T_{1j} + \sigma_2 T_{2j} = H_{1j}(x, y, s) \tag{4.91}$$

$$\frac{\partial T_{2j}}{\partial y} - \varphi_1 T_{1j} + \varphi_2 T_{2j} = H_{2j}(x, y, s) \tag{4.92}$$

where

$$\sigma_1 = \left[U_{1j}\left(a_{1j} + b_{1j}\right) - U_{1j} - s\right], \qquad \sigma_2 = U_{1j}\left(a_{2j} + b_{2j}\right),$$

$$H_{1j}(x, y, s) = g_{1j}(x, y) - U_{1j} H_{wj}(x, y, 0, s),$$

$$H_{2j}(x, y, s) = \frac{R_{cj}}{R_{2j}} g_{2j}(x, y) + U_{2j} H_{wj}(x, y, 1, s),$$

$$\varphi_1 = U_{2j}\left(a_{1j}\,\mathrm{e}^{\gamma_{1j}} + b_{1j}\,\mathrm{e}^{\gamma_{2j}}\right),$$

$$\varphi_2 = \left[\frac{R_{cj}}{R_{2j}} s + U_{2j} - U_{2j}\left(a_{2j}\,\mathrm{e}^{\gamma_{1j}} + b_{2j}\,\mathrm{e}^{\gamma_{2j}}\right)\right].$$

By applying the Laplace transform variable q against the dimensionless variable y and the dual Laplace transform, these two equations become

$$\frac{1}{Pe_{1j}} \frac{\mathrm{d}^2 \widehat{T}_{1j}}{\mathrm{d}x^2} + sign \frac{\mathrm{d}\widehat{T}_{1j}}{\mathrm{d}x} + \omega_1 \widehat{T}_{1j} = \widehat{E}_j(x, q, s) \tag{4.93}$$

$$\widehat{T}_{2j} = \frac{\varphi_1}{\varphi_2 + q} \widehat{T}_{1j} + \frac{\widehat{H}_{2j}(x, q, s) + T_{2j}(x, 0, s)}{\varphi_2 + q} \tag{4.94}$$

where

$$\omega_1 = \sigma_1 + \frac{\sigma_2 \varphi_1}{\varphi_2 + q},$$

$$\widehat{E}_j(x, q, s) = \widehat{H}_{1j}(x, q, s) - \frac{\sigma_2\left[\widehat{H}_{2j}(x, q, s) + T_{2j}(x, 0, s)\right]}{\varphi_2 + q}.$$

The general solution to eqn (4.93) is found as

$$\widehat{T}_{1j} = D_{1j}\,\mathrm{e}^{m_{1j}x} + D_{2j}\,\mathrm{e}^{m_{2j}x} + \widehat{I}_j(x,q,s) \tag{4.95}$$

where

$$m_{1j,2j} = \frac{-\,signPe_{1j} \pm \sqrt{Pe_{1j}^2 - 4Pe_{1j}\omega_j}}{2},$$

$$\widehat{I}_j(x,q,s) = \frac{Pe_{1j}}{m_{2j} - m_{1j}}$$

$$\times\left[\mathrm{e}^{m_{2j}x}\int_0^x \mathrm{e}^{-m_{2j}t}\,\widehat{E}_j(t,q,s)\,\mathrm{d}t - \mathrm{e}^{m_{1j}x}\int_0^x \mathrm{e}^{-m_{1j}t}\,\widehat{E}_j(t,q,s)\,\mathrm{d}t\right].$$

For dynamic simulation of a cross-flow exchanger with N counterdirected passes based on the dispersed plug-flow model, there are $2N$ unknown coefficients which should be determined according to the inlet, outlet and interface conditions on the side of fluid 1. Combination of these determinant conditions and eqn (4.95) results in $2N$ linear algebraic equations for these coefficients. The matrix form of these algebraic equations is expressed as

$$\mathbf{AD} = \mathbf{B} \tag{4.96}$$

where elements a_{ij} of $2N \times 2N$ matrix **A** depend upon the boundary conditions, the number of passes and the coupling order between two adjacent passes, **D** is the unknown coefficient vector $\mathbf{D} = \{d_{11},\, d_{21},\, d_{12},\, d_{22},\, \ldots,\, d_{1N},\, d_{2N}\}^{\mathrm{T}}$ and vector **B** consists of the transformed inlet temperatures such as $\widehat{F}_1(q,s)$, $F_{21}(x,s)$, $\widehat{T}_{1j}(0,q,s)$, $\widehat{T}_{1j}(1,q,s)$, and $\widehat{I}_j(1,q,s)$ as well as $\widehat{I}_j'(1,q,s)$. Clearly, the unknown $T_{2j}(x,0,s)$ is still included in eqn (4.96) for $j \geq 2$, if the dynamic simulation is aimed at a multipass cross-flow exchanger with more than one pass. In this case, iteration is needed. With an increasing number of tube passes, the iterative process becomes troublesome. To avoid such a mathematical difficulty, one may take the following recursive approach: each pass is considered as a single cross-flow exchanger, and then the real exchanger with N passes is a composite of such N single pass fictitious exchangers. The advantage of such artificial division is that the transient simulation for each unit can be done recursively. Thus, the corresponding boundary conditions for the first fictitious exchanger are:

$$t_{11} - \frac{1}{Pe_{11}} \frac{\partial t_{11}}{\partial x} = f_1(z) \qquad \text{at } x = 0 \tag{4.97a}$$

$$\frac{\partial t_{11}}{\partial x} = 0 \qquad \text{at } x = 1 \tag{4.97b}$$

$$t_{21} = f_2(x, z) \qquad \text{at } y = 0 \tag{4.97c}$$

and for the other fictitious exchangers $(1 < j < N)$

$$t_{1j} + \frac{sign}{Pe_{1j}} \frac{\partial t_{1j}}{\partial x} = f_{1j}(z) \qquad \text{at } x = \delta_j \tag{4.98a}$$

$$\frac{\partial t_{1j}}{\partial x} = 0 \qquad \text{at } x = 1 - \delta_j \tag{4.98b}$$

$$t_{2j} = f_{2j}(x, z) \qquad \text{at } y = 0 \tag{4.98c}$$

where $\delta_j = 0$ or 1 which depends upon the coupling order between two adjacent passes. Once the temperatures $f_1(y, z)$ and $f_2(x, z)$ of the incoming fluids are given as the inlet boundary conditions, $f_{1j}(y, z)$ and $f_{2j}(x, z)$ $(j > 2)$ are obtained from the outlet temperature profiles of the preceding fictitious exchanger. This approach is feasible because two adjacent passes are coupled in such a way that the inlet temperature of a fluid related to the consequent unit is equal to the outlet temperature of the same fluid pertinent to the preceding unit. For the traditional plug-flow model, the performance at the upward position (e.g., upstream) is affected by any disturbance at the backward position (e.g., downstream) because the fluid flow is assumed to be incompressible and subsonic. For the dispersed plug-flow model, the diffusion term is introduced and performances in two adjacent units affect each other. If the U-type tubes are used, the interfacial conditions similar to those listed in Table 4.1 are suitable. But for the types of headers and coupling manners between two adjacent passes, the approach of dividing a multipass cross-flow exchanger with N passes into N fictitious units may be suitable and practicable. Further computation has shown that such treatment is reliable since the energy balance between the hot and cold fluids is sustained. Therefore, eqns (4.93) and (4.94) are valid for each fictitious unit. With the boundary conditions (4.97), the transformed solution related to the first pass can be found. From its outlet results and the coupling order between the first and second pass, $\widehat{F}_{12}(q,s)$ and $\widehat{F}_{22}(x,s)$ which are respectively pertinent to $f_{12}(y,z)$ and $f_{22}(x,z)$ are determined, and then the transform temperatures $\widehat{T}_{12}(x,q,s)$ and $\widehat{T}_{22}(x,q,s)$ for the second pass are obtained. In such a recursive way, the solutions in the dual-transform image-domain for the whole cross-flow exchanger can be derived. Thus, one obtains D_{1j} and D_{2j} in expression (4.95)

$$D_{1j} = \frac{\widehat{F}_{1j}^{*}(q,s)B_{2j} + A_{2j}\widehat{T}_j'(1-\delta_j,q,s)}{A_{1j}B_{2j} - A_{2j}B_{1j}} \tag{4.99a}$$

$$D_{2j} = -\frac{\widehat{F}_{1j}^{*}(q,s)B_{1j} + A_{1j}\widehat{T}_j'(1-\delta_j,q,s)}{A_{1j}B_{2j} - A_{2j}B_{1j}} \tag{4.99b}$$

where

$$A_{1j}=\left(1+\frac{sign\, m_{1j}}{Pe_{1j}}\right)\mathrm{e}^{m_{1j}\delta_j},\; A_{2j}=\left(1+\frac{sign\, m_{2j}}{Pe_{1j}}\right)\mathrm{e}^{m_{2j}\delta_j},$$

$$B_{1j}=m_{1j}\,\mathrm{e}^{m_{1j}(1-\delta_j)},\; B_{2j}=m_{2j}\,\mathrm{e}^{m_{2j}(1-\delta_j)},$$

$$\widehat{F}^*_{1j}(q,s)=\widehat{F}_{1j}(q,s)-\widehat{I}_j(\delta_j,q,s)-\frac{sign\,\widehat{I}'_j(\delta_j,q,s)}{Pe_{1j}},\; \widehat{I}'_j(x,q,s)=\frac{\mathrm{d}\widehat{I}_j}{\mathrm{d}x}.$$

This procedure can be extended to the discrete approach to find the unknown coefficients in expression (4.76). For the special case of uniform initial conditions and $t_{21}(x,0,z)=0$, $\widehat{F}^*_{1j}(q,s)=\widehat{F}_{1j}(q,s)$ and $\widehat{I}_1(x,q,s)=0$. If no flow maldistribution occurs in fluid 1 ($Pe_{1j}\rightarrow\infty$), eqn (4.65) is reduced to the form related to the conventional plug-flow model. In this case, the solution procedure is simpler. In the dual Laplace transform domain, the expression for $\widehat{T}_{2j}(x,q,s)$ is formally the same as (4.94), but $\widehat{T}_{1j}(x,q,s)$ takes a simpler form

$$\widehat{T}_{1j}=d_j\,\mathrm{e}^{-\overline{\omega}_1 x}+\mathrm{e}^{-\overline{\omega}_1 x}\int_0^x \mathrm{e}^{\overline{\omega}_1 x}\,\widehat{E}^*_j(t,q,s)\,\mathrm{d}t \qquad (4.100)$$

where $\overline{\omega}_1=sign\omega_1$ and $\widehat{E}^*_j(t,q,s)=sign\,\widehat{E}_j(t,q,s)$. For the solution from the plug-flow model, only the inlet conditions are needed. Thus, $d_1=\widehat{F}_1(q,s)$ and $\widehat{T}_{11}$ is obtained from (4.100)

$$\widehat{T}_{11}(x,q,s)=\widehat{F}_1(q,s)\,\mathrm{e}^{-\overline{\omega}_1 x}+\mathrm{e}^{-\overline{\omega}_1 x}\int_0^x \mathrm{e}^{\overline{\omega}_1 x}\,\widehat{E}^*_1(t,q,s)\,\mathrm{d}t \qquad (4.101)$$

In a multipass cross-flow heat exchanger with N $(N>1)$ tube passes, the transformed temperatures related to the other passes $(j>1)$ can be stepwise found since c_{1j} and $T_{2j}(x,0,s)$ are obtained from the transformed solution pertinent to the preceding pass and the coupling order between two adjacent passes. It is clear that once the transformed solution has been found for each fictitious unit, the inverse transform must correspondingly be performed. For this purpose, the numerical inverse algorithm (4.24) can be used here to obtain the transient temperature distributions. The exit temperatures of both fluids are calculated according to

$$t_{1e} = \int_0^1 t_{1N}\left(x = 1 - \delta_j, y, z\right) \mathrm{d}\, y \tag{4.102a}$$

$$t_{2e} = \int_0^1 t_{2N}\left(x, y = 1, z\right) \mathrm{d}\, y \tag{4.102b}$$

One can similarly calculate the exit temperatures of both fluids at each fictitious unit by

$$t_{1ej} = \int_0^1 t_{1j}\left(x = 1 - \delta_j, y, z\right) \mathrm{d}\, y \tag{4.103a}$$

$$t_{2ej} = \int_0^1 t_{2j}\left(x, y = 1, z\right) \mathrm{d}\, y \tag{4.103b}$$

4.3.3 Examples and discussions

By both the discrete and continuous approaches, the dynamic behaviour of multipass cross-flow heat exchangers have been simulated in the above two sections. The numerical inversion of the Laplace or dual Laplace transform yields transient responses to arbitrary forms of inlet temperature disturbance or step change of flow rate, i.e., temperatures of both fluids and the core wall in the time-domain. By assuming $f_1(y,z) = 1$ and $f_2(x,z) = 0$, some examples are calculated. The curves in Fig. 4.9 illustrate the effect of flow maldistribution on dynamics of a single pass cross-flow heat exchanger with different values of Pe_1. A smaller value of Pe_1 corresponds to stronger flow maldistribution and the case of $Pe_1 \to \infty$ means the conventional plug-flow model. Flow maldistribution may remarkably degrade performance of cross-flow

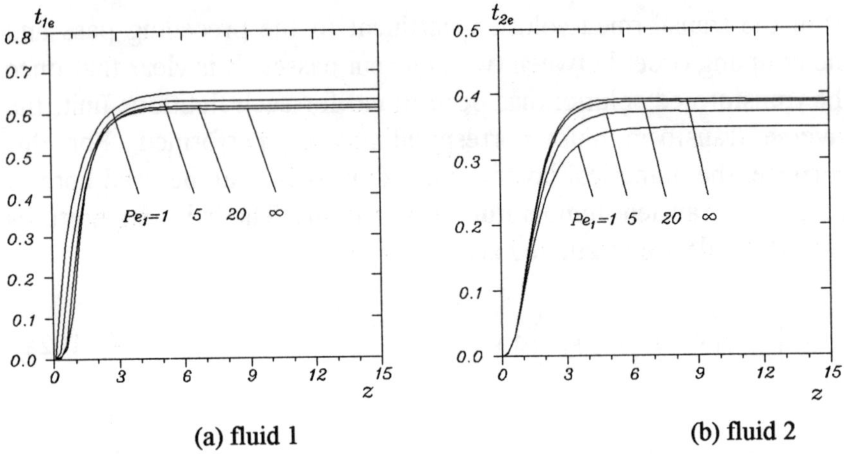

Figure 4.9: Exit temperature profiles in a single pass cross-flow heat exchanger ($R_1 = R_c = R_w = 1$, $Bi_1 = 0.2$, $U_1 = U_2 = 1.5$)

exchangers for $Pe_1 < 5$. In such cases, flow maldistribution reduces thermal effectiveness of exchangers. For $Pe_1 > 20$, however, the dispersed plug-flow model approaches the conventional plug-flow model and the effect of flow maldistribution may be neglected.

As mentioned before, there are two different ways of simulating multipass cross-flow heat exchangers: the so-called discrete and the continuous approach. The former provides more detailed information about the transient behaviour from row to row; the latter treats all rows in a pass as a continuous surface and fails to consider the real flow pattern (for example, mixing or no mixing) and different coupling orders between two adjacent rows. Generally, the continuous approach is the limit of the discrete one. The deviation between them decreases with increasing number of tube rows per pass. Under the assumption of fluid 2 mixing between adjacent rows, some curves pertinent to a single pass cross-flow exchanger are plotted in Fig. 4.10. They indicate that both approaches lead to different results for $N_R = 1$ and the deviation becomes negligible for $N_R > 4$. Furthermore, both approaches reveal almost the same dynamic behaviour within the span $z < 2$ for any row number N_R. But the discrete approach makes it possible to obtain transient temperature distributions along different rows.

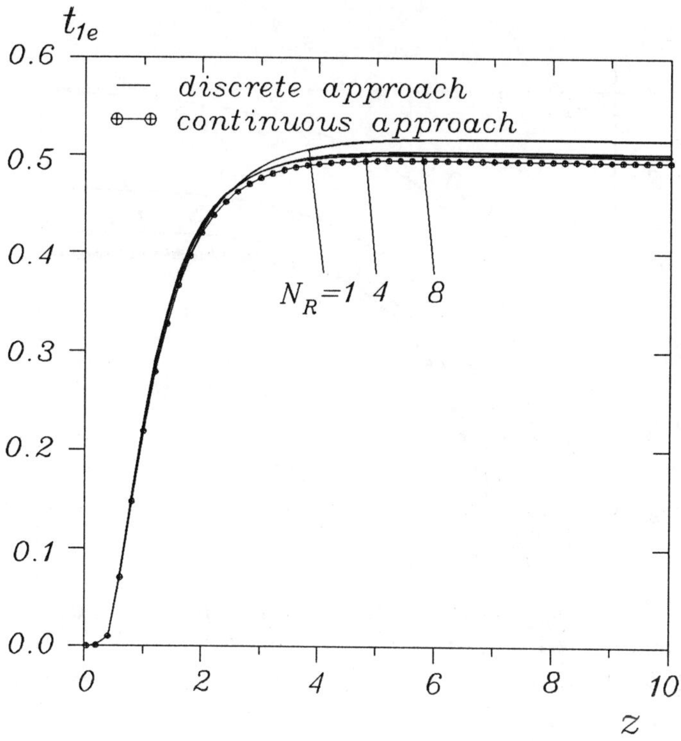

Figure 4.10: Comparison between the discrete and continuous approach ($Pe_1 = 10$, $R_1 = 0.8$, $R_c = R_w = 1$, $Bi_1 = 0.2$, $U_1 = U_2 = 2$)

Figures 4.11 and 4.12 present the influcncc of the parameters U_{1j}, U_{2j} and R_{1j} on transient responses in a two-pass cross-flow heat exchanger in which fluid 1 is coupled in the identical order and fluid 2 in the inverted order, respectively. In these figures U means $U_{11} = U_{12} = U_{21} = U_{22}$ and R means $R_{11} = R_{12}$. As expected, the effectiveness increases with U. But the tendency becomes flatter for $U > 2$. A greater value of R can damp transient response of fluid 1, but contrarily promote temperature change of fluid 2. In addition, the examples in these figures show that the algorithm can be used either for dynamic simulation or for steady-state rating and design of multipass cross-flow heat exchangers. For example, the curves in Fig. 4.9 as well as the detailed results reveal the steady-state energy balance between both fluids.

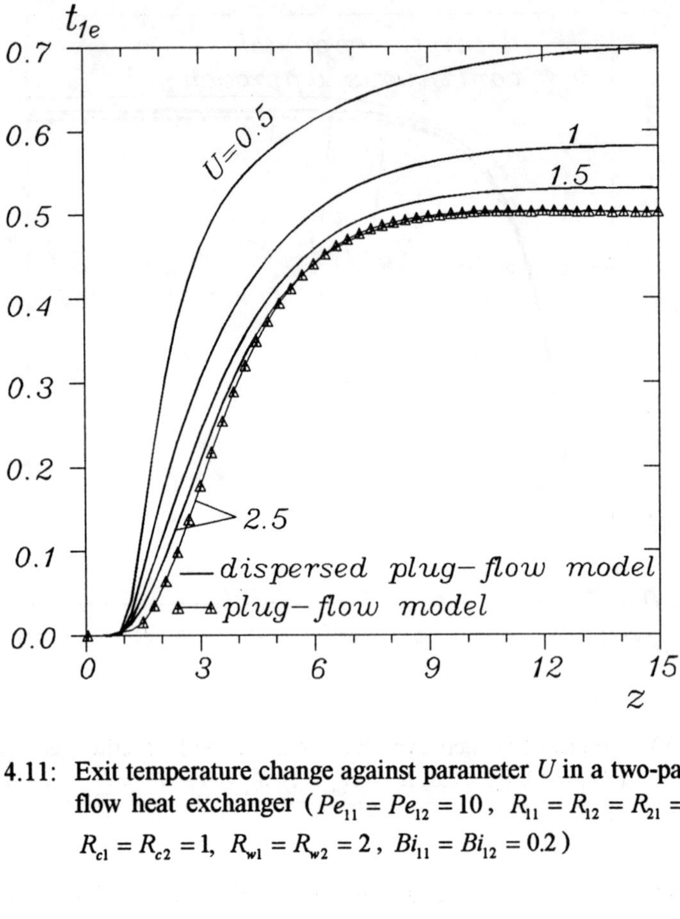

Figure 4.11: Exit temperature change against parameter U in a two-pass cross-flow heat exchanger ($Pe_{11} = Pe_{12} = 10$, $R_{11} = R_{12} = R_{21} = R_{22} = 1$, $R_{c1} = R_{c2} = 1$, $R_{w1} = R_{w2} = 2$, $Bi_{11} = Bi_{12} = 0.2$)

4.4 Transient analysis of finned cross-flow heat exchangers

In cross-flow heat exchangers, various types of extended surfaces — fins are often bonded on the side with a low heat transfer coefficient to compensate heat transfer resistances on both sides. The fin performance is discussed in detail by Incropera and DeWitt (1985). Generally, the transient analysis of cross-flow exchangers pays little attention to the effect of the fins to avoid a complicated mathematical derivation because dynamic simulation with at least two spatial variables besides the time variable makes the problem

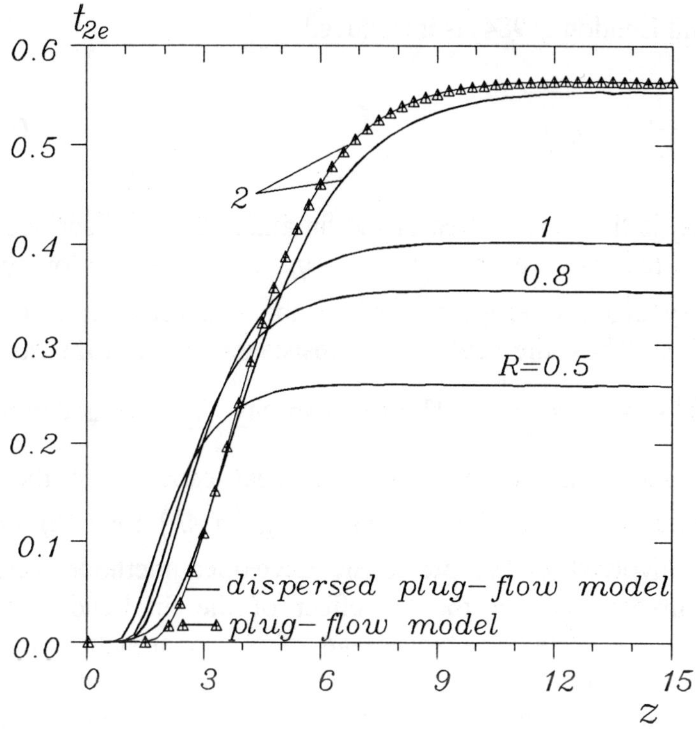

Figure 4.12: Exit temperature change against parameter R in a two-pass cross-flow heat exchanger ($Pe_{11} = Pe_{12} = 5$, $U_{11} = U_{12} = 1$, $R_{c1} = R_{c2} = 1$, $R_{w1} = R_{w2} = 1$, $Bi_{11} = Bi_{12} = 0.4$)

difficult enough. Therefore, the fins installed in most cross-flow exchangers are either neglected or modelled with lumped parameters. As shown by Müller (1976), the transient heat conduction in the fins may have a considerable influence on the dynamics of the heat exchanger in some cases. The following context introduces three ways of taking the effect of fin on dynamic characteristics of cross-flow heat exchangers into account.

4.4.1 The quasi-lumped parameter method

With this method, one separately takes the conduction resistance and the heat capacity of the fin into account. As for the conduction resistance, the total surface temperature effectiveness η_0 defined by Kays and London (1984) is introduced

$$\eta_0 = 1 - \frac{A_f}{A}\left(1 - \eta_f\right) \tag{4.104}$$

where A_f is the surface area of the fin and A the total surface area. The fin temperature effectiveness η_f is a function of the fin geometry and properties as well as the convective heat transfer coefficient. Thus, the heat transfer resistance between the fluid and the solid surface is $\frac{1}{\eta_0 Ah}$. The heat capacity C_f can be added to that of the core wall, so that the total heat capacity of the solid component is $C_w^* = C_f + C_w$. Substituting $(\eta_0 Ah)$ for (Ah) and C_w^* for C_w involved in the previously developed methods, one can approximately account for the effect of the fin bonded on the preliminary surface. The advantage of this approach is that the existing simulation algorithm can be extended to finned cross-flow heat exchangers with only a slight modification of some parameters. The disadvantage is that it cannot provide the exact description of the transient temperature distribution within the fin and may lead to greater deviation in some cases (for example, for thick fins). According to the assumption that the isotherms within the fin are in concentric circular patterns, van Aken (1994) proposed a discrete quasi-lumped parameter method to handle the effect of the fin on dynamics of finned cross-flow exchangers. He divided the fin into a number of concentric rings and applied the conventional lumped parameter model to each ring.

4.4.2 The basic element method

To account for the considerable influence of the transient heat conduction in the fin on the dynamics of an exchanger in some cases, Müller (1976) proposed the so-called basic element method. According to this method, the heat exchanger is divided into small geometric parts. The transient mathematical model for each element is derived and recombined for all elements to describe the transient behaviour of the entire heat exchanger. Ober (1978) and Schmachtenberg (1981) applied this method to the special heat exchangers involved in air-conditioning systems.

With help of the perturbation principle, Kabelac (1987) used the basic element method to simulate the transient response of cross-flow heat exchangers initiated by arbitrary inlet temperature or mass flow rate disturbances. Under the following assumptions:

(1) heat conduction in the fin is allowed only in the direction perpendicular to gas flow;

(2) the temperature profile of the gas in this direction is approximated by a linear function;

(3) there is no temperature dependence of material properties;

(4) there is no energy transport in the gas except in the direction of flow;

(5) liquid temperature and velocity are constant in the radial and circumferential direction as well as throughout the basic element;

(6) energy transport in the liquid due to the deviation from plug-flow is considered by means of a dispersion factor.

Kabelac (1989) proposed a set of transient differential equations for the fin, the gas, the tube and the liquid in each element. A similar procedure is applied to all of the basic elements which make up the heat exchanger. These governing equations are conjugated by the heat transfer relationship among the components included in the element. By means of the Laplace transform and the numerical inverse algorithm, the final results are given analytically as a transfer function in the frequency-domain or numerically as transient temperature distributions in the time-domain.

4.4.3 The numerical approach

For the design and real-time control of plate-fin heat exchangers used in cryogenic industry and gas processing, Pingaud *et al.* (1989) proposed an algorithm for steady-state and transient simulation of multistream multipass plate-fin heat exchangers. Modelling is based on the mass, momentum and energy balances. An implicit integration scheme was adapted to treat countercurrent flow. The following assumptions were introduced:

(1) phase change may occur at any point of the exchanger;

(2) temperatures in any channel are considered as uniform over any cross-section perpendicular to the flow direction;

(3) temperature drop across parting sheets is negligible but the temperature of the metal in each layer is different;

(4) flow of all streams is parallel to each other in either cocurrent or countercurrent directions;

(5) energy and momentum balances can be solved separately;

(6) heat loss is negligible;

(7) thermodynamic equilibrium exists between vapour and liquid phases;

(8) liquid hold-up and vapour retention are considered with the representation of phase slip ratio and void fraction.

On the basis of these idealizations, Pingaud *et al.* derived differential balance equations of material, momentum, and energy for fluids and solid sheets, respectively. This coupled system of nonlinear equations was solved with the finite-difference method. The successful application of the model to solve different types of industrial case studies showed its wide applicability and flexibility.

4.5 References

Chen, H. T. and Chen, K. C. (1991) Simple Method for Transient Response of Gas-to-Gas Cross-Flow Heat Exchangers with Neither Gas Mixed, *International Journal of Heat and Mass Transfer,* **34**, 2891–2898.

Dusinberre, G. M. (1959) Calculation of Transients in a Cross-Flow Heat Exchanger, *Journal of Heat Transfer,* **81**, 61–67.

Gvozdenac, D. D. (1986) Analytical Solution of the Transient Response of Gas-to-Gas Crossflow Heat Exchanger with Both Fluids Unmixed, *Journal of Heat Transfer,* **108**, 723–727.

Gvozdenac, D. D. (1991) Dynamic Response of the Crossflow Heat

Exchanger with Finite Wall Capacitance, *Wärme- und Stoffübertragung*, **26**, 207–221.

Incropera, F. P. and DeWitt, D. P. (1985) *Fundamentals of Heat and Mass Transfer*, Wiley, New York.

Kabelac, S. (1987) *Die Berechnung des Zeitverhaltens berippter Kreuzstromwärmeübertrager*, Dissertation, Universität Hannover.

Kabelac, S. (1989) The Transient Response of Finned Crossflow Heat Exchangers, *International Journal of Heat and Mass Transfer*, **32**, 1183–1198.

Kays, W. M. and London, A. L. (1984) *Compact Heat Exchangers*, McGraw-Hill Book Company, New York.

Luo, X. (1998) *Das axiale Dispersionsmodell für Kreuzstromwärmeübertrager*, Fortschritt-Berichte VDI, Reihe 19, Wärmetechnik/Kältetechnik, Nr. 109, VDI-Verlag GmbH, Düsseldorf.

Müller, K. (1976) *Das Dynamische Verhalten berippter Kreuzstromwärmeübertrager*, Dissertation, Universität Karlsruhe.

Myers, G. E., Mitchell, J. W. and Norman, R. F. (1967) The Transient Response of Crossflow Heat Exchangers, Evaporators, and Condensers, *Journal of Heat Transfer*, **89**, 75–80.

Ober, Ch. (1978) *Verhalten von Kreuzstromwärmeübertragern beim Kühlen und Entfeuchten von Luft*, Dissertation, Universität Karlsruhe.

Pingaud, H., Le Lann, J. M., and Koehret, B. (1989) Steady-Sate and Dynamic Simulation of Plate Fin Heat Exchangers, *Computers in Chemical Engineering*, **13**, 577–585.

Romie, F.E. (1983) Transient Response of Gas-to-Gas Crossflow

Heat Exchangers with Neither Gas Mixed, *Journal of Heat Transfer*, **105**, 563–570.

Roetzel, W. (1996) Transient Analysis in Heat Exchangers, in *New Development in Heat Exchangers*, (ed. by Afgan, N. *et al.*), pp. 547–575, Gordon and Breach Publishers, OPA Amsterdam, B. V.

Roetzel, W. and Luo, X. (1993) Pure Cross-Flow with Longitudinal Dispersion in One Fluid, in *Aerospace Heat Exchanger Technology 1993*, (ed. by Shah, R. K. and Hashemi, A.), Elsevier, Amsterdam.

Schmachtenberg, H. (1981) *Zum Dynamischen Verhalten großer Kreuzstromwärmeübertrager,* Dissertation, Technische Hochschule Aachen.

Schlünder, E. U. (ed.) (1986) *Heat Exchanger Design Handbook, vol. 1, Heat Exchanger Theory,* VDI-Verlag GmbH.

Spiga, M. and Spiga, G. (1988) Transient Temperature Fields in Crossflow Heat Exchangers with Finite Wall Capacitance, *Journal of Heat Transfer*, **110**, 49–53.

Spiga, M. and Spiga, G. (1992) Step Response of the Crossflow Heat Exchanger with Finite Wall Capacitance, *International Journal of Heat and Mass Transfer*, **35**(2), 559–565.

van Aken, G. J. (1994) Transient Modelling of Finned Tube Heat Exchangers, in *Proceedings of the 10th International Heat Transfer Conference* (ed. by Hewitt, G. F.), **6**, 13-HA-23.

Xuan, Y. (1994) Three-Dimensional Dynamic Simulation of Crossflow Heat Exchangers, in *Proceedings of the 10th International Heat Transfer Conference* (ed. by Hewitt, G. F.), **4**, He-24.

Xuan, Y. (1996) Transient Analysis of Multipass Crossflow Heat

Exchangers, *International Journal of Heat and Mass Transfer,* **31**, 223–230.

Xuan, Y. and Roetzel, W. (1993) Dynamics of Shell-and-Tube Heat Exchangers to Arbitrary Temperature and Step Flow Variations, *AIChE Journal*, **38**, 413–421.

Yamashita, H., Izumi, R. and Yamagushi, S. (1978) Analysis of the Dynamic Characteristics of Crossflow Heat Exchangers with Both Fluids Unmixed, *JSME Bulletin,* **21**, 479–485.

Chapter 5

Transient analysis of plate heat exchangers

Plate heat exchangers find more and more application in a variety of industrial branches due to some of their unique advantages such as high compactness, high performance of heat transfer, low pressure drop, flexibility of flow arrangements, anti-fouling ability and easy cleaning as well as maintenance, although they are less used than tubular exchangers. Plate heat exchangers have been commercially available for some decades. There appeared a number of publications on heat transfer performance and pressure drop in plate heat exchangers. Most of them are aimed at the design and steady-state performance problems. On the contrary, few works have been contributed to the dynamic behaviour of plate heat exchangers. By using the lumped parameter model, Khan *et al.* (1988) analysed the dynamic characteristics of a countercurrent plate heat exchanger and the corresponding frequency response. Based on a numerical approach, the *cinematic* model was applied by Lakshmanan and Potter (1990) to simulating the dynamic behaviour of plate heat exchangers. The so-called *cinematic* model is essentially the cell model which divides a plate heat exchanger into a number of segments along the flow direction. One common feature of these papers is that heat capacitance of plates is neglected. Instead of using the conventional plug-flow model, Das and Roetzel (1995) applied the dispersed plug-flow model to account for the effect of flow maldistribution on transient behaviour of plate heat exchangers, with including heat capacitance of plates.

5.1 Description and types of plate heat exchangers

Similar to the steady-state performance simulation and design of plate heat exchangers, dynamic simulation of this type of units is generally based on the following assumptions (Khan *et al.* 1988):

(1) A fluid is transversely mixed in each channel formed by two plates.

(2) The fluid is even distributed among channels and plug flow is assumed due to the high turbulence in the channels.

(3) Heat conduction in the flow direction is negligible.

(4) Thermal properties of fluids are constant.

(5) The heat capacity of plates is negligible.

(6) The two end plates of the exchanger are adiabatic.

Evidently, the above-mentioned idealizations are in concordance with the classical plug-flow model. In fact, some of them can be relaxed according to the specific dynamic problem and the selected simulation method. For example, assumption 4 may not be necessary for a numerical approach. Additionally, some other assumptions should be complemented, such as the idealization that no phase change occurs within the exchanger for transient simulation of single-phase heat transfer process. If the effect of flow maldistribution becomes severe, assumption 2 must be modified and the dispersed plug-flow model should be used.

Different gasketed or brazed assembly methods of the plates make it possible to put plate heat exchangers into a great diversity of applications, which is one of their main advantages. However, this flexibility of flow arrangements complicates the transient analysis of plate exchangers and no general simulation algorithm can be built for them. One should develop the simulation program for a given flow arrangement, which could not be applied to plate exchangers with other flow arrangements. The simplest flow arrangement in plate heat exchangers is the countercurrent flow (as shown in Fig. 5.1). Figure 5.1(a) represents a U-type of plate exchanger in which both the inlet and outlet ports of a fluid are located at the same end; Figure 5.1(b) shows a Z-type plate apparatus. Bassiouny and Martin (1984) gave detailed description of flow distribution and pressure drop in these types of plate units.

The dynamic behaviour of such types of countercurrent plate heat exchangers has been simulated by Khan *et al.* (1988) and Lakshmanan and Potter (1990). Furthermore, Khan *et al.* derived the first- and second-order transfer functions of the exchanger system, respectively, by defining the transfer function as the ratio of the outlet temperature of the cold stream (the controlled variable in their work) to the mass flow rate of the hot stream (the manipulated variable in their work). According to the experimental and theoretical results, they concluded that the transfer function is best represented by an overdamped second-order lag coupled with a first-order lead with dead time. Naturally, this conclusion cannot be extended to other types of plate heat exchangers.

In some types such as U- and Z-type (Fig. 5.1) of plate heat exchangers, a fluid flows through the main intake conduit and enters

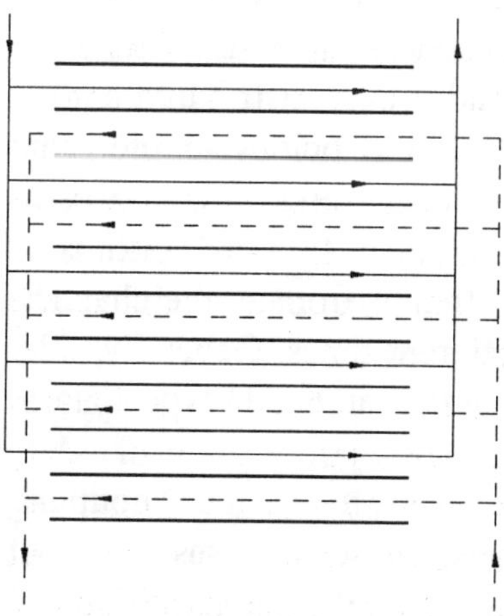

(a) U-type

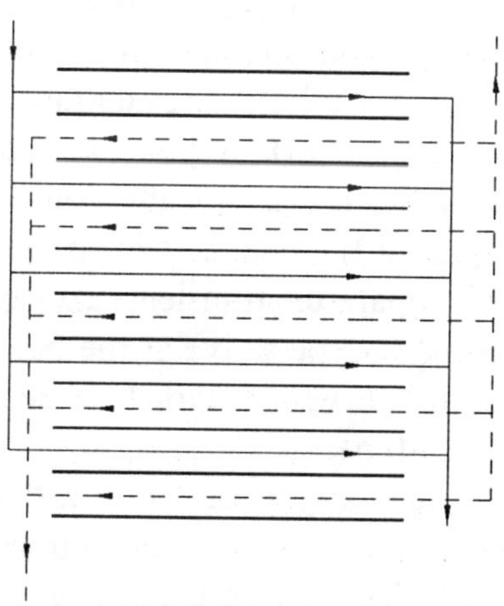

(b) Z-type

Figure 5.1: Schematic diagram of plate heat exchangers with countercurrent flow of two fluids

each channel in a pass in order of precedence, instead of entering simultaneously all channels per pass. There exists the so-called *phase lag effect* that the streams in all channels per pass separately respond to the inlet disturbance at different time spans, which makes the transient performance of a plate heat exchanger different from that of a shell-and-tube heat exchanger. This lag effect depends upon the number of plates and the flow arrangement assumed in the plate exchanger. As pointed out by Das and Roetzel (1995), the phase lag effect may become more remarkable in the Z-type.

In company with the phase lag effect at inlets, the transient response information of the stream in each channel i within the pass nearest to the outlets of a plate exchanger reaches the outlets at a different time. The transient signals which have reached the exits from these channels are integrated to form the exit dynamic characteristics of the plate exchanger. Even if a plate heat exchanger is assembled with a given number of plates but with different flow arrangements, different dynamic features appear. The differences among the flow arrangements increase with the number of channels per pass. A typical example is the comparison between U- and Z-type configurations discussed by Das and Roetzel (1995). In a Z-type plate exchanger with one pass, the stream in each channel has the same flow length from the inlet to the outlet of the exchanger; in the U-type streams from different channels have different flow lengths. Under the assumption that a fluid is even distributed among all channels in a pass, and that the flow velocities in the main conduits remain locally constant, transient features of the streams from all the channels in the Z-type simultaneously reach outlets of the apparatus, transient features of the streams from all the channels in the U-type arrive at the outlets at different times. Generally, inlet disturbances are earlier sensed at the outlets in the U-type than in the Z-type.

A variety of flow arrangements lead to various coupling conditions of the fluid among channels. In some cases, the exit conditions of the preceding channel are the entrance conditions of the subsequent channel. Although the examples in Fig. 5.1 show no coupling relationships among the streams from different channels (there is one pass for each fluid), they are mixed in the main exit conduit on each fluid side and then leave the exchanger. The

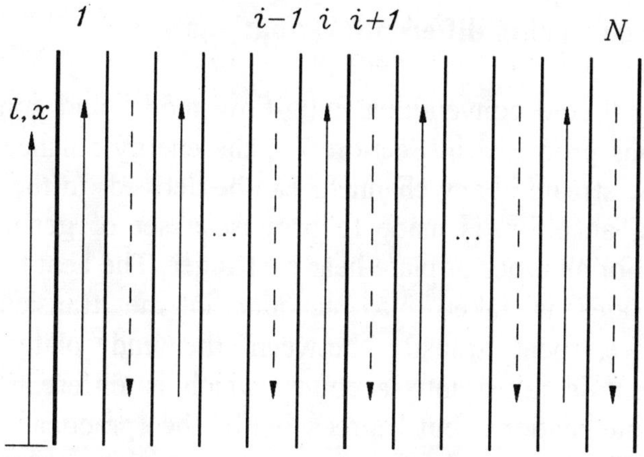

Figure 5.2: Schematic diagram of a plate heat exchanger

coupling of streams among channels controls the heat transfer performance and transient characteristics of plate heat exchangers. The end effect of a plate heat exchanger is another factor that exerts an influence on the transient heat transfer process. The differential equations describing the dynamics of the streams in both the end channels are quite different from those for the other channels. In addition, one should pay attention to the exchanger whether an even or odd number of plates are assembled because the even number of plates may cause an uneven channel distribution of the fluids.

5.2 Dynamic simulation of plate heat exchangers

In general, dynamic simulation of plate heat exchangers is considered as a one-dimensional transient problem. The temperature of a fluid changes along its flow direction. The general flow pattern in a plate heat exchanger is shown in Fig. 5.2, without paying attention to the coupling of streams among channels. Additionally, cocurrent flow and the combination of cocurrent and countercurrent flow in plate heat exchangers are also common, although these flow arrangements are not illustrated in Fig. 5.2.

5.2.1 The governing differential equations

According to the conventional plug-flow model and some of the assumptions proposed in Section 5.1, the energy balance equation of the flow stream in any channel i can be derived. In the following derivation, an effort is made to propose a set of general energy equations for all types of plate heat exchanger. The heat capacitance of the plates is taken into account in the transient model. Furthermore, heat transfer between the end plate and the environment is taken into account, which is different from the existing publications but agrees with the practical operating situation. The heat conduction resistance of a plate is neglected. In this case, a general form of transient heat transfer equation which is valid for the stream in each channel including the end channel is derived as follows

$$\frac{C_i}{L}\frac{\partial \theta_i}{\partial \tau}+(-1)^{\delta_i}\dot{W}_i\frac{\partial \theta_i}{\partial l}=\frac{(hA)_{il}}{L}(\theta_{wi}-\theta_i)+\frac{(hA)_{ir}}{L}(\theta_{wi+1}-\theta_i)$$

$$(i=1,\ 2,\ \ldots\ ,\ N) \qquad (5.1)$$

where C_i is the heat capacity of the fluid in channel i, $\dot{W}_i$ the thermal flow rate of the stream in channel i, L the length of the channel. $\delta_i=0$ if the flow direction of the stream in channel i is identical to that of the coordinate and $\delta_i=1$ if the flow direction of the stream in channel i is opposite from that of the coordinate. θ_{wi} is the temperature of the plate on the left-hand side channel i and θ_{wi+1} is the temperature of the plate on the right-hand side of channel i. h_{il} is the heat transfer coefficient between the stream and the left-hand side plate in channel i and h_{ir} the heat transfer coefficient between the stream and the right-hand side plate in channel i. Heat transfer coefficients are allowed to vary with channels and sides of a plate. If a plate heat exchanger has N_p and each pass has N_c channels, the unit contains $N=N_p\times N_c$ channels. Here the end plate on the left-hand side is counted as the first plate and the end plate on the right-hand side is plate $N+1$. Application of the lumped parameter model results in the following energy equation of plate i

$$C_{wi}\frac{\partial\theta_{wi}}{\partial l}=(hA)_{i-1,r}(\theta_{i-1}-\theta_{wi})+(hA)_{il}(\theta_i-\theta_{wi})$$

$$(i=2,\ 3,\ \ldots,\ N). \qquad (5.2a)$$

For the first and $(N+1)$th plate, one has

$$C_{w1}\frac{\partial\theta_{w1}}{\partial\tau}=(hA)_{f1}(\theta_{f1}-\theta_{w1})+(hA)_{1l}(\theta_1-\theta_{w1}) \qquad (5.2b)$$

$$C_{w,N+1}\frac{\partial\theta_{w,N+1}}{\partial\tau}=(hA)_{f,N+1}(\theta_{f,N+1}-\theta_{w,N+1})+(hA)_{Nr}(\theta_N-\theta_{w,N+1}) \qquad (5.2c)$$

where h_{f1} is heat transfer coefficient between the left-hand side of the first plate and the environmental fluid with temperature θ_{f1} and $h_{f,N+1}$ heat transfer coefficient between the right-hand side of the $(N+1)$th plate and the environmental fluid with temperature $\theta_{f,N+1}$. If h_{f1} and $h_{f,N+1}$ are set to be zero, the end plates are considered as adiabatic. By means of the following dimensionless variables and parameters

$$R_{\tau i}=\frac{\tau_{ri}}{\tau_{r1}},\ R_{wi}=\frac{C_{wi}}{C_{i-1}+C_i},\ R_{ci}=\frac{C_i}{C_1},\ R_i=\frac{\dot{W}_i}{\dot{W}_1},$$

$$U_{il}=\frac{(hA)_{il}}{\dot{W}_i},\ U_{ir}=\frac{(hA)_{ir}}{\dot{W}_i},\ (1\le i\le N)$$

$$U_{f1}=\frac{(hA)_{f1}}{\dot{W}_1},\ U_{f,N+1}=\frac{(hA)_{f,N+1}}{\dot{W}_N},\ x=\frac{l}{L},\ z=\frac{\tau}{\tau_{r1}}.$$

Equations (5.1) and (5.2) are rewritten into the following dimensionless forms:

$$R_{\tau i}\frac{\partial t_i}{\partial z}+(-1)^{\delta_i}\frac{\partial t_i}{\partial x}=U_{il}(t_{wi}-t_i)+U_{ir}(t_{wi+1}-t_i)$$

$$(i=1,\ 2,\ \dots,\ N) \qquad (5.3)$$

$$R_{wi}\frac{\partial t_{wi}}{\partial z}=\alpha_i(t_{i-1}-t_{wi})+\beta_i(t_i-t_{wi}) \qquad (i=2,\ 3,\ \dots,\ N) \qquad (5.4a)$$

$$R_{w1}\frac{\partial t_{w1}}{\partial z}=U_{f1}(t_{f1}-t_{w1})+U_{1l}(t_1-t_{w1}) \qquad (5.4b)$$

$$R_{w,N+1}\frac{\partial t_{w,N+1}}{\partial z}=\frac{U_{f,N+1}R_N}{R_{cN}}(t_{f,N+1}-t_{w,N+1})+\frac{U_{Nr}R_N}{R_{cN}}(t_N-t_{w,N+1}) \qquad (5.4c)$$

where $\alpha_i=\dfrac{U_{i-1,r}R_{i-1}}{R_{c,i-1}+R_{ci}}$ and $\beta_i=\dfrac{U_{il}R_i}{R_{c,i-1}+R_{ci}}$. Note that $C_0=0$, $R_{c0}=0$ and $R_{w,N+1}=\dfrac{C_{w,N+1}}{C_N}$. In the above equations the dimensionless dwelling time ratio $R_{\tau i}$ is defined as the ratio of the residence time of a fluid in channel i to that of the fluid in the first channel. Since the flow arrangement and the number of fluids are not involved in the derivation of the aforementioned governing equations, they are applicable to plate heat exchangers with any flow arrangements which are used to facilitate heat exchange between two fluids or among more than two fluids.

5.2.2 Initial conditions and boundary conditions

The initial conditions depend upon the state at the instant when the transient process starts. The simplest are uniform initial conditions, which corresponds to start-up phase. The general forms of initial conditions can be given as

$$t_i(x,\ 0)=g_i(x) \qquad (i=1,\ 2,\ \dots,\ N) \qquad (5.5a)$$

$$t_{wi}(x, 0) = g_{wi}(x) \qquad (i = 1, 2, \dots, N+1) \tag{5.5b}$$

No general forms of boundary conditions can be given. They are determined by the inlet conditions of the fluids, the flow arrangement, the number of passes, the number of channels per pass, and the coupling order of a fluid between two adjacent passes. One has no alternative but to find the necessary boundary and interfacial conditions according to the given plate heat exchanger with a given flow arrangement. The simplest examples may be the flow arrangements shown in Fig. 5.1, in which both fluids have only one pass and no interfacial conditions are needed. If the solid line represents fluid 1 and the dashed line fluid 2, the boundary conditions can be described as

$$t_{1i}(0, z) = f_1(z) \qquad (i = 1, 3, 5, \dots) \tag{5.6a}$$

$$t_{2i}(1, z) = f_2(z) \qquad (i = 2, 4, 6, \dots). \tag{5.6b}$$

As mentioned previously, the phase lag effect exists because a fluid is distributed among all channels in a pass but cannot simultaneously enter these channels through the main intake. It should be emphasized that the boundary conditions (5.6) are only valid if the phase lag effect is negligible. In other words, they should be modified in the case that this effect is severe. If the number of channels in a pass is relatively small or the time which is needed for a fluid to enter all channels through the main intake from the inlet of a plate heat exchanger is smaller than the residence time of the fluid stream in each channel, this lag effect may be negligible, otherwise it should be taken into account. If the dimensionless time lag at which a stream enters channel i is represented by $z_{i,lag}$, the boundary conditions (5.6) can be modified to

$$t_{1i}(0, z) = f_1\left(z - z_{i,lag}\right) u\left(z - z_{i,lag}\right) \qquad (i = 1, 3, 5, \dots) \tag{5.7a}$$

$$t_{2i}(1,z) = f_2\left(z - z_{i,lag}\right) u\left(z - z_{i,lag}\right) \qquad (i = 2,\ 4,\ 6,\ \dots\ ) \tag{5.7b}$$

where $u\left(z - z_{i,\,lag}\right)$ is the unit step function defined by

$$u\left(z - z_{i,\,lag}\right) = \begin{cases} 1, & z - z_{i,\,lag} \geq 0 \\ 0, & z - z_{i,\,lag} < 0\ . \end{cases} \tag{5.8}$$

Obviously, the dimensionless time lag $z_{i,lag}$ is different for each channel and it increases with the distance between the main inlet of a fluid into a plate heat exchanger and the entrance of a channel. It can be determined according to the number of channels per pass, the length of the channel entrance from the main inlet and the changing velocity of a fluid in the main intake. If a fluid enters only one channel pass (the so-called series-flow) through the exchanger, the phase lag effect can be neglected.

5.2.3 General solutions

A set of $(2N+1)$ differential equations forms the basis for analysis of transient behaviour of any plate heat exchangers. Combined with the given initial and boundary conditions, these equations should be solved to describe dynamic feature of a given plate unit. For this purpose, some different methods can be used. If the flow rates of the fluids change arbitrarily at the inlets or the thermal properties of the fluids vary considerably with temperatures, dynamic simulation becomes a nonlinear problem and a numerical method like the finite-difference may be preferable. If arbitrary variations of fluid temperatures and/or step disturbance of flow rates occur at inlets, dynamic simulation is considered to be linear under the assumption of constant thermal properties and the Laplace transform is applicable.

Application of the Laplace transform to eqns (5.3) and (5.4) yields

$$(-1)^{\delta_i}\frac{\mathrm{d}T_i}{\mathrm{d}x}=U_{il}T_{wi}+U_{ir}T_{w,i+1}-\left(R_{\tau i}s+U_{il}+U_{ir}\right)T_i+R_{\tau i}g_i(x)$$

$$\left(i=1,\ 2,\ \ldots,\ N\right) \qquad (5.9)$$

$$T_{wi}=\frac{\alpha_i T_{i-1}+\beta_i T_i}{\alpha_i+\beta_i+R_{wi}s}+\frac{R_{wi}g_{wi}(x)}{\alpha_i+\beta_i+R_{wi}s} \qquad \left(i=2,\ 3,\ \ldots,\ N\right) \qquad (5.10a)$$

$$T_{w1}=\frac{\alpha_1 T_{f1}+\beta_1 T_1}{\alpha_1+\beta_1+R_{w1}s}+\frac{R_{w1}g_{w1}(x)}{\alpha_1+\beta_1+R_{w1}s} \qquad (5.10b)$$

$$T_{w,N+1}=\frac{\alpha_{N+1}T_{f,N+1}+\beta_{N+1}T_N}{\alpha_{N+1}+\beta_{N+1}+R_{w,N+1}s}+\frac{R_{w,N+1}g_{w,N+1}(x)}{\alpha_{N+1}+\beta_{N+1}+R_{w,N+1}s} \qquad (5.10c)$$

where $\alpha_1=U_{f1}$, $\beta_1=U_{1l}$, $\alpha_{N+1}=\dfrac{U_{f,N+1}R_N}{R_{cN}}$, $\beta_{N+1}=\dfrac{U_{Nr}R_N}{R_{cN}}$. Thus, eqn (5.9) becomes

$$\frac{\mathrm{d}T_i}{\mathrm{d}x}=a_iT_{i-1}+b_iT_i+c_iT_{i+1}+h_i(x) \qquad \left(i=1,\ 2,\ \ldots,\ N\right) \qquad (5.11)$$

where T_0 and $T_{f,N+1}$ are the temperatures of the environmental fluids,

$T_0=T_{f1}$, $T_{N+1}=T_{f,N+1}$,

$$a_i=\frac{(-1)^{\delta_i}U_{il}\alpha_i}{\alpha_i+\beta_i+R_{wi}s},$$

$$b_i=(-1)^{\delta_i}\left[\frac{U_{il}b_i}{\alpha_i+\beta_i+R_{wi}s}+\frac{U_{ir}a_{i+1}}{\alpha_{i+1}+\beta_{i+1}+R_{w,i+1}s}-\left(R_{\tau i}s+U_{il}+U_{ir}\right)\right],$$

$$c_i = \frac{(-1)^{\delta_i} U_{ir} b_{i+1}}{\alpha_{i+1} + \beta_{i+1} + R_{w,i+1} s},$$

$$h_i(x) = (-1)^{\delta_i} \left[\frac{U_{il}}{\alpha_i + \beta_i + R_{wi} s} g_{wi}(x) \right.$$

$$\left. + \frac{U_{ir}}{\alpha_{i+1} + \beta_{i+1} + R_{w,i+1} s} g_{w,i+1} + R_{ti} g_i(x) \right].$$

Once the N conjugated differential equations given by (5.11) are solved, the dynamic behaviour of the plate heat exchanger in the transform domain or the frequency domain is determined. These equations can be represented in the following matrix form

$$\frac{\mathrm{d}\mathbf{T}}{\mathrm{d}x} = \mathbf{AT} + \mathbf{H}(x) \tag{5.12}$$

where **A** is a tridiagonal matrix whose elements are as follows:

$$\begin{bmatrix} b_1 & c_1 & & & & \\ a_2 & b_2 & c_2 & & 0 & \\ & a_3 & b_3 & \mathbf{c}_3 & & \\ & & & \cdots & & \\ & 0 & & a_{N-1} & b_{N-1} & c_{N-1} \\ & & & & a_N & b_N \end{bmatrix}$$

and vector **T** as well as **H**(x) is given as

$$\mathbf{T} = \{T_1,\ T_2,\ \ldots,\ T_N\}^{\mathrm{T}},$$

$$\mathbf{H}(x) = \{a_1 T_{f1} + h_1(x),\ h_2(x),\ \ldots,\ h_{N-1}(x),\ c_N T_{f,N+1} + h_N(x)\}^{\mathrm{T}}.$$

Only if the initial conditions are constant and uniform and heat exchange between the end plates and the environment is negligible, the system of differential equations is homogeneous. The eigenvalues and eigenvectors can be found by different algorithms (Steinberg, 1974). If the elements of the tridiagonal matrix satisfy the following conditions

$$a_2 = a_3 = \ \ldots \ = a_N, \ b_1 = b_2 = \ \ldots \ = b_N, \ c_1 = c_2 = \ \ldots \ = c_{N-1} \tag{5.13}$$

the eigenvalues of **A** are given by Bellman (1970)

$$\lambda_j = b - 2\sqrt{ac}\cos j\varphi, \ \varphi = \pi/(N+1) \ \left(j = 1, \ 2, \ \ldots \ , \ N\right). \tag{5.14}$$

The solution to the system of differential equations (5.12) is familiar and a general form is

$$\mathbf{T} = \mathrm{e}^{\mathbf{A}x}\,\mathbf{D}^{*} + \mathrm{e}^{\mathbf{A}x}\int \mathrm{e}^{-\mathbf{A}x'}\,\mathbf{H}(x')\,\mathrm{d}\,x' \tag{5.15}$$

if all eigenvalues of **A** are distinct. This solution can be represented in another form

$$\mathbf{T} = \mathbf{U}\,\mathrm{e}^{\mathbf{L}x}\,\mathbf{D} + \int \mathbf{U}\,\mathrm{e}^{\mathbf{L}(x-x')}\,\mathbf{U}^{-1}\mathbf{H}(x')\,\mathrm{d}\,x' \tag{5.15$'$}$$

where the diagonal matrix $\mathrm{e}^{\mathbf{L}x} = diag\left\{\mathrm{e}^{\lambda_1 x}, \ \mathrm{e}^{\lambda_2 x}, \ \ldots \ , \ \mathrm{e}^{\lambda_N x}\right\}$, **U** is a $N \times N$ matrix whose elements are the corresponding eigenvectors, i.e., $\mathbf{U} = \left\{\mathbf{u}_1, \ \mathbf{u}_2, \ \ldots \ , \ \mathbf{u}_N\right\}$. This solution is applicable to all types of plate heat exchangers. **D** is an unknown coefficient vector. For a given plate heat exchanger, the flow arrangement and boundary conditions are known, and **D** can be determined according to the given boundary and interfacial conditions in order to obtain the dynamic behaviour of the given apparatus. To do this, the given boundary and interfacial conditions should simultaneously be transformed. For example, application of the Laplace transform to boundary conditions (5.7) yields

$$\mathrm{L}\left[t_{1i}(0,z)\right] = \mathrm{e}^{-z_{i,lag}\,s} F_1(s) \qquad (i = 1,\ 3,\ 5\ ,\ \dots) \tag{5.16a}$$

$$\mathrm{L}\left[t_{2i}(0,z)\right] = \mathrm{e}^{-z_{i,lag}\,s} F_2(s) \qquad (i = 2,\ 4,\ 6\ ,\ \cdots). \tag{5.16b}$$

Insertion of the transformed forms of the given boundary and interfacial conditions related to a selected flow arrangement into the general solution (5.15) or (5.15′) leads to a matrix equation for the unknown coefficients $d_1,\ d_2,\ \dots\ ,\ d_N$

$$\mathbf{WD} = \mathbf{F} + \mathbf{S} \tag{5.17}$$

where **W** is a matrix whose elements depend upon the number of channels per pass, the number of passes and the flow arrangement. Besides these factors, the inlet conditions of a plate heat exchanger are involved in vector **F**. Corresponding to the given boundary and interfacial conditions as well as arrangement order of matrix **W**, vector **S** is determined according to the calculated results from the second part on the right-hand side of the solution (5.15′). For the transient simulation problem of an actual exchanger, the inverse matrix of **W** always exists, and then the coefficient vector **D** follows

$$\mathbf{D} = \mathbf{W}^{-1}(\mathbf{F} + \mathbf{S})\,. \tag{5.18}$$

Once the coefficient vector **D** has been determined, one finds the dynamic feature of the streams in all channels and plates of a given plate heat exchanger in the image-domain. For the convenience of analysis, the transformed temperature $T_i(x,s)$ in channel i is expressed as

$$T_i(x,s) = \sum_{j=1}^{N} d_j u_{ij}\,\mathrm{e}^{\lambda_j x} + \sum_{j=1}^{N}\sum_{k=1}^{N} u_{ij} v_{jk} \int \mathrm{e}^{\lambda_j (x-x')} h_k(x')\,\mathrm{d}\,x'$$

$$(i = 1,\ 2,\ \dots\ ,\ N) \tag{5.19}$$

where d_j, u_{ij}, v_{jk} and $h_k(x)$ are elements of **D**, **U**, $\mathbf{U}^{-1}$ and **H**(x), respectively. Application of the aforementioned numerical algorithms of the inverse transform to solution (5.19) and (5.10) yields the transient temperature profiles of all streams and plates in the time-domain.

5.2.4 Transient responses to inlet disturbances

As shown, the inverse transform of the solution (5.19) gives the real-time responses of plate heat exchangers to arbitrary temperature variations and step disturbances of the flow rates at the inlets. Strictly speaking, one can obtain only the transient behaviour at the exit of each channel from these results, not the transient behaviour at the exits of a plate exchanger. If there is only one channel per pass (such as a series-flow arrangement), the dynamic behaviour at the exit of the channel is approximately identical to that at the exit of an exchanger for a fluid. For the cases in which the phase lag effect is negligible and a fluid is evenly distributed among N_c channels in a pass, the temperature of a fluid leaving the exchanger may be expressed as

$$t_e = \frac{\sum_{k=1}^{N_c} t_{ke}\dot{m}_k}{\sum_{k=1}^{N_c} \dot{m}_k} \tag{5.20}$$

where $\dot{m}_k$ and t_{ke} are the mass flow rate in channel k and the stream temperature at the exit of channel k, respectively. The mean exit temperature is applicable to each fluid in a plate exchanger.

If the phase lag effect is severe, however, one cannot calculate the transient temperatures at the exits of a plate heat exchanger by expression (5.20). In this case, one should pay attention to the fact that the streams from different channels reach the exit of the exchanger at different times. Note the U-type and Z-type shown in Fig. 5.3, which are taken as examples to explain how to calculate the transient temperatures at the exits of plate heat exchangers.

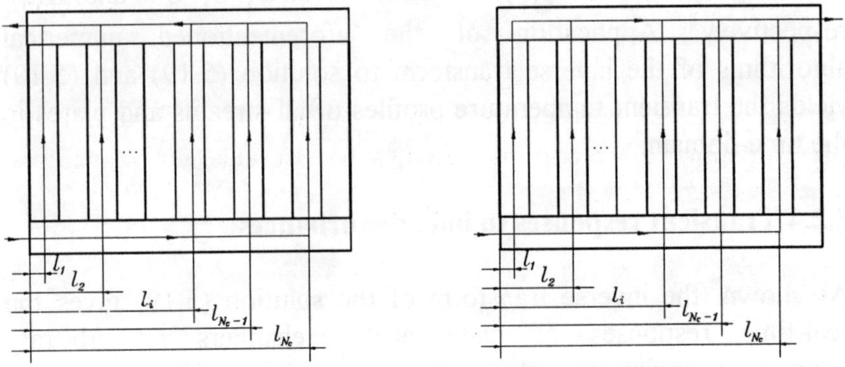

(a) U-type (b) Z-type

Figure 5.3: Flow of a fluid in plate heat exchangers

As shown, a fluid flows through the intake conduit and successively enters each channel in a pass. The streams from the different channels converge in the exhaust conduit and leave the exchanger. Evidently, the fluid velocities decrease with increasing distance from the inlet in both intake conduits and the exhaust conduit of the Z-type exchanger (Fig. 5.3(b)), while in the exhaust conduit of the U-type exchanger the velocity decreases with growing distance from the inlet. If the flow velocity entering channel i on the side of a fluid is assumed as v_i, the flow velocity in the intake conduit after channel i is expressed as

$$_i = V_0 - \sum_{k=1}^{i} v_k \frac{A_k}{A} \tag{5.21}$$

where V_0 is the frontal velocity of a fluid at the entrance of a plate heat exchanger. A and A_k are the flow area of the main intake conduit and that of channel k, respectively. The time required for the fluid to travel the distance between the channels can be calculated as

$$\Delta\tau_1 = \frac{l_1}{V_0} \tag{5.22a}$$

$$\Delta\tau_i = \frac{l_i - l_{i-1}}{V_i} \qquad (1 < i \le N_c) \tag{5.22b}$$

where N_c is the number of channels on the side of fluid 1. Hence the dimensionless phase lag between the entrances of consecutive channels is expressed as

$$\Delta\phi_i = \frac{\Delta\tau_i}{\tau_{r1}}. \tag{5.23}$$

The total phase lag at the entrance of channel i is given by

$$\phi_i = \sum_{k=1}^{i} \Delta\phi_k . \tag{5.24}$$

The phase lag $\phi_{i,exit}$ from the exit of the channel to the exit of the exchanger can be calculated in a similar way. In most of plate heat exchangers, there is the following relationship:

$$\phi_{i,exit} = \phi_i \qquad \text{for the U - type, and} \tag{5.25a}$$

$$\phi_{i,exit} = \phi_{n_1 - i + 1} \qquad \text{for the Z - type.} \tag{5.25b}$$

The temperature response at the exits of a plate heat exchanger is affected by the aforementioned lag time and expressed as

$$t_e = \frac{\sum_{k=1}^{N_c} t_{ke}(\tau^*)\dot{m}_k}{\sum_{k=1}^{N_c} \dot{m}_k} \tag{5.26}$$

where $\tau^* = \tau - \tau_{r1}\phi_{i,exit}$. If $\tau^* \le 0$, t_{ke} is equal to the initial temperature. With this expression, one can calculate the exit temperatures in a plate heat exchanger in which the phase lag effect may be severe.

5.3 Dynamic analysis of plate heat exchangers with dispersion

The above-mentioned dynamic analysis of plate heat exchangers is based on the conventional plug-flow model, without paying attention to the flow maldistribution or backmixing. In fact, the effect of the flow maldistribution in plate heat exchangers is rather prominent as observed by Amooie-Foumeny (1975) and Haseler *et al.* (1992). The flow maldistribution in plate heat exchangers may include uneven distribution of a fluid among channels in a pass, stagnant flow and other flow patterns deviating from the plug-flow. In the previous chapters, the dispersed plug-flow model is used to account for the effect of flow maldistribution on transient behaviour of exchangers. Similarly, the dispersed term is introduced into the differential equations describing the transient characteristics of plate heat exchangers to modify the deviation of the real flow pattern from the idealized plug-flow in plate exchangers.

5.3.1 Governing differential equations

Some of the assumptions involved here are the same as those mentioned in Section 5.2. The additional assumption arising from the dispersion effect is that the flow maldistribution in plate exchangers can be described by a dispersion term, i.e., the apparent conduction term. According to the dispersed plug-flow model, a set of governing differential equations can be derived. For the stream in channel i $(1 \le i \le N)$

$$\frac{C_i}{L}\frac{\partial \theta_i}{\partial \tau} + (-1)^{\delta_i}\dot{W}_i\frac{\partial \theta_i}{\partial l} = A_q D_i \frac{\partial^2 \theta_i}{\partial l^2} + \frac{(hA)_{il}}{L}\left(\theta_{wi} - \theta_i\right)$$

$$+\frac{(hA)_{ir}}{L}\left(\theta_{wi+1} - \theta_i\right) \qquad (5.27)$$

The only difference between eqns (5.27) and (5.1) is that a dispersion term is inserted on the right-hand side of eqn (5.27). The other differential equations for the plates remain the same as those derived in Section 5.2.1, if axial heat conduction is neglected in the plates. Thus,

$$C_{w1}\frac{\partial \theta_{w1}}{\partial \tau} = (hA)_{f1}\left(\theta_{f1} - \theta_{w1}\right) + (hA)_{1l}\left(\theta_1 - \theta_{w1}\right) \tag{5.28a}$$

$$C_{wi}\frac{\partial \theta_{wi}}{\partial l} = (hA)_{i-1,r}\left(\theta_{i-1} - \theta_{wi}\right) + (hA)_{il}\left(\theta_i - \theta_{wi}\right) \qquad (i = 2,\ 3,\ \dots,\ N) \tag{5.28b}$$

$$C_{w,N+1}\frac{\partial \theta_{w,N+1}}{\partial \tau} = (hA)_{f,N+1}\left(\theta_{f,N+1} - \theta_{w,N+1}\right) + (hA)_{Nr}\left(\theta_N - \theta_{w,N+1}\right). \tag{5.28c}$$

Equations (5.27) and (5.28) describe the transient behaviour of plate heat exchangers, which account for the effect of flow maldistribution in the channels between the plates. It should be emphasized that all variables and parameters in these equations, like those in eqns (5.1) and (5.2), conform to their corresponding values in the channels and not the values of the incoming fluids. The dimensionless forms of these equations are as follows:

$$R_{\tau i}\frac{\partial t_i}{\partial z} + (-1)^{\delta_i}\frac{\partial t_i}{\partial x} = \frac{1}{Pe_i}\frac{\partial^2 t_i}{\partial x^2} + U_{il}\left(t_{wi} - t_i\right) + U_{ir}\left(t_{wi+1} - t_i\right)$$

$$(i = 1,\ 2,\ \dots,\ N) \tag{5.29}$$

$$R_{wi}\frac{\partial t_{wi}}{\partial z} = \alpha_i\left(t_{i-1} - t_{wi}\right) + \beta_i\left(t_i - t_{wi}\right) \qquad (i = 2,\ 3,\ \dots,\ N) \tag{5.4a}$$

$$R_{w1}\frac{\partial t_{w1}}{\partial z} = U_{f1}\left(t_{f1} - t_{w1}\right) + U_{1l}\left(t_1 - t_{w1}\right) \tag{5.4b}$$

$$R_{w,N+1}\frac{\partial t_{w,N+1}}{\partial z}=\frac{U_{f,N+1}R_N}{R_{cN}}\left(t_{f,N+1}-t_{w,N+1}\right)+\frac{U_{Nr}R_N}{R_{cN}}\left(t_N-t_{w,N+1}\right) \tag{5.4c}$$

where $Pe_i=\left(\dot{W}_iL\right)/\left(A_qD_i\right)$ and the other parameters are defined in Section 5.2.1. The initial conditions are generally be given by

$$t_i(x,0)=g_i(x) \qquad (i=1,\ 2,\ \ldots,\ N) \tag{5.5a}$$

$$t_{wi}(x,0)=g_{wi}(x) \qquad (i=1,\ 2,\ \ldots,\ N+1). \tag{5.5b}$$

The boundary conditions depend upon the alignment among streams from channels between two adjacent passes. No general forms can be given. The simplest case is the one-pass U- and Z-type exchangers. For an application in which two fluids with inlet temperatures $f_1(x)$ and $f_2(x)$ are involved in a U-type or Z-type plate exchanger, the boundary conditions are as follows:

at $x=0$:

$$t_i-\frac{1}{Pe_i}\frac{\partial t_i}{\partial x}=f_1\left(z-z_{i,lag}\right)u\left(z-z_{i,lag}\right) \quad (i=1,\ 3,\ 5,\ \ldots,\ N-1) \tag{5.30a}$$

$$\frac{\partial t_i}{\partial x}=0 \qquad (i=2,\ 4,\ 6,\ \ldots,\ N), \tag{5.30b}$$

at $x=1$:

$$t_i+\frac{1}{Pe_i}\frac{\partial t_i}{\partial x}=f_2\left(z-z_{i,lag}\right)u\left(z-z_{i,lag}\right) \quad (i=2,\ 4,\ 6,\ \ldots,\ N) \tag{5.30c}$$

$$\frac{\partial t_i}{\partial x}=0 \qquad (i=1,\ 3,\ 5,\ \ldots,\ N-1). \tag{5.30d}$$

No interfacial conditions are needed. To derive the above boundary conditions, it is assumed that the channels are symmetrically distributed between two fluids which flow in countercurrent direction, i.e., there is an even number N of channels and each fluid occupies one half of the channels. Under steady-state conditions, the inlet temperature of a stream entering an odd channel is equal to the incoming temperature of fluid 1 entering the exchanger; the inlet temperature of a stream entering an even channel is equal to that of incoming fluid 2 entering the exchanger.

5.3.2 Solution to transient temperature

Once $2N$ boundary and interfacial (if necessary) conditions as well as $2N+1$ initial conditions are given, one can solve eqns (5.29) and (5.4) to find the transient temperatures in a plate heat exchanger. If the flow rate of a fluid changes arbitrarily or the thermal properties of a fluid vary considerably with temperature, one should resort to numerical methods. If arbitrary temperature variations and/or step flow disturbances occur at inlets of the exchanger, the Laplace transform method can be used to find a solution for transient temperatures under the assumption of constant properties. The solution procedure is similar to that described in the previous section.

By using the Laplace transform with respect to the dimensionless time variable, the system of eqns (5.29) and (5.4) is transformed into:

$$\frac{\mathrm{d}^2 T_i}{\mathrm{d}x^2} = (-1)^{\delta_i} Pe_i \frac{\mathrm{d}T_i}{\mathrm{d}x} + a_i T_{i-1} + b_i T_i + c_i T_{i+1} + h_i(x) \qquad (i = 1,\ 2,\ \ldots,\ N) \qquad (5.31)$$

$$T_{wi} = \frac{\alpha_i T_{i-1} + \beta_i T_i}{\alpha_i + \beta_i + R_{wi} s} + \frac{R_{wi} g_{wi}(x)}{\alpha_i + \beta_i + R_{wi} s} \qquad (i = 1,\ 2,\ \ldots,\ N+1) \qquad (5.32)$$

where

$$a_i = -\frac{Pe_i U_{il} \alpha_i}{\Delta_i}, \quad b_i = \left(R_i s + U_{il} + U_{ir} - \frac{U_{il} \beta_i}{\Delta_i} - \frac{U_{ir} \alpha_{i+1}}{\Delta_{i+1}} \right) Pe_i,$$

$$c_i = -\frac{Pe_i U_{ir} \beta_{i+1}}{\Delta_{i+1}},$$

$$h_i(x) = \left[R_i g_i(x) - \frac{U_{il} R_{wi}}{\Delta_i} g_{wi}(x) - \frac{U_{ir} R_{w,i+1}}{\Delta_{i+1}} g_{w,i+1}(x) \right] Pe_i,$$

$$\Delta_i = \alpha_i + \beta_i + R_{wi} s, \; T_0 = T_{f1}, \quad T_{N+1} = T_{f,N+1}.$$

Equation (5.31) is expressed as the matrix form:

$$\frac{\mathrm{d}\mathbf{T}}{\mathrm{d}x} = \mathbf{AT} + \mathbf{H}(x) \tag{5.33}$$

where

$$\mathbf{T} = \left\{ T_1, \; T_2, \; \ldots, \; T_N, \; \frac{\mathrm{d}T_1}{\mathrm{d}x}, \; \frac{\mathrm{d}T_2}{\mathrm{d}x}, \; \ldots, \; \frac{\mathrm{d}T_N}{\mathrm{d}x} \right\}^{\mathrm{T}},$$

$$\mathbf{H}(x) = \left\{ 0, \; 0, \; \ldots, \; h_1(x) + a_1 T_{f1}, \; h_2(x), \; h_3(x), \; \ldots, \right.$$

$$\left. h_{N-1}(x), \; h_N(x) + c_N T_{f,N+1} \right\}^{\mathrm{T}}.$$

The elements of matrix **A** is as follows:

For $1 \le i \le N$, $1 \le j \le 2N$, $\quad a_{ij} = \begin{cases} 1, & i = j \\ 0, & i \ne j \end{cases}$

For $i = N+1$, $1 \le j \le 2N$, $a_{ij} = \begin{cases} b_{i-N}, & j = i-N \\ c_{i-N}, & j = i-N+1 \\ (-1)^{\delta_{i-N}} Pe_{i-N}, & j = i \\ 0, & \text{other } j \end{cases}$

For $N+2 \le i \le 2N-1$, $1 \le j \le 2N$,

$$a_{ij} = \begin{cases} a_{i-N}, & j = i-(N+1) \\ b_{i-N}, & j = i-N \\ c_{i-N}, & j = i-N+1 \\ (-1)^{\delta_{i-N}} Pe_{i-N}, & j = i \\ 0, & \text{other } j \end{cases}$$

For $i = 2N$, $1 \le j \le 2N$, $a_{ij} = \begin{cases} a_{i-N}, & j = i-(N+1) \\ b_{i-N}, & j = i-N \\ (-1)^{\delta_{i-N}} Pe_{i-N}, & j = i \\ 0, & \text{other } j \end{cases}$

The solution form to the matrix equation (5.33) can be obtained by deriving the eigenvalues $\{\lambda_j\}$ and eigenvector $\{\mathbf{u}_j\}$ $(j = 1, 2, \ldots, 2N)$ of the coefficient matrix **A**, which is proposed in the previous sections. The simplest case corresponds to the uniform initial conditions. In this case the solution yields

$$T_i = \sum_{j=1}^{2N} d_j u_{ij} \, \mathrm{e}^{\lambda_j x} \qquad (i = 1, 2, \ldots, N) \qquad (5.34a)$$

$$\frac{dT_i}{dx} = \sum_{j=1}^{2N} d_j u_{ij}\, e^{\lambda_j x} \qquad (i = N+1,\ N+2,\ \ldots,\ 2N) \qquad (5.34b)$$

where the coefficient vector **D** should be determined according to the given boundary and interfacial conditions. The matrix equation to find **D** can be expressed as

$$\mathbf{WD} = \mathbf{S} \qquad (5.35)$$

where the elements of matrix **W** and vector **D** depend upon the boundary conditions and the composition order of the matrix equation (5.35). In the light of practical application, the inverse matrix of **W** always exists, and then **D** can be obtained from

$$\mathbf{D} = \mathbf{W}^{-1}\mathbf{S}. \qquad (5.36)$$

Thus, the solution for the transformed temperatures in the frequency-domain is found. To obtain the response in the time-domain, a numerical inversion of the Laplace transform is performed, which has already been described in the previous sections.

Some numerical examples for the U- and Z-type have been computed and illustrated by Das and Roetzel (1995). They found that the outlet temperatures of hot and cold fluids differ for U- and Z-type configurations, and the difference between the two types increases with the number of channels. The time lag of the temperature response in the U-type plate heat exchangers is less than that in the Z-type. As pointed out previously, flow maldistribution always results in degradation of thermal performance of a heat exchanger. The Péclet number is the quantitative indicator for the estimation of such an effect, which has been illustrated in the examples.

5.4 Transient analysis of spiral plate heat exchangers

Another type of plate heat exchangers is the spiral plate heat exchanger which consists of spiral or spiral/axial flow passages. Because of its advantages (Gupta, 1986), the spiral plate heat exchanger finds more and more application in a variety of industrial applications. According to whether one or both of the fluids are in spiral flow, there are three principal types of flow arrangements (Gupta, 1986):

(1) Spiral-spiral flow: both of the fluids are in spiral flow.

(2) Spiral-cross flow: one fluid is in spiral flow and the other is in cross-flow parallel to the axis of the spiral element.

(3) Spiral flow-combination cross-and-spiral flow: this is used when one of the fluids is a mixture of vapour and noncondensables.

Because of complexity of the flow conduit, it is difficult to simulate the thermal performance in spiral plate heat exchangers. Bes and Roetzel (1992a, b, 1993) proposed an analytical algorithm for the rating and design of spiral heat exchangers. With and without heat losses to the surroundings, Targett *et al.* (1992) proposed solutions in closed form for a double-spiral heat exchanger. Zaleski and Lachowski (1987) developed a model for the unsteady behaviour of a parallel-flow spiral plate heat exchanger, which is one of the few papers devoted to the transient simulation of spiral plate heat exchangers.

To set up a system of governing differential equations for the transient performance of spiral plate heat exchangers, Zaleski and Lachowski made the following assumptions:

(1) the heat transfer coefficient and thermal properties of the fluids are constant along the entire flow paths;

(2) the flow is rotary plug flow (velocity proportional to the radius);

(3) heat transfer within the fluid is negligible;

(4) heat transfer along the wall is negligible.

Obviously, these idealizations are similar to those for transient simulation of other types of heat exchangers.

In a countercurrent or cocurrent spiral plate heat exchanger the two fluids involved in the heat transfer process flow through conduits which are b_1 and b_2 in width. For a $2n$-channel spiral plate heat exchanger in which n channels contain fluid 1 and n channels contain fluid 2, Zaleski and Lachowski (1987) derived the following matrix forms of the governing equations to describe the transient temperature distributions of both the fluids and the plates:

$$\mathbf{B}\frac{\partial \mathbf{t}}{\partial \varphi}+\mathbf{D}\frac{\partial \mathbf{t}}{\partial \tau}=-\mathbf{FCt}+\mathbf{FHt}_s \tag{5.37}$$

$$\mathbf{E}\frac{\partial \mathbf{t}_s}{\partial \tau}=\mathbf{Kt}+\mathbf{Gt}_s \tag{5.38}$$

where $\mathbf{t}=\left\{t_1,\, t_2,\, \ldots\, ,\, t_{2n}\right\}^{\mathrm{T}}$ and $\mathbf{t}_s=\left\{t_{0,s},\, t_{1,s},\, t_{2,s},\, \ldots\, ,\, t_{2n,s}\right\}^{\mathrm{T}}$ are vectors of the temperatures of the fluids and the plates. The elements of coefficient vectors **B, C, D, E, F, G, H** and **K** are functions of the geometrical dimensions of the channels, heat transfer coefficients, thermal flow rates and heat capacities of the fluids and the plates. With the given initial, boundary and interfacial conditions, the system of aforementioned differential equations is solved by the method of characteristics. In this method, the system of partial differential equations is transformed into a system of ordinary equations, by expressing them in terms of curves called characteristics. In some simpler cases, the system of transformed ordinary equations may be solved analytically; in more complex cases, a numerical method is necessary.

In the light of their computational results, Zaleski and Lachowski pointed out that a considerable amount of heat is accumulated in the thick plates, which affects the temperature profiles of the fluids. The models used to account for the heat capacity of the plates are accurate but very time-consuming. Therefore, a simplified model, originating from eqns (5.37) and (5.38) is defined as follows:

$$\mathbf{B}\frac{\partial \mathbf{t}}{\partial \varphi}+\mathbf{D}\frac{\partial \mathbf{t}}{\partial \tau}=\mathbf{Mt} \tag{5.39}$$

where $\mathbf{M} = -(\mathbf{FC} + \mathbf{FHG}^{-1}\mathbf{K})$. Equation (5.39) neglects the heat capacity of the plates and has sufficient accuracy.

5.5 References

Amooie-Foumeny, M. M. (1975) *Flow Distribution in Plate Heat Exchangers*, Ph.D. Thesis, University of Bradford.

Bassiouny, M.K. and Martin, H. (1984) Flow Distribution and Pressure Drop in Plate Heat Exchangers, *Chemical Engineering Science*, **39**, 693–704.

Bellman, R. (1970) *Introduction to Matrix Analysis*, McGraw-Hill, Inc., New York.

Bes, Th. and Roetzel, W. (1992a) Distribution of Heat Flux Density in Spiral Heat Exchangers, *International Journal of Heat and Mass Transfer*, **35**, 1331–1347.

Bes, Th. and Roetzel, W. (1992b) Approximate Theory of Spiral Heat Exchangers, in *Design and Operation of Heat Exchangers* (ed. by Roetzel, W., Heggs, P. J., and Butterworth, D.), pp. 223–232, Springer-Verlag, Berlin.

Bes, Th. and Roetzel, W. (1993) Thermal Theory of the Spiral Heat Exchanger, *International Journal of Heat and Mass Transfer*, **36**, 765–773.

Das, S. K. and Roetzel, W. (1995) Dynamic Analysis of Plate Heat Exchangers with Dispersion in Both Fluids, *International Journal of Heat and Mass Transfer*, **38**, 1127–1140.

Das, S. K., Spang, B. and Roetzel, W. (1995) Dynamic Behaviour of Plate Heat Exchangers — Experiments and Modeling, *Journal of Heat Transfer*, **117**, 859–864.

Gupta, J. P. (1986) *Fundamentals of Heat Exchanger and Pressure Vessel Technology*, Hemisphere Publishing Corporation, Washington.

Haseler, L. E., Wadekar, V. V. and Clarke, R. H. (1992), Flow Distribution Effects in A Plate and Frame Heat Exchanger, in *Proceedings of 1992 ICHMT International Symposium*, pp. 361–367.

Lakshmanan, C. C. and Potter, O. E. (1990) Dynamic Simulation of Plate Heat Exchangers, *International Journal of Heat and Mass Transfer*, **33**, 995–1002.

Khan, A. R., Baker, N. S. and Wardle, A. P. (1988) The Dynamic Characteristics of a Counter-current Plate Heat Exchanger, *International Journal of Heat and Mass Transfer*, **31**, 1269–1278.

Roetzel, W. and Das, S. K. (1995) Hyperbolic Axial Dispersion Model: Concept and Its Application to a Plate Heat Exchanger, *International Journal of Heat and Mass Transfer*, **38**, 3065–3076.

Steinberg, D. I. (1974) *Computational Matrix Algebra*, McGraw-Hill, Inc., New York.

Targett, M. J., Retallick, W. B., and Churchill, S. W. (1992) Solutions in Closed Form for a Double-Spiral Heat Exchanger, *Industrial and Engineering Chemistry Research*, **31**, 658–669.

Zaleski, T. and Lachowski, A. (1987) Unsteady Temperature Profiles in Parallel-Flow Spiral Heat Exchangers, *International Chemical Engineering*, **27**, 556–565.

Chapter 6

Dynamic simulation of other types of heat exchangers

There are a variety of types of commercially available heat exchangers. According to the operation conditions and parameters, e.g., temperatures, pressures, states of fluids, the user selects a particular type of heat exchanger. In addition to the aforementioned heat exchangers, other types of heat exchangers such as double-pipe heat exchangers, heat-pipe exchangers and regenerators are widely used in industrial applications. The geometric structure and dimensions and the flow arrangements affect the thermal performance and transient behaviour of heat exchangers. For each type of heat exchanger, it is necessary to develop a special simulation algorithm, although it may be similar to those for other types of heat exchangers. In the following context, some other types of heat exchangers are considered and suitable simulation methods are suggested for the specific geometry and operating conditions.

6.1 Dynamic analysis of double-pipe heat exchangers

A double-pipe heat exchanger consists of two concentric pipes of different diameters, usually in the form of a hair-pin. The double-pipe exchanger is used for low heat duties requiring surface areas of up to about 50 m^2. A number of double-pipe heat exchangers can be connected in series or parallel as necessary. It has the advantages of flexibility since units can be added or removed as required, and it is easy to service and requires a low inventory of spares because of its standardization (Gupta, 1986). Although the double-pipe exchanger can be considered as a type of shell-and-tube exchanger, the flow pattern in a double-pipe exchanger is quite different from that in a conventional shell-and-tube exchanger. The flow in the annulus of a double-pipe exchanger is longitudinal, the shellside flow in a baffled shell-and-tube exchanger is a combination of cross-flow between the baffles and longitudinal flow through the baffle windows. Furthermore, all the fluid flows through each of the passes in a multipass shell-and-tube exchanger, unlike the double-pipe exchangers where the stream may split into substreams which enter different units connected in series–parallel forms.

The idealizations required to derive a system of governing differential equations for the double-pipe heat exchanger are similar to those for shell-and-tube heat exchangers. Matulla and Orlicek (1971) applied the plug-flow model to deriving a set of governing differential equations describing the temperatures of both fluids and the inner pipe. By using the method of frequency analysis, they determined the heat transfer coefficients in a double-pipe exchanger.

6.1.1 Dynamics of single double-pipe heat exchanger

In fact, the dynamic simulation of double-pipe heat exchangers is similar to that of shell-and-tube heat exchangers. Based on some of the assumptions applied to shell-and-tube heat exchangers, the following governing equations are derived

$$\frac{C_1}{L}\frac{\partial \theta_1}{\partial \tau} + \dot{W}_1 \frac{\partial \theta_1}{\partial l} = \frac{h_1 A_{1i}}{L}\left(\theta_{1w} - \theta_1\right) \tag{6.1}$$

$$\frac{C_{1w}}{L}\frac{\partial \theta_{1w}}{\partial \tau} = \frac{h_1 A_{1i}}{L}\left(\theta_1 - \theta_{1w}\right) + \frac{h_2 A_{1o}}{L}\left(\theta_2 - \theta_{1w}\right) \tag{6.2}$$

$$\frac{C_2}{L}\frac{\partial\theta_2}{\partial\tau} \pm \dot{W}_2\frac{\partial\theta_2}{\partial l} = \frac{h_2 A_{1o}}{L}(\theta_{1w} - \theta_2) + \frac{h_2 A_{2r}}{L}(\theta_{2w} - \theta_2) \tag{6.3}$$

$$\frac{C_{2w}}{L}\frac{\partial\theta_{2w}}{\partial\tau} = \frac{h_2 A_{2i}}{L}(\theta_2 - \theta_{2w}) + \frac{h_a A_{2o}}{L}(\theta_a - \theta_{2w}) \tag{6.4}$$

where $C_1 = \rho_1 c_{p1} r_{1i}^2 \pi L$, $C_2 = \rho_2 c_{p2}\pi(r_{2i}^2 - r_{1o}^2)L$, $A_{1i} = 2\pi r_{1i} L$, $A_{1o} = 2\pi r_{1o} L$, $A_{2i} = 2\pi r_{2i} L$, $A_{2o} = 2\pi r_{2o} L$. r_{1i} and r_{1o} are the inner and outer radiuses of the core pipe, r_{2i} and r_{2o} the inner and outer radiuses of the outer pipe. θ_a is the temperature of the environmental fluid. The negative sign of the second term on the left-hand side of eqn (6.3) is valid for the case of countercurrent flow. The lumped parameter model is applied to both the inner and outer pipes. Since flow maldistribution in double-pipe exchangers is not severe, the conventional plug-flow model is adopted for both fluids. These equations are rewritten in the following dimensionless form

$$\frac{\partial t_1}{\partial z} + \frac{\partial t_1}{\partial x} + U_{1i}(t_1 - t_{1w}) = 0 \tag{6.5}$$

$$R_{1w}\frac{\partial t_{1w}}{\partial z} = U_{1i}(t_1 - t_{1w}) + U_{1o}R_2(t_2 - t_{2w}) \tag{6.6}$$

$$R_\tau R_1\frac{\partial t_2}{\partial z} \pm \frac{\partial t_2}{\partial x} + U_{1o}(t_2 - t_{1w}) + U_{2i}(t_2 - t_{1w}) = 0 \tag{6.7}$$

$$R_1 R_{2w}\frac{\partial t_{2w}}{\partial z} + U_{2i}(t_{2w} - t_2) + U_{2o}(t_{2w} - t_a) = 0 \tag{6.8}$$

where

$$U_{1i} = \frac{h_1 2\pi r_{1i} L}{\dot{W}_1},\ U_{1o} = \frac{h_2 2\pi r_{1o} L}{\dot{W}_2},\ U_{2i} = \frac{h_2 2\pi r_{2i} L}{\dot{W}_2},$$

$$U_a = \frac{h_a 2\pi r_{2o} L}{\dot{W}_2}, \quad R_{1w} = \frac{\rho_{1w} c_{p1w}}{\rho_1 c_{p1}} \left[\left(\frac{r_{1o}}{r_{1i}} \right)^2 - 1 \right],$$

$$R_{2w} = \frac{\rho_{2w} c_{p2w} \left(r_{2o}^2 - r_{2i}^2 \right)}{\rho_1 c_{p1} r_{1i}^2}, \quad R_c = \frac{\rho_{2w} c_{p2} \left(r_{2i}^2 - r_{1o}^2 \right)}{\rho_1 c_{p1} r_{1i}^2}.$$

These differential equations are similar to those for shell-and-tube heat exchangers. The general forms of the initial conditions are

$$t_1(x, 0) = g_1(x) \tag{6.9a}$$

$$t_{1w}(x, 0) = g_{1w}(x) \tag{6.9b}$$

$$t_2(x, 0) = g_2(x) \tag{6.9c}$$

$$t_{2w}(x, 0) = g_{2w}(x). \tag{6.9d}$$

The boundary conditions at the entrances yield

$$t_1(0, z) = f_1(z) \tag{6.10a}$$

$$t_2(0, z) = f_2(z) \qquad \text{for the cocurrent flow} \tag{6.10b}$$

$$t_2(1, z) = f_2(z) \qquad \text{for the countercurrent flow.} \tag{6.10c}$$

The system of differential equations (6.5)–(6.8) with the related initial and boundary conditions can be used to describe the dynamic simulation of double-pipe exchangers subject to arbitrary disturbances of temperature and flow rates at entrances. If the flow rates vary arbitrarily at the entrances, one should resort to numerical methods, such as the finite-difference and the finite-element methods. In the other cases, the Laplace transform is applied to the above-mentioned system. By using s with respect to the dimensionless time variable z, the transformation yields

$$\frac{\mathrm{d}T_1}{\mathrm{d}x} = a_{11}T_1 + a_{12}T_2 + H_1(x,s) \tag{6.11}$$

$$\frac{\mathrm{d}T_2}{\mathrm{d}x} = a_{21}T_1 + a_{22}T_2 + H_2(x,s) \tag{6.12}$$

$$T_{1w} = \frac{U_{1i}T_1 + U_{1o}R_2T_2}{\Delta_1} + G_{1w}(x,s) \tag{6.13}$$

$$T_{2w} = \frac{U_{2i}T_2 + U_{2o}T_a}{\Delta_2} + G_{2w}(x,s) \tag{6.14}$$

where

$$a_{11} = \frac{U_{1i}^2}{\Delta_1} - s,\ a_{12} = \frac{U_{1i}U_{1o}R_2}{\Delta_1},\ a_{21} = sign\frac{U_{1o}U_{1i}}{\Delta_1},$$

$$a_{22} = sign\left(\frac{U_{1o}^2R_2}{\Delta_1} + \frac{U_{2i}^2}{\Delta_2} - R_1R_cs + U_{1o} + U_{2i}\right),\ sign = \pm$$

$$\Delta_1 = U_{1i} + U_{1o}R_2 + R_{1w}s,\ \Delta_2 = U_{2i} + U_{2o}R_2 + R_1R_{2w}s$$

$$H_1(x,s) = g_1(x) + U_{1i}G_{1w}(x,s),$$

$$H_2(x) = sign\left[U_{2i}G_{2w}(x,s) + U_{1o}G_{1w}(x,s) + R_1R_cg_2(x) + \frac{U_{2i}U_{2o}}{\Delta_2}T_a\right]$$

$$G_{1w}(x,s) = \frac{R_{1w}g_{1w}(x)}{\Delta_1},\ G_{2w}(x,s) = \frac{R_1R_{2w}g_{2w}(x)}{\Delta_2}$$

Nonhomogeneous differential equations (6.11) and (6.12) can be solved with the method described by Xuan and Roetzel (1993a). If the initial conditions are uniform, the solution becomes simpler. In this case, the combination of these two equations leads to

$$\frac{\mathrm{d}^2 T_1}{\mathrm{d}x^2} - \left(a_{11} + a_{22}\right)\frac{\mathrm{d}T_1}{\mathrm{d}x} + \left(a_{11}a_{22} - a_{12}a_{21}\right)T_1 = 0 \tag{6.15}$$

Its solution is

$$T_1(x,s) = B_1\,\mathrm{e}^{\lambda_1 x} + B_2\,\mathrm{e}^{\lambda_2 x} \tag{6.16}$$

and the solution for T_2 is

$$T_2(x,s) = \frac{\lambda_1 - a_{11}}{a_{12}} B_1\,\mathrm{e}^{\lambda_1 x} + \frac{\lambda_2 - a_{11}}{a_{12}} B_2\,\mathrm{e}^{\lambda_2 x} \tag{6.17}$$

where $\lambda_{1,2} = \dfrac{\left(a_{11} + a_{22}\right) \pm \sqrt{\left(a_{11} - a_{22}\right)^2 + 4a_{12}a_{21}}}{2}$. According to the given boundary conditions, one can determine two coefficients B_1 and B_2. For cocurrent flow,

$$B_1 = \frac{1}{\lambda_2 - \lambda_1}\left[\left(\lambda_2 - a_{11}\right)F_1(s) - a_{12}F_2(s)\right],$$

$$B_2 = \frac{1}{\lambda_1 - \lambda_2}\left[\left(\lambda_1 - a_{11}\right)F_1(s) - a_{12}F_2(s)\right]$$

For countercurrent flow

$$B_1 = \frac{\left(\lambda_2 - a_{11}\right)\mathrm{e}^{\lambda_2} F_1(s) - a_{12}F_2(s)}{\left(\lambda_2 - a_{11}\right)\mathrm{e}^{\lambda_2} - \left(\lambda_1 - a_{11}\right)\mathrm{e}^{\lambda_1}},$$

$$B_2 = \frac{\left(\lambda_1 - a_{11}\right)\mathrm{e}^{\lambda_1} F_1(s) - a_{12}F_2(s)}{\left(\lambda_1 - a_{11}\right)\mathrm{e}^{\lambda_1} - \left(\lambda_2 - a_{11}\right)\mathrm{e}^{\lambda_2}}.$$

Thus, the transformed solutions have been found. Application of the numerical inversion algorithm results in transient temperature distributions. If the periodic oscillation of fluid temperature occurs at the entrance of fluid 1 or fluid 2, the heat transfer coefficients can experimentally be determined by estimating the amplitude attenuation and phase delay of the oscillatory wave during propagation. For this purpose, the above-derived solution is needed.

An alternative lumped-differential formulation of dynamic response of double-pipe heat exchangers was developed by Neto and Cotta (1993). They proposed a set of transient energy equations for fully developed laminar–laminar or laminar–turbulent flow situations in double-pipe exchangers. In their model, the temperature distribution in the outer annular channel is radially lumped. Instead of using the conventional plug-flow model, they investigated the effect of flow distribution in the inner pipe. But the heat capacities of the pipes were neglected. The generalized integral transform technique is used to find the solution.

6.1.2 Dynamics of a double-pipe heat exchanger group

A number of double-pipe heat exchangers can be connected in series or parallel for a particular purpose. If one of the two fluid streams is very large in volume compared with the other, for example, the double-pipe exchangers are set up in a series–parallel arrangement (Gupta, 1986). According to the heat duty, the number of double-pipe exchangers can be increased or decreased as required. Dynamic behaviour of a double-pipe exchanger group is related with both characteristics of single exchanger and the connection manner.

The system of differential equations (6.5)–(6.8) or (6.11)–(6.14) can be used to describe the dynamic behaviour of an exchanger group consisting of N double-pipe exchangers. These equations are valid for each exchanger in the group. For such an exchanger group, therefore, one has $2N$ ordinary differential equations in the image-domain. The transformed equations are expressed in the following matrix form:

$$\frac{d\mathbf{T}}{dx} = \mathbf{AT} + \mathbf{H} \tag{6.18}$$

where

$$\mathbf{T} = \{T_{11},\ T_{21},\ T_{21},\ T_{22},\ \dots,\ T_{N1},\ T_{N2}\}^{\mathrm{T}},$$

$$\mathbf{H} = \{h_{11}(x,s),\ h_{21}(x,s),\ h_{12}(x,s),\ h_{22}(x,s),\ \dots,h_{N1}(x,s),\ h_{N2}(x,s)\}^{T}$$

and elements of the coefficient matrix **A** depend upon heat transfer coefficients, thermal flow rates, heat capacities of fluids and the solid components. With the given initial and boundary conditions as well as the interfacial conditions determined by the series arrangement, the system (6.18) can be solved with the method described in Chapter 2. Thus, the transient response of a double-pipe exchanger group can be found.

6.2 Transient characteristics of multi-fluid heat exchangers

Although most heat exchangers are applied to the heat transfer process between two fluids, multi-fluid heat exchangers are put into operation in some cases, for example, in chemical engineering and cryogenic systems (separation and liquefaction of oxygen and nitrogen, liquefaction of natural gas, etc.). Compared with two-fluid exchangers, multi-fluid heat exchangers conveniently accomplish heat transfer among three or more than three fluids. Sekulic and Shah (1995) described examples of a variety of types of multi-fluid exchangers. Generally, there are shell-and-tube multi-fluid exchangers, plate-fin multi-fluid exchangers and Hampson multi-fluid exchangers (Kao, 1965).

One of the features of multi-fluid exchangers that makes them different from those of two-fluid ones is that one fluid may be in thermal contact with two or more than two fluids. There exists some published literature, although not voluminous, which contains the necessary data regarding the design theory and the steady-state operation mode (Aulds and Barron, 1967, Sekulic and Shah, 1995). The dynamic behaviour of a multi-fluid heat exchanger is a topic to be further tackled. In principal, the dynamic behaviour of a multi-fluid exchanger may be described by a set of conjugated, linear partial differential equations derived from energy balance under some assumptions. The number and forms of these equations depend upon the number and sorts of fluids as well as the geometrical structure. The differential equations (5.1) and (5.2) derived in Chapter 5 are suitable for predicting the transient behaviour of plate exchangers involved in the heat transfer process between more than two fluids. In the following the transient model of a three-fluid heat exchanger is investigated.

6.2.1 Transient model of a tubular three-fluid heat exchanger

Figure 6.1 schematically represents a special tubular three-fluid heat exchanger, in which both fluid 1 and fluid 3 are in respective thermal contact with fluid 2. There exists no direct thermal contact between fluid 1 and fluid 3. Introducing some of assumptions, Sekulic and Herman (1987) derived three differential equations for this type of tubular three-fluid heat exchanger.

These three differential equations are as follows

$$\frac{\partial t_1}{\partial \tau} + \delta_1 a_1 \frac{\partial t_1}{\partial l} + b_{1,2}(t_1 - t_2) = 0 \tag{6.19}$$

$$\frac{\partial t_2}{\partial \tau} + \delta_2 a_2 \frac{\partial t_2}{\partial l} + b_{2,1}(t_2 - t_1) + b_{2,3}(t_2 - t_3) = 0 \tag{6.20}$$

$$\frac{\partial t_3}{\partial \tau} + \delta_3 a_3 \frac{\partial t_3}{\partial l} + b_{3,2}(t_3 - t_2) = 0 \tag{6.21}$$

where $\delta_i = \pm 1$ $(i = 1,\ 2,\ 3)$ and whether δ_i takes the positive or negative value depends upon the flow direction and the coordinate direction. Coefficients a_i and $b_{i,j}$ $(i = 1,\ 2,\ 3;\quad j = 1,\ 2,\ 3)$ are

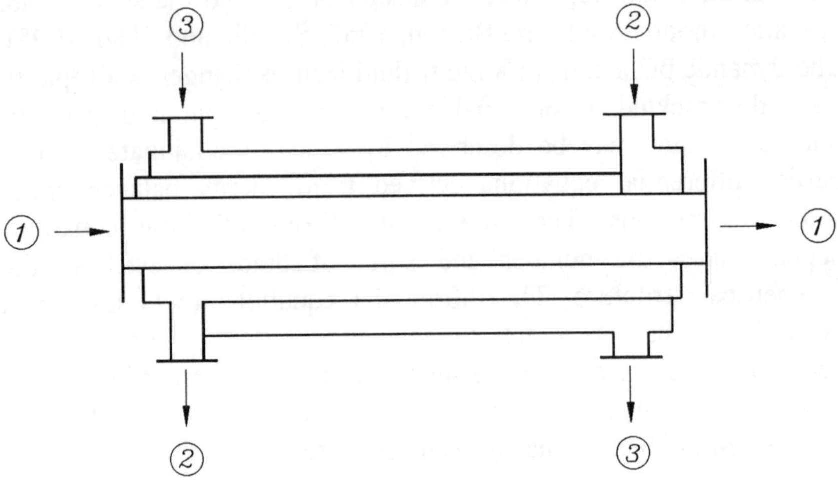

Figure 6.1: Schematic representation of a tubular three-fluid heat exchanger

functions of flow rates, heat transfer coefficients and heat capacities of fluid streams. Evidently, the heat capacitance of solid components is neglected and no equations are introduced for solid components in the above model.

With regard to the given boundary and initial conditions, Sekulic and Herman obtained the numerical solution. In another paper, Sekulic *et al.* (1991) made comparison between experimental and theoretical results about the dynamic behaviour of a tubular three-fluid heat exchanger. Alternatively, eqns (6.19)–(6.21) can be solved with the methods developed in the previous chapters without any further effort.

6.2.2 Modified transient model of a three-fluid heat exchanger

If the heat capacitance of solid components and heat loss to the environment should be taken into account, the system of eqns (6.19)–(6.21) needs to be modified. In this case, a set of differential equations is derived as:

$$\frac{\partial t_1}{\partial \tau} + \delta_1 a_1^* \frac{\partial t_1}{\partial l} + b_{1,2}^* \left(t_1 - t_{w1} \right) = 0 \tag{6.22}$$

$$\frac{\partial t_{w1}}{\partial \tau} + c_{w1,1} \left(t_{w1} - t_1 \right) + c_{w1,2} \left(t_{w1} - t_2 \right) = 0 \tag{6.23}$$

$$\frac{\partial t_2}{\partial \tau} + \delta_2 a_2^* \frac{\partial t_2}{\partial l} + b_{2,w2}^* \left(t_2 - t_{w2} \right) + b_{2,w1}^* \left(t_2 - t_{w1} \right) = 0 \tag{6.24}$$

$$\frac{\partial t_{w2}}{\partial \tau} + c_{w2,2} \left(t_{w2} - t_2 \right) + c_{w2,3} \left(t_{w2} - t_3 \right) = 0 \tag{6.25}$$

$$\frac{\partial t_3}{\partial \tau} + \delta_2 a_3^* \frac{\partial t_3}{\partial l} + b_{3,w3}^* \left(t_3 - t_{w3} \right) + b_{3,w2}^* \left(t_3 - t_{w2} \right) = 0 \tag{6.26}$$

$$\frac{\partial t_{w3}}{\partial \tau} + c_{w3,3} \left(t_{w3} - t_3 \right) + c_{w3,f} \left(t_{w3} - t_f \right) = 0 \tag{6.27}$$

where all coefficients depend upon the heat transfer coefficients, flow rates, heat capacities of fluid streams and solid components. The forms of these equations are similar to those derived in the previous chapters and the solution methods used previously can be applied to finding the transient response subject to inlet temperature and flow rate disturbances.

According to the aforementioned procedure, simulation of dynamic process of a multi-fluid heat exchanger is similar to that of a two-fluid apparatus. Various multi-fluid heat exchangers have many different applications. The geometrical structure of units, the number of fluid streams, the heat transfer coefficients and the flow rate as well as the flow arrangement vary from case to case. One should grasp the characteristics of transient heat transfer process in a multi-fluid exchanger and set up a set of differential equations. The solution methods in the previous chapters are applicable.

6.3 Simulation methods of thermal regenerators

Besides recuperators, in which heat is transferred between fluid streams through a wall, regenerative heat exchangers (regenerators) find wide use in various fields, such as industrial waste heat recovery, air conditioning and cryogenic processes. There are two typical types of regenerators, i.e., a fixed-matrix regenerator and a rotary regenerator. The design theory of regenerators has been studied for many years; nevertheless, much research interest is still aimed at regenerator simulation. The first complete and scientifically correct theory on thermal regenerators was developed by Hausen (1929): Shah (1981) and Heggs (1986) provided an excellent description of thermal models of a regenerator. Various methods such as the Laplace transform, integral transform and finite-difference are applied to finding solutions to thermal simulation of regenerators. Discretizing the space variable, Razelos (1979) derived a solution to a steady periodic-flow regenerator. An accurate and rapid method was proposed by Hill and Willmott (1989) by means of the trapezoidal discretization rule. Li (1983) presented a finite-difference algorithm for performance evaluation of a periodic-flow exchanger, by extending the conventional TDMA numerical method to a set of the four-knot algebraic equations. With the possible axial gas conduction, Kuo *et al.* (1990) made non-Darcian flow analysis of cryogenic packed-bed regenerators.

Although the operating principle of regenerative exchangers is relatively simple, the thermal design and rating calculation of regenerators are quite complicated because of the complex thermal and hydraulic phenomena, for example, flow maldistribution, flow stagnation, axial and lateral heat conduction, carryover of gases in the matrices and extreme operation conditions, etc. The thermal simulation of regenerators has so far been based on a lot of assumptions (Shah, 1981), some of which are necessary for engineering design, but some may greatly deviate from reality. For example, the heat capacities of fluids are negligible compared to the capacities of the matrices because a regenerator is exclusively used for gas-to-gas heat transfer process. But axial heat conduction in the matrices may not be neglected, especially for metal matrices. Therefore, the theory of regenerators still needs to be further improved, which is the impetus to further research work on the simulation of regenerators.

In fact, the forced convection is not the sole heat transfer mechanism in regenerators, as described in most publications on thermal regenerator simulation. Axial and lateral heat conduction in the matrices always exists and carryover as well as leakage of fluids also exerts an effect on the performance of regenerators. The mathematical difficulty makes it impractical to present a thermal simulation model including all heat transfer modes in a regenerator. Some idealizations are essential for forming the governing differential equations. Among the conventional assumptions involved in regenerator simulation (Shah, 1981), some were separately investigated. With respect to accuracy, neglecting axial heat conduction in matrices and maldistribution (for example, carryover and leakage) of fluids may result in a considerable error. Transverse heat conduction may not be negligible, either, particularly for regenerators whose matrices are constructed from the ceramic materials used in the glassmaking industry. However, it is usually either neglected or separated (as shown by Hausen) to avoid a two-dimensional simulation problem.

6.3.1 Governing differential equations

By means of the fundamental principle of heat transfer in porous media, a versatile simulation scheme of regenerators is now developed. Axial conduction in matrices is involved in the energy balance equation. Being different from most of the existing publications, a macroscopic dispersion term is introduced into the energy equation to include the effect of maldistribution of fluids such as carryover, leakage, bypassing and backmixing.

In either fixed-matrix or rotary regenerators, the governing differential equations confining matrix and fluid temperatures can be reduced to one-dimensional forms, i.e., the temperatures depend upon the spatial coordinate l and the time variable τ, if the transverse conduction resistance of the matrices is neglected. Although the fluid temperatures in a rotary regenerator varies with the axial coordinate l and the angular one θ for a stationary observer on the ground, they can be considered as functions of l and τ for an observer riding on the matrix (Hausen, 1976; Shah, 1981), which means the following derivation is valid for both the fixed-matrix and rotary regenerators under the suitable reference systems. On the basis of most conventional idealizations involved in regenerator simulation (Hausen, 1976; Shah, 1981; Heggs, 1986), the governing differential equations for the fluid and the matrix during the hot period are derived as follows (Xuan and Han, 1996):

$$\rho_h c_{pfh} \varphi \frac{\partial \theta_h}{\partial \tau} + \rho_h c_{pfh} \varphi u_h \frac{\partial \theta_h}{\partial l} = h_h \sigma(\theta_{sh} - \theta_h) + \mathcal{D}_{fh} \varphi \frac{\partial^2 \theta_h}{\partial l^2} \tag{6.28}$$

$$\rho_s c_{ps} (1-\varphi) \frac{\partial \theta_{sh}}{\partial \tau} = h_h \sigma(\theta_{sh} - \theta_{sh}) + k_s (1-\varphi) \frac{\partial^2 \theta_{sh}}{\partial l^2} \tag{6.29}$$

where the entrance of hot fluid is located at $l = 0$. These two equations are reduced as:

$$\frac{L}{u_h P_h} \frac{\partial t_h}{\partial \eta} + \frac{\partial t_h}{\partial \xi} = \Lambda_h (t_{sh} - t_h) + \frac{1}{Pe_h} \frac{\partial^2 t_h}{\partial \xi^2} \tag{6.30}$$

$$\frac{\partial t_{sh}}{\partial \eta} = \Pi_h (t_h - t_{sh}) + \lambda_h \frac{\partial^2 t_{sh}}{\partial \xi^2} . \tag{6.31}$$

Similarly, the reduced equations during the cold-period yield:

$$\frac{L}{u_c P_c} \frac{\partial t_c}{\partial \eta} \pm \frac{\partial t_c}{\partial \xi} = \Lambda_c (t_{sc} - t_c) + \frac{1}{Pe_c} \frac{\partial^2 t_c}{\partial \xi^2} \tag{6.32}$$

$$\frac{\partial t_{sc}}{\partial \eta} = \Pi_c (t_c - t_{sc}) + \lambda_c \frac{\partial^2 t_{sc}}{\partial \xi^2} \tag{6.33}$$

where the positive sign (+) and the negative (–) of ± are valid for the cocurrent and countercurrent flow, respectively. Axial conduction in matrices and flow maldistribution may be two main factors of degrading thermal performance of regenerators. The coefficient Pe appearing in both equations (6.30) and (6.32) describes the effect of fluid maldistribution on thermal performance of regenerators. The parameter λ_h or λ_c in equation (6.31) or (6.33) represents the effect of axial heat conduction. In general, the hot (or cold) period P_h (or P_c) is much greater than the dwelling time $\tau_{dh} = L/u_h$ or $\tau_{dc} = L/u_c$ of a fluid in the regenerator. Thus, eqns (6.30) and (6.32) are reduced to

$$\frac{\partial t_h}{\partial \xi} = \Lambda_h\left(t_{sh} - t_h\right) + \frac{1}{Pe_h}\frac{\partial^2 t_h}{\partial \xi^2} \tag{6.34}$$

$$\pm\frac{\partial t_c}{\partial \xi} = \Lambda_c\left(t_{sc} - t_c\right) + \frac{1}{Pe_c}\frac{\partial^2 t_c}{\partial \xi^2}. \tag{6.35}$$

These parabolic equations require two boundary conditions. Since the dispersed plug-flow model is introduced to describe flow maldistribution in regenerators, the boundary conditions proposed by Danckwerts (1953) is used to construct the necessary conditions for fluid temperatures at the entrance and exit of a regenerator. These conditions are as follows:

$$t_h - \frac{1}{Pe_h}\frac{\partial t_h}{\partial \xi} = f_1(\eta) \text{ at } \xi = 0; \quad \frac{\partial t_h}{\partial \xi} = 0 \text{ at } \xi = 1 \tag{6.36}$$

$$t_c - \frac{1}{Pe_c}\frac{\partial t_c}{\partial \xi} = f_2(\eta) \text{ at } \xi = 0; \quad \frac{\partial t_c}{\partial \xi} = 0 \text{ at } \xi = 1$$

(for the cocurrent flow) (6.37a)

$$t_c + \frac{1}{Pe_c}\frac{\partial t_c}{\partial \xi} = f_2(\eta) \text{ at } \xi = 1; \quad \frac{\partial t_c}{\partial \xi} = 0 \text{ at } \xi = 0$$

(for the countercurrent flow). (6.37b)

Correspondingly, one can give the two boundary conditions for the matrix temperature during each period. For simplicity, it is assumed that the matrix is of the adiabatic boundary condition. The initial conditions of the matrix for the hot-period yield:

$$t_{sh}(\xi,0)=g_{gh}(\xi) \qquad \text{for the transient response.} \tag{6.38}$$

For the steady operation, the initial conditions are expressed by the inter-period conditions:

$$t_{sh}(\xi,0)=t_{sc}(\xi,1) \text{ and } t_{sc}(\xi,0)=t_{sh}(\xi,1). \tag{6.39}$$

A set of the equations describing the heat transfer process in regenerators has been so far established. In the foregoing equations there are some parameters that still have to be explained. As mentioned before, the axial dispersion (or effective conduction) term introduced into the energy equations of fluids consists of the effective conduction and the axial macroscopic dispersion of fluid, and then the equivalent conductivity $\mathcal{D}$ of a fluid is usually expressed as $\mathcal{D}=k_f+\mathcal{D}_{dis}$. The dispersion coefficient $\mathcal{D}_{dis}$ depends upon the inherent construction of regenerator matrices, manufacturing clearances, backmixing, leakage, uneven distribution of fluid streams, etc. It should generally be determined by experiment. In cases that are short of necessary experimental data, the following dispersion conductivity for packed-sphere beds given by Wakao and Kaguei (1982)

$$\mathcal{D}_{dis}=0.5\rho_f c_{pf} u d_e \tag{6.40}$$

may be applied to regenerator simulation. Here an equivalent particle diameter $d_e=6/\sigma$ of the general matrix is introduced as the diameter of a hypothetical sphere with the same specific surface area σ based on the solid matrix (Kaviany, 1991). The interstitial convective coefficient h_h (or h_c) varies with the Reynolds number *Re* and the geometry of the regenerator matrix. If the matrix consists of thin plates, the suitable formula of calculating $h_{h,c}$ can be found in a book by Kays and London (1984). For the matrix consisting of spherical particles, the interstitial heat transfer coefficient h can be calculated with the following equation:

$$\frac{hd}{k_f} = 2 + 1.1 Re^{0.6} Pr^{1/3}. \tag{6.41}$$

The next task is to determine the temperature distributions of the fluid and matrix from the foregoing closed system of differential equations with the corresponding initial and boundary conditions. Because of the complexity of the above equations, it is almost impossible to derive analytical solution forms. Finite-difference and the Laplace transform methods are two reasonable choices. The latter is selected here to avoid time-consuming calculation along both the spatial and time coordinates step by step with the finite-difference method. In the light of the methods developed in the previous chapters, there exists the Laplace transform pair:

$$F(\xi, s) = \int_0^\infty f(\xi, \eta)\, \mathrm{e}^{-s\eta}\, \mathrm{d}\eta$$

$$f(\xi, \eta) = \frac{1}{2\pi i} \int_{c_1 - i\infty}^{c_1 + i\infty} F(\xi, s)\, \mathrm{e}^{s\eta}\, \mathrm{d}s \tag{6.42}$$

and then the numerical inverse algorithm is expressed as

$$f(\xi, \eta) = \sum_{j=1}^{M} \chi_j F[\xi,\ a_j(\eta)] \tag{6.43}$$

where χ_j and $a_j(\eta)$ may take different forms according to the selected inverse algorithm and M depends upon the desired accuracy. From the previously derived differential equations, one can determine the response and simulate the steady periodic-flow performance of a regenerator.

6.3.2 Transient response

If the inlet temperature or flow rate of a fluid abruptly changes, a dynamic process takes place in a regenerator, which is different from the steady periodic performance. Such a transient process is governed by eqns (6.31) and (6.34) or (6.33) and (6.35). Applying the Laplace transform to eqns (6.31) and (6.34), one obtains the differential equations as follows:

$$\frac{\mathrm{d}\,T_h}{\mathrm{d}\,\xi} = \Lambda_h\left(T_{sh} - T_h\right) + \frac{1}{Pe_h}\frac{\mathrm{d}^2\,T_h}{\mathrm{d}\,\xi^2} \tag{6.44}$$

$$\Pi_h\left(T_h - T_{sh}\right) + \lambda_h\frac{\mathrm{d}^2\,T_{sh}}{\mathrm{d}\,\xi^2} = sT_{sh} - g_{sh}(\xi)\,. \tag{6.45}$$

For a general form of the initial condition $g_{sh}(\xi)$, the parameter variation principle should be employed (Xuan and Roetzel, 1993b) to find the solution to these two equations. For the uniform initial conditions, one is able to derive the following general solutions in the transformed domain without much effort:

$$T_h(\xi, s) = \sum_{i=1}^{4} c_i\,\mathrm{e}^{m\,\xi_i} \tag{6.46}$$

$$T_{sh}(\xi, s) = \sum_{i=1}^{4} c_i\alpha_{hi}\,\mathrm{e}^{m\,\xi_i} \tag{6.47}$$

where $\alpha_{hi} = 1 + \dfrac{m_i}{\Lambda_h} - \dfrac{m_i^2}{Pe_h}$ and m_i $(i = 1,\ 2,\ 3,\ 4)$ are roots of the polynomial below:

$$\frac{\lambda_h}{Pe_h}m^4 - \lambda_h m^3 - \left(\lambda_h\Lambda_h + \frac{s}{Pe_h}\right)m^2 + \left(s + \Pi_h\right)m + s\Lambda_h = 0\,. \tag{6.48}$$

Obviously, solutions (6.46) and (6.47) are valid for the case that the four roots of polynomial (6.48) are distinct. If the multiplicity of roots is greater than 1, the transformed solutions take different forms. According to the given boundary conditions, the coefficients c_i can be determined. Thus, the dynamic response of the hot fluid and matrices subject to abrupt inlet disturbances have been obtained in the image domain. By means of the numerical inverse algorithm, one can obtain transient temperature distributions in the real-time domain. It should be emphasized that this solution is valid only for the start-up period.

6.3.3 Steady periodic-flow operation

Without taking flow maldistribution or axial conduction into account, Nahavandi and Weinstein (1961), Hausen (1927, 1942, 1976) and Hill and Willmott (1989) respectively derived the solution to the steady periodic-flow regenerative heat exchanger problem. A general feature of these solutions is that the reversal (cyclic equilibrium) conditions between the hot and cold periods and the related power series have been introduced to deduce the solution. Without question, the accuracy of such a solution depends upon the term number and form of the selected series. A similar approach is used to simulate steady periodic-flow operation of regenerators, taking axial conduction and flow maldistribution into account.

(1) Flow maldistribution and no axial conduction in the matrix

In this case $\lambda_{h,c} = 0$, the transformed solutions corresponding to the hot-period is as follows:

$$T_h(\xi,s) = c_{h1}\,\mathrm{e}^{m_{h1}\xi} + c_{h2}\,\mathrm{e}^{m_{h2}\xi} + \beta_h G_h(\xi,s) \tag{6.49}$$

$$T_{sh}(\xi,s) = \frac{\Pi_h}{\Pi_s + s}\left[c_{h1}\,\mathrm{e}^{m_{h1}\xi} + c_{h2}\,\mathrm{e}^{m_{h2}\xi} + \beta_h G_h(\xi,s)\right] + \frac{g_{sh}(\xi)}{\Pi_h + s} \tag{6.50}$$

where

$$c_{h1} = \frac{\left(1 - m_{h2}/Pe_h\right)\beta_h G_h'(1,s) + m_{h2}\,\mathrm{e}^{m_{h2}}\,F_h(s)}{\Delta_h},$$

$$c_{h2} = \frac{c_{h1} m_{h1}\,\mathrm{e}^{m_{h1}} - \beta_h G_h'(1,s)}{m_{h2}\,\mathrm{e}^{m_{h2}}}$$

$$\Delta_h = m_{h2}\,\mathrm{e}^{m_{h2}}\left(1 - \frac{m_{h1}}{Pe_h}\right) - m_{h1}\,\mathrm{e}^{m_{h1}}\left(1 - \frac{m_{h2}}{Pe_h}\right),$$

$$G_h(\xi,s) = \int_0^{\xi}\left[\mathrm{e}^{m_{h2}(\xi-\xi')} - \mathrm{e}^{m_{h1}(\xi-\xi')}\right] g_{sh}(\xi')\,\mathrm{d}\xi'$$

$$m_{h1,2} = \frac{Pe_h \pm \sqrt{Pe_h^2 + 4Pe_h\Lambda_h s/\left(\Pi_h + s\right)}}{2},$$

$$\beta_h = \frac{\Lambda_h Pe_h}{\left(m_{h1} - m_{h2}\right)\left(\Pi_h + s\right)}.$$

For the cocurrent arrangement, substitution of subscript "*c*" for "*h*" in the above solutions directly yields the transformed temperature for the cold-period. For the countercurrent flow, the following solutions related to the cold-period are derived,

$$T_c(\xi,s) = c_{c1}\,\mathrm{e}^{m_{c1}\xi} + c_{c2}\,\mathrm{e}^{m_{c2}\xi} + \beta_c G_c(\xi,s) \tag{6.51}$$

$$T_{sc}(\xi,s) = \frac{\Pi_c}{\Pi_c + s}\left[c_{c1}\,\mathrm{e}^{m_{c1}\xi} + c_{c2}\,\mathrm{e}^{m_{c2}\xi} + \beta_c G_c(\xi,s)\right] + \frac{g_{sc}(\xi)}{\Pi_c + s} \tag{6.52}$$

where

$$c_{c1} = \frac{F_2(s) - \beta_c G_c(1,s) - \beta_c G_c'(1,s)/Pe_c}{\Delta_c}, \quad c_{c2} = \frac{m_{c1}}{m_{c2}} c_{c1}$$

$$\Delta_c = e^{m_{c2}} - \frac{m_{c1}\, e^{m_{c2}}}{m_{c2}} - \frac{m_{c1}\, e^{m_{c1}} - m_{c1}\, e^{m_{c2}}}{Pe_c},$$

$$G_c(\xi,s) = \int_0^{\xi} \left[e^{m_{c2}(\xi-\xi')} - e^{m_{c1}(\xi-\xi')} \right] g_{sc}(\xi')\, d\xi'$$

$$m_{c1,2} = \frac{-Pe_c \pm \sqrt{Pe_c^2 + 4Pe_c \Lambda_c s/(\Pi_c + s)}}{2},$$

$$\beta_c = \frac{\Lambda_c Pe_c}{(m_{c1} - m_{c2})(\Pi_c + s)}.$$

The expression $g_{sh}(\xi)$ in $G_h(\xi,s)$ and $g_{sc}(\xi)$ in $G_c(\xi,s)$ are still unknown. They should be determined with the reversal conditions (6.39). Thus, these functions are assumed as

$$g_{sh}(\xi) = \sum_{n=0}^{\infty} a_n \xi^n, \quad g_{scf}(\xi^*) = \sum_{n-0}^{\infty} b_n (\xi^*)^n. \tag{6.53}$$

Consider a truncated power series up to power N. One should find $2N+2$ coefficients a_n and b_n $(n = 0, 1, 2, \dots, N)$ of the above expansions. Integration of $G_h(\xi,s)$ and $G_c(\xi,s)$ yields

$$G_h(\xi,s) = \sum_{n=0}^{N} \sum_{j=1}^{n} a_n \left[D_j(m_{h1}, \xi, n) - D_j(m_{h2}, \xi, n) \right] \tag{6.54a}$$

$$G_c(\xi,s) = \sum_{n=0}^{N} \sum_{j=1}^{n} b_n \left[D_j(m_{c1}, \xi^*, n) - D_j(m_{c2}, \xi^*, n) \right] \tag{6.54b}$$

where $D_j(m,\xi,n) = \frac{n!}{m^j(n-j+1)!}\xi^{(n-j+1)} - \frac{n!}{m^{j+1}}\mathrm{e}^{m\xi}$.

So far the transformed temperatures of the hot and cold fluids as well as that of the matrix have been formally derived. Application of the numerical inverse algorithm to these transformed solutions results in the transient temperature profiles in the real time domain, although the coefficients a_n and b_n $(n = 0,\ 1,\ 2,\ \ldots,\ N)$ are still unknown. According to the transient temperatures $t_{sh}(\xi,\eta)$ and $t_{sc}(\xi,\eta)$, the reversal condition (6.39) yields $2N+2$ algebraic equations for these coefficients by selecting $2N+2$ discrete points within the intervals $0 \le \xi \le 1$ and $0 \le \xi^* \le 1$. The matrix form of these algebraic equations is:

$$\mathbf{DY} = \mathbf{B} \tag{6.55}$$

where $\mathbf{Y} = \{a_0,\ a_1,\ \ldots,\ a_N,\ b_0,\ b_1,\ \ldots,\ b_N\}^{\mathrm{T}}$ and matrix $\mathbf{D}$ and vector $\mathbf{B}$ are confined by the reversal conditions (6.39), (6.50) and (6.52). Once these coefficients, $G_{sh}(\xi,s)$ and $G_{sc}(\xi,s)$ are determined, one finds the transformed temperatures of hot and cold fluids as well as that of the matrix. Application of the inverse algorithm results in the transient temperature profiles for the steady periodic-flow. It is clear that the accuracy of the final solution depends upon the polynomial forms (6.53) which may be Lagrangian polynomial, Chebyshev series or simple power series expansions.

(2) Axial conduction and no flow maldistribution

In this case $Pe \to \infty$ and $\lambda_{h,c} > 0$, and then the traditional plug-flow model is applied to the fluid. Thus, the following transformed solutions pertinent to the hot period are derived as:

$$T_h = c_{h1}\,\mathrm{e}^{m_{h1}\xi} + c_{h2}\,\mathrm{e}^{m_{h2}\xi} + c_{h3}\,\mathrm{e}^{m_{h3}\xi} + G_h(\xi,s) \tag{6.56}$$

$$T_{sh} = \frac{1}{\Lambda_h}\frac{\mathrm{d}T_h}{\mathrm{d}\xi} + T_h \tag{6.57}$$

where m_{hi} $(i = 1,\ 2,\ 3)$ are the roots of polynomial $\lambda_h m_h^3 + \lambda_h \Lambda_h m_h^2 - (\Pi_h + s)m_h - s\Lambda_h = 0$,

$$G_h(\xi, s) = \alpha_1 \int_0^{\xi} \mathrm{e}^{m_{h1}(\xi-\xi')}\, g_{sh}(\xi')\,\mathrm{d}\,\xi' + \alpha_2 \int_0^{\xi} \mathrm{e}^{m_{h2}(\xi-\xi')}\, g_{sh}(\xi')\,\mathrm{d}\,\xi'$$

$$+\alpha_3 \int_0^{\xi} \mathrm{e}^{m_{h3}(\xi-\xi')}\, g_{sh}(\xi')\,\mathrm{d}\,\xi',$$

$$\alpha_1 = -\frac{\Lambda_h}{(m_{h1} - m_{h3})(m_{h1} - m_{h2})},\quad \alpha_2 = -\frac{\Lambda_h}{(m_{h2} - m_{h3})(m_{h2} - m_{h1})},$$

$$\alpha_3 = \frac{\Lambda_h}{(m_{h2} - m_{h3})(m_{h1} - m_{h3})}.$$

Three coefficients c_{hi} $(i = 1,\ 2,\ 3)$ are determined according to the boundary conditions of the fluid and solid matrix. For the simplest case of the adiabatic boundary conditions, one has

$$c_{h1} = \frac{\mu_6\mu_2 - \mu_3\mu_5}{\mu_1\mu_5 - \mu_4\mu_2},\quad c_{h2} = -\frac{c_{h1}\mu_1 + \mu_3}{\mu_2},\quad c_{h3} = F_h(s) - c_{h1} - c_{h2} \tag{6.58}$$

where

$$\mu_1 = \frac{m_{h1}^2}{\Lambda_h} + m_{h1} - \frac{m_{h3}^2}{\Lambda_h} - m_{h3},\qquad \mu_2 = \frac{m_{h2}^2}{\Lambda_h} + m_{h2} - \frac{m_{h3}^2}{\Lambda_h} - m_{h3}$$

$$\mu_3 = \frac{m_{h3}^2}{\Lambda_h} + m_{h3},\qquad \mu_4 = \left(\frac{m_{h1}^2}{\Lambda_h} + m_{h1}\right)\mathrm{e}^{m_{h1}} - \left(\frac{m_{h3}^2}{\Lambda_h} + m_{h3}\right)\mathrm{e}^{m_{h3}}$$

$$\mu_5 = \left(\frac{m_{h2}^2}{\Lambda_h} + m_{h2}\right) e^{m_{h2}} - \left(\frac{m_{h3}^2}{\Lambda_h} + m_{h3}\right) e^{m_{h3}},$$

$$\mu_6 = \frac{G_h''(1,s)}{\Lambda_h} + G_h'(1,s) + \left(\frac{m_{h3}^2}{\Lambda_h} + m_{h3}\right) e^{m_{h3}} F_h(s).$$

The above solutions can be extended to the cold period for the cocurrent flow arrangement by substituting the subscript "*c*" for "*h*". For the countercurrent flow, the transformed solutions in the cold period are derived as

$$T_c = c_{c1} e^{m_{c1}\xi} + c_{c2} e^{m_{c2}\xi} + c_{c3} e^{m_{c3}\xi} + G_c(\xi, s) \tag{6.59}$$

$$T_{sc} = T_c - \frac{1}{\Lambda_c}\frac{dT_c}{d\xi} \tag{6.60}$$

where m_{ci} $(i = 1, 2, 3)$ are the roots of polynomial $\lambda_c m_c^3 - \lambda_c \Lambda_c m_c^2 - (\Pi_c + s) m_c + s\Lambda_c = 0$.

$$G_c(\xi, s) = \beta_1 \int_0^\xi e^{m_{c1}(\xi-\xi')} g_{sc}(\xi') d\xi' + \beta_2 \int_0^\xi e^{m_{c2}(\xi-\xi')} g_{sc}(\xi') d\xi'$$

$$+ \beta_3 \int_0^\xi e^{m_{c3}(\xi-\xi')} g_{sc}(\xi') d\xi',$$

$$\beta_1 = -\frac{\Lambda_c}{(m_{c1} - m_{c3})(m_{c1} - m_{c2})}, \quad \beta_2 = \frac{\Lambda_c}{(m_{c2} - m_{c3})(m_{c2} - m_{c1})},$$

$$\beta_3 = \frac{\Lambda_c}{(m_{c2} - m_{c3})(m_{c1} - m_{c3})}.$$

For the adiabatic boundary conditions of the solid matrix, one obtains the coefficients c_{ci}

$$c_{c1} = \frac{v_6 v_2 - v_3 v_5}{v_1 v_5 - v_4 v_2}, \quad c_{c2} = -\frac{v_3 - c_{c1} v_1}{v_2},$$

$$c_{c3} = \mathrm{e}^{-m_{c3}} \left[F_c(s) - G_c(1,s) - c_{h1}\, \mathrm{e}^{m_{c1}} - c_{h2}\, \mathrm{e}^{m_{c2}} \right] \tag{6.61}$$

where

$$v_1 = m_{c1} - \frac{m_{c1}^2}{\Lambda_c} - \left(m_{c3} - \frac{m_{c3}^2}{\Lambda_c} \right) \mathrm{e}^{m_{c2} - m_{c3}},$$

$$v_2 = m_{c2} - \frac{m_{c2}^2}{\Lambda_c} - \left(m_{c3} - \frac{m_{c3}^2}{\Lambda_c} \right) \mathrm{e}^{m_{c2} - m_{c3}},$$

$$v_3 = \left(m_{c3} - \frac{m_{c3}^2}{\Lambda_c} \right) \frac{F_c(s) - G_c(1,s)}{\mathrm{e}^{m_{c3}}},$$

$$v_4 = \mathrm{e}^{m_{c1}} \left(m_{c1} - \frac{m_{c1}^2}{\Lambda_c} - m_{c3} + \frac{m_{c3}^2}{\Lambda_c} \right),$$

$$v_5 = \mathrm{e}^{m_{c2}} \left(m_{c2} - \frac{m_{c2}^2}{\Lambda_c} - m_{c3} + \frac{m_{c3}^2}{\Lambda_c} \right),$$

$$v_6 = G_c'(1,s) - \frac{G_c''(1,s)}{\Lambda_c} - \left(m_{c3} - \frac{m_{c3}^2}{\Lambda_c} \right) \left[F_c(s) - G_c(1,s) \right].$$

Obviously the functions $g_{sh}(\xi)$ and $g_{sc}(\xi)$ are still unknown. Using the procedure described previously, one is able to find both the initial distributions from the inter-periodic relationship (6.39). Thus, the transformed temperatures are determined.

6.3.4 Performance of regenerators

Having found the transformed temperatures, one applies the inverse algorithm to obtain the transient temperature profiles. The average fluid outlet temperatures are useful for estimating the overall performance of regenerators. The time-average temperatures of both the fluids at the outlets are defined as

$$t_{ho} = \int_0^1 t_h(\xi = 1, \eta)\,\mathrm{d}\,\eta \tag{6.62a}$$

$$t_{co} = \int_0^1 t_c(\xi = 1, \eta)\,\mathrm{d}\,\eta \qquad \text{(for cocurrent flow)} \tag{6.62b}$$

$$t_{co} = \int_0^1 t_c(\xi = 0, \eta)\,\mathrm{d}\,\eta \qquad \text{(for countercurrent flow).} \tag{6.62c}$$

The regenerator effectiveness ε is defined as the ratio of the actual heat transfer rate to the thermodynamically limited maximum obtainable heat transfer rate in a countercurrent flow exchanger with infinite transfer surface area per cycle

$$\varepsilon = \frac{\left(\dot{m}c_{pf}P\right)_h\left(t_{hi} - h_{ho}\right)}{\left(\dot{m}c_{pf}P\right)_{\min}\left(t_{hi} - h_{ci}\right)} = \frac{\left(\dot{m}c_{pf}P\right)_c\left(t_{co} - h_{ci}\right)}{\left(\dot{m}c_{pf}P\right)_{\min}\left(t_{hi} - h_{ci}\right)}. \tag{6.63}$$

The previously described algorithm has been programmed in the FORTRAN language. A number of cases have been calculated and some results are illustrated in Table 6.1. The results corresponding to $Pe = \infty$ and $\lambda = 0$ agree with Razelos' work based on the conventional model of simulating regenerators, which reveals that the algorithm is reliable and accurate. The values listed in Table 6.1 show the effect of flow maldistribution on the performance of regenerators. The case of $Pe = \infty$ means no flow maldistribution so the plug-flow exists in a regenerator. This problem was solved by Hausen (1929) and Razelos (1979) as well as other researchers. With a decreasing value of *Pe*, the function of

Table 6.1: Dimensionless matrix temperature t_{sc} for a countercurrent regenerator with $\Lambda_h = \Lambda_c = 10$, $\Pi_h = \Pi_c = \pi$

η	$\xi = 0.0$	$\xi = 0.2$	$\xi = 0.4$	$\xi = 0.5$	$\xi = 0.6$	$\xi = 0.8$	$\xi = 1.0$
Results from Razelos's work (1979)							
0.00	0.9901	0.8799	0.7219	0.6366	0.5501	0.2788	0.2227
0.25	0.9446	0.8173	0.6540	0.5676	0.4750	0.3004	0.1023
0.50	0.8930	0.7523	0.5852	0.4978	0.4092	0.2283	0.0470
0.75	0.8368	0.6867	0.5163	0.4286	0.3405	0.1674	0.0216
1.00	0.7771	0.6199	0.4486	0.3621	0.2769	0.1192	0.0099
Calculated results ($Pe = \infty$, $\lambda = 0$)							
0.00	0.9906	0.8810	0.7232	0.6379	0.5513	0.3797	0.2224
0.25	0.9456	0.8186	0.6553	0.5689	0.4811	0.3013	0.1014
0.50	0.8941	0.7540	0.5865	0.4989	0.4102	0.2289	0.0470
0.75	0.8379	0.6879	0.5174	0.4296	0.3413	0.1678	0.0211
1.00	0.7782	0.6210	0.4495	0.3628	0.2774	0.1194	0.0096
($Pe = 15$, $\lambda = 0$)							
0.00	0.9531	0.8322	0.6915	0.6182	0.5444	0.3922	0.2968
0.25	0.8862	0.7753	0.6325	0.5585	0.4830	0.3302	0.1789
0.50	0.8236	0.7175	0.5729	0.4980	0.4218	0.2674	0.1111
0.75	0.7631	0.6593	0.5138	0.4388	0.3633	0.2132	0.0713
1.00	0.7038	0.6015	0.4562	0.3824	0.3091	0.1683	0.0470
($Pe = \infty$, $\lambda = 0$)							
0.00	0.8755	0.7643	0.6504	0.5939	0.5385	0.4386	0.3847
0.25	0.7968	0.7133	0.6039	0.5473	0.4906	0.3788	0.2734
0.50	0.7303	0.6621	0.5562	0.4994	0.4417	0.3245	0.2034
0.75	0.6708	0.6115	0.5085	0.4521	0.3994	0.2769	0.1571
1.00	0.6159	0.5620	0.4621	0.4067	0.3501	0.2362	0.1248

flow diffusion is strengthened and the temperature distributions against the spatial variable ξ becomes flatter, and then the thermal effectiveness of a regenerator is reduced.

Calculation has shown that in some cases reverse heat transfer may take place in a regenerator with cocurrent flow, i.e., heat is transferred from the matrix to the fluid during the hot period, and vice versa during the cold period. This phenomenon is related to such parameters as time, period, *Pe*, Λ and Π. A greater value of *Pe* (it refers to a steeper temperature profile against ξ) or a longer period may induce *reverse heat transfer*. With increasing Λ, the position ξ_r where reverse heat transfer occurs is shifted towards the location $\xi = 0$. In the case $Pe_h = Pe_c = 10$, $\Pi_h = \Pi_c = 0$ and $\eta = 0.1$, for example, if $\Lambda_h = \Lambda_c = 7$, $\xi_r = 0.85$; if $\Lambda_h = \Lambda_c = 8$, $\xi_r = 0.75$; if $\Lambda_h = \Lambda_c = 9$, $\xi_r = 0.68$. Reverse heat transfer degrades the performance of a regenerator and it should be avoided. In fact, regenerators with cocurrent flow are seldom applied.

6.3.5 Hybrid Laplace transform-finite-difference algorithm

The conventional finite-difference method can be applied to the above differential equations to simulate regenerator performance, while the time derivative is tackled by using the Crank–Nicolson difference formula. Since computation is carried out step by step along the time coordinate to reach the specified time, it costs much time to obtain a solution for a long time process. This drawback may be overcome by combining the Laplace transform and the finite-difference to form the hybrid/finite-difference algorithm. The procedure is as follows:

(1) The Laplace transform is applied to the original differential equations.

(2) The transformed equations and boundary conditions are discretized against the other variable.

(3) The derived algebraic equations are solved.

(4) Application of the numerical inverse algorithm to the transformed solutions yields the final results in the real time-domain.

By means of this hybrid method, one can directly calculate temperature at any specified time without integration along each time step.

The transformed forms of eqns (6.31) and (6.34) belong to the convection–diffusion type. The upwind difference should be applied to discretize the convective terms in these equations in order to obtain convergent and real solutions. Discretization of the transformed equations pertinent to the hot period yields

$$\frac{T_{h,i} - T_{h,i-1}}{\Delta\xi} = \Lambda_h \left(T_{sh,i} - T_{h,i}\right) + \frac{1}{Pe_h} \frac{T_{h,i+1} - 2T_{h,i} + T_{h,i-1}}{\left(\Delta\xi\right)^2} \tag{6.64}$$

$$sT_{sh,i} - g_{sh}(\xi) = \Pi_h\left(T_{h,i} - T_{sh,i}\right) + \lambda_h \frac{T_{sh,i+1} - 2T_{sh,i} + T_{sh,i-1}}{(\Delta\xi)^2} \tag{6.65}$$

The boundary conditions are discretized as

$$\left(Pe_h \Delta\xi + 1\right)T_{h,1} - T_{h,2} = Pe_h \Delta\xi\, F_h(s) \tag{6.66a}$$

$$T_{h,N+1} = T_{h,N} \tag{6.66b}$$

$$T_{sh,1} = T_{sh,2} \tag{6.66c}$$

$$T_{sh,N+1} = T_{sh,N} \tag{6.66d}$$

The above discrete equations for all grid points can be written in the matrix form

$$\mathbf{A}_h \mathbf{T}_h = \mathbf{a}_h + \mathbf{D}_{sh} \mathbf{T}_{sh} \tag{6.67}$$

$$\mathbf{B}_{sh} \mathbf{T}_{sh} = g_{sh} + E_h \mathbf{T}_h \tag{6.68}$$

The solutions are found as

$$\mathbf{T}_h = \mathbf{A}_h^{-1} \mathbf{a}_h + \mathbf{A}_h^{-1} \mathbf{D}_{sh} \mathbf{T}_{sh} \tag{6.69}$$

$$\mathbf{T}_{sh} = \mathbf{P}_{sh} \mathbf{g}_{sh} + \mathbf{q}_{sh} \tag{6.70}$$

where matrix $\mathbf{P}_{sh} = \left(\mathbf{B}_{sh} - \mathbf{E}_h \mathbf{A}_h^{-1} \mathbf{D}_{sh}\right)^{-1}$ and vector $\mathbf{q}_{sh} = \mathbf{P}_{sh} \mathbf{E}_h \mathbf{A}_h^{-1} \mathbf{a}_h$. The transformed temperature of the matrix at the discrete point *i* is

$$T_{sh,i} = \sum_{j=1}^{N+1} p_{sh,ij}\, g_{sh,j} + q_{sh,i}\,. \tag{6.71}$$

For the cold period, a similar solution yields

$$T_{sc,i} = \sum_{j=1}^{N+1} p_{sc,ij} g_{sc,j} + q_{sc,i} . \tag{6.72}$$

This solution form can directly be obtained for the cold period by replacing subscript "*h*" with "*c*" in the solutions (6.69) and (6.70). With the inverse algorithm, one can find the transient temperature of the matrix and that of the fluid from these expressions. Keeping in mind that the initial conditions $g_{sh}(\xi_j)$ and $g_{sc}(\xi_j^*)$ of the hot and cold periods are still unknown, one should use the reversal condition to determine them in the time-domain. If the interval $0 \le \xi \le 1$ is discretized into N elements, one obtains $2N+2$ algebraic equations as

$$\mathbf{Wg} = \mathbf{b} \tag{6.73}$$

where

$$\mathbf{g} = \left\{ g_{sh}(\xi_1),\ g_{sh}(\xi_2),\ \ldots,\ g_{sh}(\xi_{N+1}),\ g_{sc}(\xi_1^*),\ g_{sc}(\xi_2^*),\ \ldots,\ g_{sc}(\xi_{N+1}^*) \right\}^{\mathrm{T}} .$$

Thus, one has found the transient temperature distribution of the fluids and of the matrix for the cocurrent flow. If a regenerator operates with countercurrent flow, the difference is only the discretization of eqn (6.35) and the related boundary conditions. For example, eqn (6.35) is discretized as

$$-\frac{T_{c,i+1} - T_{c,i}}{\Delta\xi} = \Lambda_c \left(T_{sc,i} - T_{c,i} \right) + \frac{1}{Pe_c} \frac{T_{c,i+1} - 2T_{c,i} + T_{c,i-1}}{(\Delta\xi)^2} \tag{6.74}$$

The consequent procedure is similar to that mentioned before.

6.4 Transient behaviour of heat pipes

Heat pipes find wide use in a variety of other scientific research and industrial activities, although they were first applied to space missions. Because of the unique advantages such as high performance of heat transfer, excellently isothermal characteristic and anti-corrosion ability, heat pipes have become one of the most effective and economic devices in waste heat recovery. A few published papers exist on steady-state thermal performance of heat-pipe exchangers (Huang and Tsuei, 1985; Wadowski *et al*., 1991). Recently, researchers have been looking at the transient behaviour of heat pipes. Tilton *et al*. (1988) developed a simple one-dimensional model for transient response of a liquid metal heat pipe under the assumption that the heat pipe is completely isothermal at any given time. Using a two-dimensional compressible flow model and considering the vapour core as a channel flow, Issacci *et al*. (1990) analysed the vapour flow pattern during a start-up transient in heat pipes. Chang *et al*. (1992) proposed a three-dimensional alternating-direction-implicit (ADI) finite-difference numerical model to simulate transient behaviour of heat pipes with thermal energy storage under reverse-pulsed heat loads.

The heat transfer process in a heat pipe with wick includes (Bowman, 1991): heat conduction through an evaporator wall, evaporation of the working fluid at the evaporating section and movement of the vapour from the evaporator through the heat pipe vapour space and into the condenser, condensation of the vapour at the condensing section and heat conduction through the condenser wall. According to this analysis, Bowman (1991) proposed a transient model of a heat pipe with wick. His model consists of two parts: the vapour and wall model.

6.4.1 The vapour model

The vapour flow is assumed to be one-dimensional, compressible and unsteady. The continuity equation is

$$\frac{\partial \rho}{\partial \tau}+\frac{\partial(\rho u)}{\partial x}+\frac{2\rho_w V}{R}=0 \tag{6.75}$$

where V is the vapour evaporation/condensation velocity, ρ the density of vapour, ρ_w the density of the working fluid and R the vapour space radius. The subscript w refers to fluid properties evaluated at the vapour/wick interface condition.

The momentum equation is obtained as

$$\frac{\partial(\rho u)}{\partial \tau}+\frac{\partial}{\partial x}\left(\rho u^2+p-\frac{4}{3}\mu\frac{\partial u}{\partial x}\right)+\frac{2\tau_w}{R}=0. \tag{6.76}$$

The shear stress τ_w in eqn (6.76) is expressed as

$$\tau_w=f\frac{\rho u^2}{2}. \tag{6.77}$$

Bowman (1991) used the simplest friction model to determine the friction coefficient f. If the local Reynolds number (based on pipe diameter) is below 2×10^3, this coefficient yields

$$f=\frac{16}{Re}. \tag{6.78}$$

If the Reynolds number is greater than 2×10^3, the flow is turbulent and the coefficient is

$$f=\frac{0.046}{Re^{1/5}}. \tag{6.79}$$

The energy equation is given as

$$\frac{\partial(\rho E)}{\partial \tau}+\frac{\partial}{\partial x}\left[\rho u\left(E+\frac{p}{\rho}\right)-k\frac{\partial t}{\partial x}\right]+\frac{2\rho_w V(E+p/\rho)_w}{R}=0 \tag{6.80}$$

where $E = e + u^2/2$ and e is the specific internal energy. In this equation the axial conduction term is not ignored, and this has been shown to be very important for the numerical stability of the solution (Bowman and Sweeten, 1989).

The complete set of differential equations for describing the vapour flow can be given in the general matrix form

$$\frac{\partial U}{\partial \tau} + \frac{\partial F}{\partial x} + H = 0 \tag{6.81}$$

where

$$U = \begin{bmatrix} \rho \\ \rho u \\ \rho E \end{bmatrix}, \quad F = \begin{bmatrix} \rho u \\ \rho u^2 + p - \frac{4}{3}\mu \frac{\partial u}{\partial x} \\ \rho u\left(E + \frac{p}{\rho}\right) - k\frac{\partial t}{\partial x} \end{bmatrix}, \quad H = \frac{1}{R}\begin{bmatrix} 2\rho_w V \\ 2\tau_w \\ 2\rho_w V\left(E + \frac{p}{\rho}\right)_w \end{bmatrix}.$$

Bowman proposed the following boundary conditions for the vapour space. At the ends of the vapour space, the axial velocity u is zero and the axial pressure and temperature gradients are zero. Along the length of the heat pipe, the evaporation/condensation mass velocity $\rho_w V$ is specified by the wall model, based on the energy flux leaving or entering the heat pipe wall using the expression

$$\rho_w V = q/h_{fg} \tag{6.82}$$

where h_{fg} is the latent heat of vaporization. The vapour is assumed to be an ideal gas with constant isochoric specific heat capacity and the internal energy is expressed as $e = c_v T$. Besides, the ideal gas equation of state $p = \rho \mathbf{R} T$ is applicable.

6.4.2 The wall model

The form of the wall model depends upon whether the explicit or implicit algorithm is used. For the first, the heat transfer through the wall is assumed to be unsteady and one-dimensional at each axial location along the heat pipe; however, each axial location along the heat pipe is studied separately. The one-dimensional heat conduction equation solved at each axial location along the heat pipe for the explicit algorithm is

$$\frac{1}{r}\frac{\partial}{\partial r}\left(r\frac{\partial t}{\partial r}\right)=\frac{1}{a}\frac{\partial t}{\partial \tau}. \tag{6.83}$$

For the latter, the two-dimensional heat conduction equation is introduced as

$$\frac{1}{r}\frac{\partial}{\partial r}\left(r\frac{\partial t}{\partial r}\right)+\frac{\partial^2 t}{\partial z^2}=\frac{1}{a}\frac{\partial t}{\partial \tau}. \tag{6.84}$$

With regard to the wall boundary condition on the exterior surface of the evaporator, an energy flux is specified as

$$q=-k\left.\frac{\partial t}{\partial r}\right|_{\text{outside wall}} \tag{6.85}$$

On the inside wall (wall/vapour interface) an evaporation/condensation boundary condition is used. The heat conducted away from (or to) the wall is set equal to the heat absorbed (or given) off) by the evaporation (or condensing) fluid

$$-k\left.\frac{\partial t}{\partial r}\right|_{\text{inside wall}}=q=h_i\left(t_{vap}-t_w\right) \tag{6.86}$$

where t_w is the inside wall temperature and t_{vap} the saturated vapour temperature. With this vapour temperature and the wall boundary condition both the wall and vapour models are coupled. Incidentally, the energy fluxes in both these expressions are the same. On the outside surface of the condenser the heat conducted out of the wall is set equal to the heat transferred to the freestream environment

$$-k\left.\frac{\partial t}{\partial r}\right|_{\text{outside wall}} = h_e\left(t_w - t_{env}\right) \tag{6.87}$$

where t_w is the external condenser wall temperature and t_{env} is the environment temperature around the heat pipe.

For sake of simplicity the wall, wick and liquid in the wick are treated as a homogeneous layer in Bowman's model. An equivalent conductivity for the compound layer is found by using an equivalent thermal resistance technique (Chi, 1976). A more complicated explicit method is to divide the wall into two layers: one layer for the heat pipe container and the other for the wick–liquid container. The evaporation/condensation coefficient can be calculated from some sources (Stephan, 1988). In his work, Bowman assumed the coefficients to be equal to 2500 W/m²K.

The finite-difference method is used to solve a set of the above-mentioned differential equations. To save computation time, Bowman proposed the use of the so-called time warping to combine the vapour and wall models by using different time steps, if the explicit algorithm is applied. The time warping is implemented by first marching the wall model through time one large time step and then marching the vapour model through time one small time step. The numerical results show that the water vapour's temperature, pressure and density would vary with time (increasing as the heat pipe starts) but varies slightly axially along the heat pipe at any given time. The evaporation and condensation velocity V is essentially constant axially along the heat pipe at any given time, only changing sign between the evaporated and condensed sections. The water vapour's axial velocity u would increase axially along the evaporator, reach a maximum at the evaporator/condenser interface and then decrease along the condenser.

6.5 Dynamic behaviour of multi-phase heat exchangers

Besides the heat exchangers with single phase flow on both sides, in power and chemical engineering as well as the air-conditioning systems heat exchangers are frequently used in cases where a saturated vapour is condensed or a liquid stream is evaporated on one side. Such apparatus for the heat transfer process with liquid-vapour mixture are called heat exchangers with double-phase change. The energy transport in a two-phase flow system has the following main advantages (Mayinger and Schult, 1981):

(1) Heat transfer can occur under a small temperature difference between two streams and economic heat transport is possible.

(2) The mass flow rate for the energy transport may be relatively small due to the large latent heat of evaporation.

(3) Operation may be possible also with free convection only because of the large density difference between the phases.

The heat exchangers involved in phase-change processes may generally be classified as condensers and evaporators. In the first a vapour is reduced to a denser form, generally to a liquid, when it comes into contact with a cold surface and is cooled below at saturation temperature. In the latter a liquid is vaporized as vapour by absorbing the latent heat, when it comes into contact with a hot surface. The design of a condenser or evaporator is more complicated than that of a single-phase heat exchanger, as discussed by Gupta (1986).

6.5.1 Fundamental governing equations

Heat and mass transfer occur simultaneously and are coupled in exchangers with phase change. The modelling of such apparatus is quite different from that of exchangers with single phase flow. Mayinger and Schult (1981) outlined the dynamic features of double-phase change heat exchangers and proposed a simplified transient model for condensers. Schoenberg (1966) analysed the responses of the condensing pressure and interface movement to sinusoidal variation in the inlet vapour flow. His work was aimed at space power application where transient disturbance of the condensing pressure greatly affects the efficiency of Rankine power cycle. Liao *et al*. (1988) analysed the transient flow behaviour in a single-tube condenser and obtained the responses of the vapour-void fraction and outlet flow rate to sudden changes of the wall heat flux and the inlet vapour flow. Later, Liao *et al*. (1990) proposed a two-phase two-fluid model to analyse the transient phenomena in a double-pipe condenser by introducing the momentum equation, which includes the effects of low inertia, pressure variation, vapour compressibility and coolant flow conditions, to highlight the essential mechanisms behind the transient phenomena of the condensing flow. Wang and Touber (1991) proposed a distributed and transient model of an air cooler and gave some detailed examples.

As already known, the two phase flow exists in exchangers with phase change. The transient behaviour of the thermal performance of these types of exchangers is related to heat and mass transfer processes. The flow rates, the vapour-void fraction distribution and propagation, the vaporizing or condensing pressure and the wall shear stress play a great role in the operation of these exchangers. In the following it is assumed that the two-phase flow occurs on one side and the single phase flow on the other side in evaporators or condensers, which is the common case. The model of an exchanger with phase change consists of two parts: the submodel of the two-phase flow and the submodel of the single phase flow. The first is more complicated than the second, which includes continuity equations, momentum equations and energy equations. The basic idealizations for the first submodel are as follows:

(1) The vapour and liquid are in thermodynamic equilibrium but may have different velocities;

(2) The flow is one-dimensional, i.e., the vapour and liquid each have an average cross-sectional velocity and the void fraction is used to describe the ratio of cross-sectional area occupied by the vapour to the total cross-sectional area.

(1) The submodel of the two-phase flow

Continuity equations for the liquid and vapour phase are respectively derived as

$$\frac{\partial}{\partial\tau}\left[\langle\phi\rangle\rho_v\right]+\frac{\partial}{\partial l}\left[\langle\phi\rangle\rho_v u_v\right]=\dot{m}_{lv} \tag{6.88}$$

$$\frac{\partial}{\partial\tau}\left[\left(1-\langle\phi\rangle\right)\rho_l\right]+\frac{\partial}{\partial l}\left[\left(1-\langle\phi\rangle\right)\rho_l u_l\right]=-\dot{m}_{lv} \tag{6.89}$$

where $\dot{m}_{lv}$ is the interfacial mass-transport rate from the liquid to the vapour phase, which has a positive value for vaporizing processes and a negative value for a condensing processes. $\langle\phi\rangle$ is the void fraction based on the cross-section area, $\langle\phi\rangle = A_v/A$ and $A = A_l + A_v$.

The energy equation for the vapour–liquid mixture is

$$\frac{\partial}{\partial\tau}\left[\langle\phi\rangle\rho_v h_v+\left(1-\langle\phi\rangle\right)\rho_l h_l\right]+\frac{\partial}{\partial l}\left[\langle\phi\rangle\rho_v u_v h_v+\left(1-\langle\phi\rangle\right)\rho_l u_l h_l\right]=\frac{A_s}{A}\dot{q}_{tp} \tag{6.90}$$

where $\dot{q}_{tp}$ is the heat flux from the environment to the mixture on the unit surface area and A_s is the heat transfer surface area per unit length. Obviously, the axial heat conduction in both the vapour and liquid phases is neglected. Since thermodynamic equilibrium between the phases is assumed at any instant, only one energy equation for the mixture is needed.

The momentum equations for the vapour and liquid phases are respectively as follows:

$$\frac{\partial}{\partial\tau}\left(\langle\phi\rangle\rho_v u_v\right)+\frac{\partial}{\partial l}\left(\langle\phi\rangle\rho_v u_v u_v\right)=-\langle\phi\rangle\frac{\partial p}{\partial l}-\langle\phi\rangle\rho_v \mathbf{g}\sin\vartheta-F_{inter}$$

$$-F_v+\dot{M}_v \tag{6.91}$$

$$\frac{\partial}{\partial \tau}\left[\left(1-\langle\phi\rangle\right)\rho_l u_l\right]+\frac{\partial}{\partial l}\left[\left(1-\langle\phi\rangle\right)\rho_l u_l u_l\right]$$

$$=-\left(1-\langle\phi\rangle\right)\frac{\partial p}{\partial l}-\left(1-\langle\phi\rangle\right)\rho_l \mathrm{g}\sin\vartheta+F_{inter}+F_v-\dot{M}_v-F_w . \quad (6.92)$$

In both these momentum equations, the second terms on the right-hand side account for the effect of gravity. But they are equal to zero if the angle of evaporator or condenser inclination from the horizontal is zero. The third terms represent the interfacial shear stress. For condensation inside a tube with an inner diameter D_i, F_{inter} is given as (Liao and Wang, 1990)

$$F_{inter}=\frac{4\sqrt{\langle\phi\rangle}}{D_i}\tau_i . \quad (6.93)$$

The shear stress τ_i at the liquid–vapour interface is defined as

$$\tau_i=f_i\frac{\rho_i}{2}\left(u_v-u_l\right)^2 \quad (6.94)$$

where f_i is the interfacial friction factor. To account for the effect of interfacial roughness, a correlation is proposed by Whalley and Hewitt (1977)

$$f_i=f_v\left[1+24\left(\frac{\rho_l}{\rho_v}\right)^{1/3}\frac{\delta}{D_i}\right] \quad (6.95)$$

where f_v can be evaluated from

$$f=\begin{cases}0.079Re^{-0.25}, & 4000<Re<30\,000\\ 0.046Re^{-0.2}, & Re>30\,000\end{cases} \quad (6.96)$$

and δ is the thickness of the liquid film. In the annular flow model, f_i is written as

$$f_i = f_v\left[1+12\left(\frac{\rho_l}{\rho_v}\right)^{1/3}\left(1-\sqrt{\langle\phi\rangle}\right)\right]. \tag{6.97}$$

The fourth terms are the virtual mass force. The general form of the virtual mass force, as derived by Drew *et al.* (1979), is expressed as

$$F_v = \langle\phi\rangle\rho_l C_v\left\{\frac{\partial u_v}{\partial\tau}-\frac{\partial u_l}{\partial\tau}+u_v\frac{\partial(u_v-u_l)}{\partial l}\right.$$

$$\left.+(u_v-u_l)\left[(\lambda-2)\frac{\partial u_v}{\partial l}+(1-\lambda)\frac{\partial u_l}{\partial l}\right]\right\} \tag{6.98}$$

where C_v is the coefficient of virtual mass and λ is a constant. No and Kazimi (1985) derived the quantitative bounds on C_v and λ from consideration of the second law of thermodynamics and numerical stability. Lahey *et al.* (1980) pointed out that inclusion of the virtual mass force helps to overcome the difficulties from the two-fluid model of two-phase flow which is an ill-posed initial value problem and to improve the numerical efficiency. But Wang and Touber (1991) neglected such virtual mass force in their transient model of an air cooler because they assumed that the equilibrium process of momentum transport is rapid and the momentum equation was considered as time-independent.

The fifth terms represent the momentum of mass transfer between the phases and the value is equal to $\dot{M}_v = \dot{m}_{lv}u_i$. Usually, the interfacial velocity u_i is assumed to be equal to the liquid velocity u_l. For a condensing process, $\dot{M}_v = -|\dot{m}_{lv}|u_i$. The sixth term on the right-hand side of eqn (6.92) is the friction resistance on the wall surface. For a condensing process inside a tube with diameter D_i, F_w is equal to

$$F_w = \frac{4}{D_i} \tau_w \tag{6.99}$$

where τ_w is the wall shear stress. According to the two-phase flow pattern and the theoretical model, there are different formulas to calculate τ_w. Liao and Wang (1990) used a Lockhart-Martinelli type correlation in their work.

(2) The submodel of single-phase flow and wall

The second submodel is for the single-phase flow on the other side of the exchanger as well as the wall, which depends upon the geometrical structure of the exchangers and flow pattern. The differential equations in the previous chapters can be used here to describe temperature distributions of single-phase flow stream and the wall. In the analysis of condensation in a double-pipe condenser (Liao and Wang, 1990), the equations governing the temperature variations in the tube wall and coolant flow yield, respectively,

$$\frac{1}{4}\left(D_o^2 - D_i^2\right)\rho_w c_{pw} \frac{\partial \theta_w}{\partial \tau} = q_w D_i - q_c D_o \tag{6.100}$$

$$\rho_c c_{pc}\left(\frac{\partial \theta_c}{\partial \tau} + u_c \frac{\partial \theta_c}{\partial l}\right) = \frac{q_c D_o}{\left(D_2^2 - D_o^2\right)/4} \tag{6.101}$$

where D_o and D_i are the outer and inner diameters of the condenser tube and D_2 is the inner diameter of the outer shell; θ_w and θ_c are the temperatures of the tube wall and coolant, respectively; q_w and q_c are the heat fluxes passing through the tube wall and that conveyed by the coolant, respectively; u_c is the average coolant velocity; ρ_w, c_{pw}, ρ_c and c_{pc} are the corresponding density and specific heat of the tube wall and the coolant, respectively.

So far closure of the system of differential equations for phase-change exchangers has been formed. Naturally, some further modifications of the above-mentioned basic equations may be possible according to the vaporizing or condensing process and geometrical structure. The system of transient equations can be solved, with regard to the given initial and boundary conditions which may vary from case to case. At the evaporator inlet, the vaporized liquid may have the saturated temperature corresponding to a given pressure and the vapour-void fraction is zero. At the condenser inlet, the refrigerant may be considered as 100% vapour at the saturation condition. The pressure and mass flow rate of the refrigerant may be specified at the inlet while the coolant temperature may be given at the condenser outlet (coolant inlet).

6.5.2 Transient behaviour

An effective method to solve the transient two-phase flow problem is the so-called semi-implicit method for pressure-linked equations (SIMPLE) (Patankar and Spalding, 1972). Two staggered spatial meshes are used for the finite-difference equations to maintain the numerical stability. Generally, the velocity and void fraction correspond to one of the staggered meshes and the pressure, density and other state variables as well as parameters are placed on another mesh system. Wang and Touber (1991) used an advanced computer code PHOENICS, based on the SIMPLE method, to analyse the transient behaviour of the two-phase flow of a refrigerant evaporator. Liao and Wang (1990) directly applied the SIMPLE method to the transient responses of a double-pipe condenser subject to change in inlet pressure, change in refrigerant inlet flow rate and change in coolant flow rate. From some computed examples and analysis, they pointed out that a change in system input (inlet vapour pressure, flow rate or coolant flow rate) causes the vapour void in the condenser to change, and then gives rise to the flow surge phenomenon in the outlet condensate flow. The response of the condensing flow is very sensitive to a change in inlet vapour pressure. The response to a change on the coolant side is rather slow because heat capacity of the tube wall damps the response.

6.6 Transient behaviour of packed porous beds

Application of heat transfer between a flowing fluid and a packed porous medium is of great interest in a variety of fields such as waste heat recovery, solar heat storage systems, absorbers and catalytic reactors. With high compactness and a high volumetric heat transfer coefficient, heat exchangers consisting of packed particles and perforated plates as well as screens may be more effective and economic than other types of heat exchangers. In his monograph, Kaviany (1991) discussed the heat and mass transfer mechanism in porous media, theoretical models and their application. By using the concepts of volume porosity, directional surface permeability, distributed resistance and distributed heat source and sink, Sha (1984) proposed a general porous media formulation for multiphase flow with heat transfer.

According to the mode of energy transport, packed porous beds may be sorted into the sensible and the latent heat types. The first type refers to that energy which is stored as sensible energy in solids or fluids; the latter means that energy is stored as the latent heat required for a phase change of the stored materials or fluids, or as reaction heat in a reversible chemical process. The sensible heat type is widely used. For the mathematical treatment of heat and/or mass transfer process in packed porous beds through which a fluid flows, there are two general types of models: the two-phase models and the single-phase models. For the two-phase models, both phases exchange heat and/or mass and no local equilibrium between the phases exists. For the single-phase models, it is assumed that a local equilibrium exists between the phases and the solid matrix and the flowing fluids are approximated to be a quasi-homogeneous medium. Of these two types of models, the two-phase models are more realistic. The single-phase model is the limit of the two-phase model, as the volumetric heat transfer coefficient tends to infinity. Vortmeyer and Schaefer (1974) estimated the equivalence of one- and two-phase models for heat transfer processes in packed beds. To describe the dynamic behaviour of packed bed heat exchangers and absorbers, Jefferson (1968) proposed a finite stage model, with axial dispersion, interphase resistance and intraparticle diffusion in the absence of gross radial transfer.

6.6.1 Mathematical models of packed porous beds

Figure 6.2 is schematic description of a packed porous bed. The thermal performance and flow characteristics of a packed bed depend upon such factors as local velocity and distribution of a fluid, geometry, porosity and specific surface area of a packed bed, and thermal properties of the solid matrix and the fluid, etc. The axial heat conduction and flow maldistribution may affect the thermal effectiveness of a packed bed. In general, mathematical models of packed porous beds are based on the following main assumptions:

(1) The porous bed is regarded as a continuous media.

(2) The flow takes one-dimensional plug-flow pattern.

(3) Thermal properties of the solid matrix and the fluid in the bed are temperature-independent.

Additional idealizations may be introduced to simplify the mathematical treatment.

(1) The two-phase model

Since there exists no thermal equilibrium between the fluid and the solid matrix, the two-phase model includes governing differential equations for the fluid and the solid matrix, respectively. The temperature distributions of both these phases are coupled by the interfacial heat transfer $h_{sf}\sigma\left(\theta_s - \theta_f\right)$. The interstitial convection heat transfer coefficient, h_{sf}, and the specific surface, σ, are two

Figure 6.2 schematic representation of a packed porous bed

important parameters in the two-phase model. Kaviany (1991) discussed its meaning in detail. $\left(h_{sf}\sigma\right)$ has the dimension W/m³ K, so it is called the volumetric heat transfer coefficient in some cases. It should be determined experimentally. With regard to the energy balance between the fluid and the solid matrix, one is able to derive a set of coupled partial differential equations:

fluid:

$$\phi(\rho\, c_p)_f\left(\frac{\partial\theta_f}{\partial\tau}+u_f\,\frac{\partial\theta_f}{\partial l}\right)=\phi\, k_{fe}\,\frac{\partial^2\theta_f}{\partial l^2}+h_{sf}\sigma\left(\theta_s-\theta_f\right) \tag{6.102}$$

solid matrix:

$$(1-\phi)(\rho\, c_p)_s\,\frac{\partial\theta_s}{\partial\tau}=(1-\phi)\, k_{se}\,\frac{\partial^2\theta_s}{\partial l^2}+h_{sf}\sigma\left(\theta_f-\theta_s\right) \tag{6.103}$$

where ϕ is the porosity of the packed bed. u_f is the average pore velocity. The effective heat conductivities k_{fe} and k_{se} should be determined experimentally. In the case of a lack of experimental data, it may be assumed that $k_{fe}=k_f$ and $k_{se}=k_s$. If necessary, k_{fe} in eqn (6.102) can be replaced by a dispersion coefficient or apparent heat conductivity to account for the flow maldistribution (dispersion) of the fluid through the bed (Xuan and Roetzel, 1995), which is the essential of the so-called dispersed plug-flow model. Obviously, both these equations are of the same form as eqns (6.28) and (6.29) for regenerators.

By defining $\dot{w}_f=(\rho\, c_p)_f u_f\phi$ and the following dimensionless variables and parameters

$$z=\frac{\tau}{L/u_f},\quad x=\frac{l}{L},\quad Pe=\frac{(\rho\, c_p)_f u_f L}{k_{fe}},\quad U_{sf}=\frac{h_{sf}\sigma L}{\dot{w}_f},$$

$$R_{sf}=\frac{(1-\phi)(\rho\, c_p)_s}{\phi(\rho\, c_p)_f},\quad \lambda=\frac{(1-\phi)k_{se}}{\dot{w}_f\, L},$$

one can rewrite these two differential equations as follows:

$$\frac{\partial t_f}{\partial z}+\frac{\partial t_f}{\partial x}=\frac{1}{Pe}\frac{\partial^2 t_f}{\partial x^2}+U_{sf}\left(t_s-t_f\right), \tag{6.104}$$

$$R_{sf}\frac{\partial t_s}{\partial z}=\lambda\frac{\partial^2 t_s}{\partial x^2}+U_{sf}\left(t_f-t_s\right). \tag{6.105}$$

Among the previously defined dimensionless parameters, the conductance parameter λ is one useful criterion whose value determines the actual effect of longitudinal conduction of the solid matrix on the thermal performance of a packed porous bed. If axial conduction in the matrix is negligible, i.e., the matrix has zero conductivity along the flow direction, $\lambda = 0$. For large Reynolds numbers $(Re > 150)$, the axial conduction of the bed can be neglected. These differential equations reduce to the Schumann model (Schumann, 1929). The Péclet number *Pe* appearing in eqn (6.104) quantitatively describes the comprehensive effect of axial heat conduction and flow maldistribution of the fluid through the bed. In fact, the axial conduction of a gas (usually, a gas flows through a packed bed) is negligibly small and *Pe* mainly reveals the influence of flow maldistribution and transverse dispersion. If no account for the maldistribution and dispersion is included, $Pe \to \infty$, which corresponds to the conventional plug-flow model.

The solution to the above-mentioned differential equations requires initial and boundary conditions. The general forms of initial conditions are $t_f(x, 0) = g_f(x)$ and $t_s(x, 0) = g_s(x)$. The boundary conditions can be obtained from the energy balance at the entrance and exit of a packed bed. Vortmeyer and Schaefer (1974) proposed the appropriate boundary conditions related to the two-phase model. Similar forms of these boundary conditions are applied to the thermal estimation of matrix heat exchangers (Xuan and Roetzel, 1995).

(2) The single-phase model

If there exists thermal equilibrium between the solid matrix and the fluid stream flowing through the bed, i.e.,

$\theta_f(x,\tau)=\theta_s(x,\tau)=\theta(x,\tau)$, the two-phase reduces to the single-phase model. Combination of eqns (6.102) and (6.103) yields

$$\left[\phi(\rho c_p)_f+(1-\phi)(\rho c_p)_s\right]\frac{\partial\theta}{\partial\tau}+\phi(\rho c_p)_f u_f\frac{\partial\theta}{\partial l}$$

$$=\left[\phi k_{fe}+(1-\phi)k_{se}\right]\frac{\partial^2\theta}{\partial l^2}. \tag{6.106}$$

This is the so-called single-phase model. By defining

$$(\rho c_p)^*=\phi(\rho c_p)_f+(1-\phi)(\rho c_p)_s,$$

$$\dot{w}=\phi(\rho c_p)_f u_f,\quad k^*=\phi k_{fe}+(1-\phi)k_{se},$$

it is rewritten as

$$(\rho c_p)^*\frac{\partial\theta}{\partial\tau}+\dot{w}\frac{\partial\theta}{\partial l}=k^*\frac{\partial^2\theta}{\partial l^2} \tag{6.107}$$

If the dispersion in the packed bed is severe, the effective conductivity k^* should include the dispersion effect and be an equivalent conductivity of the bed. This parameter accounts for the effect of the pore-level velocity nonuniformity on the temperature distribution. Kaviany (1991) discussed the calculation of such a parameter in detail. The initial condition may be expressed as $\theta(x,0)=g(x)$. The appropriate boundary conditions corresponding to the single-phase model are the Danckwerts' type of boundary conditions:

$$\text{at } l=0:\quad -k^*\frac{\partial\theta}{\partial l}=\phi(\rho c_p)_f u_f(\theta_{in}-\theta) \tag{6.108a}$$

$$\text{at } l=L:\quad \frac{\partial\theta}{\partial l}=0 \tag{6.108b}$$

where θ_{in} is the temperature of the incoming fluid stream. The boundary conditions (6.108) are derived from the energy balance at the entrance and exit of the packed porous bed. The dimensionless forms of the single-phase model and the related boundary conditions are as:

$$\frac{\partial t}{\partial z}+\frac{\partial t}{\partial x}=\frac{1}{Pe^*}\frac{\partial^2 t}{\partial x^2} \tag{6.109}$$

$$\text{at } x=0: \quad t_{in}=t-\frac{1}{Pe^*}\frac{\partial t}{\partial x} \tag{6.110a}$$

$$\text{at } x=1: \quad \frac{\partial t}{\partial x}=0 \tag{6.110b}$$

where $Pe^* = \dot{w}L/k^*$.

The single-phase model of a packed bed is simpler than the two-phase model. The two-phase model considers the interaction between the matrix and the fluid stream through a packed bed and the temperature distributions of both the phases. Based on the local thermal equilibrium between the solid and fluid phases, the single-phase model treats a packed bed through which a fluid flows as a quasi-homogeneous medium. The deviation between these models becomes small, as the interstitial heat transfer coefficient increases.

6.6.2 Solution of the models

Generally, it is difficult to obtain an analytical solution to either the two-phase model or the single-phase model. In few cases the closed-formed solutions may be found by introducing some further idealizations (Riaz, 1977). The finite-difference method is an efficient implement to solve these models, especially in cases where both the temperature and flow rate of an incoming fluid arbitrarily change at the inlet. For arbitrary temperature variations and step disturbance of an incoming fluid, the Laplace transform can be used to find the solution.

According to the given initial and boundary conditions, the solution corresponding to the two-phase model is easily found by a similar procedure to that described in Section 6.3. If the single-phase model is used, the solution procedure is rather simpler than that related to the two-phase model. By applying the Laplace transform to eqn (6.109), one has

$$\frac{\mathrm{d}^2 T}{\mathrm{d}x^2} - Pe^* \frac{\mathrm{d}T}{\mathrm{d}x} - sPe^* T = -Pe^* g(x). \tag{6.111}$$

Application of the parameter variation method yields the solution to eqn (6.111)

$$T(x,s) = c_1\,\mathrm{e}^{m_1 x} + c_2\,\mathrm{e}^{m_2 x} + G(x,s) \tag{6.112}$$

where $G(x,s) = \dfrac{Pe^*}{m_1 - m_2}\left[-\int \mathrm{e}^{m_1(x-x')} g(x')\,\mathrm{d}x' + \int \mathrm{e}^{m_2(x-x')} g(x')\,\mathrm{d}x'\right]$

and $m_{1,2}$ are given as

$$m_{1,2} = \frac{Pe^* \pm \sqrt{(Pe^*)^2 + 4sPe^*}}{2}.$$

Insertion of the boundary conditions into the solution (6.112) leads to two coefficients:

$$c_1 = \frac{m_2\,\mathrm{e}^{m_2}\,T_{in} + (1 - 1/Pe^*)G'(1,s)}{(1 - 1/Pe^*)(m_2\,\mathrm{e}^{m_2} - m_1\,\mathrm{e}^{m_1})},$$

$$c_2 = -\frac{m_1\,\mathrm{e}^{m_1}\,T_{in} + (1 - 1/Pe^*)G'(1,s)}{(1 - 1/Pe^*)(m_2\,\mathrm{e}^{m_2} - m_1\,\mathrm{e}^{m_1})}.$$

So far the solution to the single-phase model in the transform-domain has been found. The necessary condition for this solution is that both eigenvalues $m_{1,2}$ are distinct. As mentioned in the previous sections, application of the numerical inverse transform yields the transient temperature distributions in the real time domain subject to the inlet disturbances.

By assuming zero heat capacitance and no dispersion of the fluid in the bed, Vortmeyer and Mong (1976) proposed a two-phase regenerator model and obtained the numerical solution. In this paper they also determined the solution to the single-phase model and compared the results from both these models. The comparison shows that the result from the single-phase model coincides with the temperature distribution of the solid matrix from the two-phase model. Deviation between the two models decreases with increasing the Reynolds number *Re*, whether the solid matrix is heated or cooled.

6.7 Dynamic behaviour of reactors

In the previous section, dynamic simulation of packed beds which serve for the core part of a variety of compact heat exchangers is involved. In the chemical process industries, packed beds also serve as catalyst reactors. The performance of fixed-bed catalytic reactors is affected by their operation under transient conditions. Many papers have been devoted to behaviour prediction of catalyst reactors. An excellent review of the different models was presented by Froment and Bischoff (1979) and later by Daszkowski (1991). Since the complicated interaction between heat, mass and momentum transfer coupled with chemical reaction appears in the fixed-bed catalytic reactors, reactor models must be rather complex if a realistic prediction is to be obtained. As pointed out by Elkanizi and Aly (1992), a realistic description of a fixed-bed catalytic reactor usually results in multi-dimensional models often leading to a complicated set of coupled nonlinear differential equations in which the nonlinearity and/or the dimensionality of these equations reflects the main feature of the models. In addition, the nonlinear boundary conditions may multiply the complexity of a model. Therefore, some physicochemical assumptions are necessary to simplify the problem description. Some idealizations may be as follows:

(1) The flow takes the plug-flow pattern;

(2) The reaction rate is determined by Arrhenius' law;

(3) Properties and heat and mass transfer coefficients are constant;

(4) The catalyst activity is constant.

According to the feature of the problem to be handled, some assumptions may be relaxed and/or other assumptions may be introduced.

6.7.1 Mathematical model of the reactor with switch flow

By introducing the additional assumptions:

(1) Chemical reaction occurs only on the catalyst surface;

(2) Heat conduction is negligible in the packed catalytic bed;

(3) Mass transfer resistance between the gas and the catalyst surface is neglected,

Gawdzik and Rakowski (1989) proposed a mathematical model of the adiabatic tubular reactor with switch flow.

The following governing differential equations are derived.

For the gas:

$$\frac{\partial \alpha_A}{\partial \tau}+u\frac{\partial \alpha_A}{\partial l}=\frac{r_A}{\phi\, c_{A0}} \tag{6.113}$$

$$(\rho\, c_p)_g \phi\left(\frac{\partial \theta_g}{\partial \tau}+u\frac{\partial \theta_g}{\partial l}\right)=h\sigma\left(\theta_c-\theta_g\right) \tag{6.114}$$

and for the catalyst

$$(1-\phi)(\rho\, c_p)_c\frac{\partial \theta_c}{\partial \tau}=(-\Delta h)_R \mathbf{r}_A+h\sigma\left(\theta_g-\theta_c\right) \tag{6.115}$$

where c_{A0} (mol/m^3) is the initial concentration of component A, α_A the conversion of component A, u the local pore-level velocity, $(-\Delta h)_R$ (J/mol) the reaction enthalpy. The subscripts c and g

represent the catalyst and the gas, respectively. Obviously, the exothermic reaction occurs in the reactor. The first-order reaction rate $\mathbf{r}_A$ is described by

$$\mathbf{r}_A = k_\infty \exp\left[-\frac{E}{R(\theta + 273.15)}\right]\frac{n_{A0}}{A_c u \phi}(1-\alpha_A)\left(1-\frac{\alpha_A}{1-\alpha_A}\frac{1}{K}\right) \tag{6.116}$$

where A_c is the cross-sectional area of the bed, E (J/mol) the activation energy, k_∞ (s^{-1}) a frequency, n_{A0} the initial mole number of component A, R the gas constant, and K the equilibrium constant which tends to infinity for the irreversible reaction.

By introducing the following dimensionless variables and moduli:

$$x = \frac{l}{L},\quad z = \frac{\tau}{L/u},\quad t_g = \frac{\theta_g - \theta_r}{\theta_r + 273.15},\quad t_c = \frac{\theta_c - \theta_r}{\theta_r + 273.15},\quad Da = \frac{k_\infty L}{u\phi},$$

$$\gamma = \frac{E}{R(\theta_r + 273.15)},\quad U_1 = \frac{h\sigma L}{(\rho c_p)_g u\phi},\quad U_2 = \frac{h\sigma L}{(\rho c_p)_c u(1-\phi)},$$

$$B = \frac{(-\Delta h)_R n_{A0}}{(\rho c_p)_c A_c u(\theta_r + 273.15)(1-\phi)},$$

one can transform eqns (6.113)–(6.115) into the dimensionless forms:

$$\frac{\partial \alpha_A}{\partial z} + \frac{\partial \alpha_A}{\partial x} = Da(1-\alpha_A)\exp\left(-\frac{\gamma}{1+t_c}\right)\left(1-\frac{1}{K}\frac{\alpha_A}{1-\alpha_A}\right) \tag{6.117}$$

$$\frac{\partial t_g}{\partial z} + \frac{\partial t_g}{\partial x} = U_1(t_c - t_g) \tag{6.118}$$

$$\frac{\partial t_c}{\partial z} = BDa(1-\alpha_A)\exp\left(-\frac{\gamma}{1+t_c}\right)\left(1-\frac{1}{K}\frac{\alpha_A}{1-\alpha_A}\right) - U_2(t_c - t_g) \tag{6.119}$$

with the following conditions:

$$\alpha_A(z, 0) = 0, \quad t_g(z, 0) = 0, \quad t_c(0, x) = 0. \tag{6.120}$$

After each flow reversal the inlet to the reactor becomes its outlet and the catalyst temperature profile is reversed, which is described by

$$t_{n+1}\left(z_p, x\right) = t_n\left(z_p, 1 - x\right) \tag{6.121}$$

where z_p determines the moment of the flow reversal.

It is impossible to solve the above-derived boundary value problem analytically. Gawdzik and Rakowski (1989) applied both the method of characteristics and the method of orthogonal collocation to finding the solution to the set of differential equations, respectively. The first is based on the principle proposed by Sneddon (1957), which may be highly time-consuming. The essence of the orthogonal collocation method lies in the substitution of the solutions at the selected collocation points for the values of the derivatives at these points (Finlayson, 1972). Such substitution makes it possible to reduce an ordinary differential equation to a set of algebraic equations, and a partial differential equation with two independent variables to a set of ordinary differential equations. Of course, one can use other numerical methods such as the finite-difference method to determine the solution.

The proposed model may be further improved according to the desired accuracy. If axial conduction of the bed should be taken into account, for example, the energy equation (6.115) for the catalyst is modified as

$$(1-\phi)\left(\rho c_p\right)_c \frac{\partial \theta_c}{\partial \tau} = (1-\phi)\frac{\partial^2 \theta_c}{\partial l^2}(-\Delta h)_R \mathbf{r}_A + h\sigma\left(\theta_g - \theta_c\right) \tag{6.122}$$

where k_{ce} is the effective conductivity of the catalyst in the bed.

6.7.2 Two-dimensional model of the fixed-bed reactor

A two-dimensional model of the fixed-bed reactor may be more accurate because it accounts for the spatial variation of component concentrations and temperatures in both the axial and radial directions as well as the time-dependent change. Daszkowski (1991) proposed a two-dimensional transient model for a tubular reactor with chemical reaction, considering a two-dimensional flow field. If the plug-flow pattern in the bed is assumed, the following mathematical model of the fixed-bed reactor is derived.

For the gas:

$$\frac{\partial c}{\partial \tau}+u\frac{\partial c}{\partial l}=D\left(\frac{\partial^2 c}{\partial l^2}+\frac{1}{r}\frac{\partial c}{\partial r}+\frac{\partial^2 c}{\partial r^2}\right)+\frac{h_M\sigma}{\phi}\left(c_s-c\right) \tag{6.123}$$

$$\left(\rho c_p\right)_g\phi\left(\frac{\partial \theta_g}{\partial \tau}+u\frac{\partial \theta_g}{\partial l}\right)=k_{ge}\phi\left(\frac{\partial^2 \theta_g}{\partial l^2}+\frac{1}{r}\frac{\partial \theta_g}{\partial r}+\frac{\partial^2 \theta_g}{\partial r^2}\right)+h\sigma\left(\theta_c-\theta_g\right) \tag{6.124}$$

and for the solid phase (catalyst)

$$h_M\sigma\left(c-c_s\right)-\eta f\left(c_s,\theta_s\right)=0 \tag{6.125}$$

$$\left(1-\phi\right)\left(\rho c_p\right)_g\frac{\partial \theta_s}{\partial \tau}=\left(1-\phi\right)k_{se}\left(\frac{\partial^2 \theta_s}{\partial l^2}+\frac{1}{r}\frac{\partial \theta_s}{\partial r}+\frac{\partial^2 \theta_s}{\partial r^2}\right)$$

$$+\left(-\Delta h\right)_R\eta f\left(c_s,\theta_s\right)+h\sigma\left(\theta_c-\theta_g\right) \tag{6.126}$$

where c and c_s (mol/m^3) are the mole concentration of the reacted gas in the bed and that on the surface of the catalyst, respectively. D (m^2/s) is the diffusion coefficient of the reacted gas, h_M (m/s) the interphase mass transfer coefficient, $f\left(c_s,\theta_s\right)$ (mol/m^{3}s) the reaction rate and η the dimensionless effectiveness factor.

Equations (6.123)–(6.124) describe mass and heat balance for the reacted gas phase in the bed; equations (6.125)–(6.126) are valid for mass and heat balance of the solid phase (catalyst), respectively. Such a model takes the essential transfer mechanism of the fixed-bed reactor into account. With the given initial and boundary conditions, this model can be solved numerically. By means of the finite-element method, Elkanzi and Aly (1992) applied a similar model to simulating the steady-state behaviour of a fixed-bed catalytic reactor.

6.8 Dynamic simulation of kilns and furnaces

In a variety of industrial activities, there are various types of kilns and furnaces which serve different purposes such as heating materials and elements, heat treatment, drying processes, or producing processes. In the kilns and furnaces physical and/or chemical reactions occur. Mathematical models of furnaccs and kilns may be different according to the geometrical structure and operating procedure, although they are based on the fundamental principle of heat and mass transfer. Spang (1972) developed a transient model of a cement kiln in which a complex series of exothermic and endothermic reactions take place. Yu (1994) took heat conduction in the products and in the lining bricks of kiln cars, gas flow and chemical reaction into account and proposed a dynamic model of a tunnel kiln. A heat and mass transfer model was built by Trystram *et al.* (1992) to describe a natural gas indirect-fired baking tunnel-oven. This model included heat transfer from combustion gas to the biscuits, conduction (for the band conveyor), convection and radiation fluxes. Sullivan *et al.* (1992) validated a model of a transient direct-fired batch reheating furnace.

As described by Chapman *et al.* (1990), furnace models can be broadly classified in two groups: detailed phenomenological models and simplified system models. The phenomenological models use the fundamental principles of fluid mechanics, thermodynamics, and heat transfer to provide detailed information on the gas flow pattern, heat transfer process and pollutant formation within a furnace. For all but the simplest furnaces, however, the computer resources required by such fundamental models make them impractical for furnace design and optimization study. Rather, simplified system models are typically employed to predict the thermal efficiency and fuel consumption of real furnaces. System models are based on the phenomenological models, but include substantial simplifying assumptions that allow a designer to explore a wide range of possible operating conditions and design options. Such models typically do not contain the level of detail required to predict the gas flow field or the amount of pollutant formation within a furnace.

6.8.1 Mathematical models of kilns

There are many tunnel kilns and rotary kilns operating in the refractory and ceramic industry. Such kilns have highly complex dynamic behaviour due to the chemical reactions involved. One cannot expect to find a universal model for all types of kilns. In fact, the geometrical structure, operating conditions and procedures of kilns make it impossible to establish a transient model which can be used to describe the dynamic behaviour of all kilns. A comprehensive model is always composed of mass balance equations for all of the components involved and heat balance equations with appropriate initial and boundary conditions. Yu (1994) divided a tunnel kiln into three zones: the preheating, the firing and the cooling zones. By using the principle of the finite-difference method, he developed mass and heat balance equations for each zone and proposed a dynamic model of a tunnel kiln under the condition that the gas temperatures and oxygen concentration profiles in the firing zone are given. It was assumed that the exhaust-gas components are mainly N_2 (nitrogen), O_2 (oxygen), CO_2 (carbon dioxide), H_2O (water), and SO_2 (sulphur dioxide). To set up balance equations, each zone was divided into a number of elements. The mass balance in element k is described as

$$\sum_{i=1}^{I} V_{in,i}^{k}(\tau) - \sum_{j=1}^{J} V_{out,j}^{k}(\tau) + V_s^k(\tau) = 0 \qquad (6.127)$$

where $V_{in,i}^{k}(\tau)$ and $V_{out,j}^{k}(\tau)$ are respectively the flow rates of a gas flowing into and out of the element k at the instant τ by means of mode k , $V_{s}^{k}(\tau)$ is the generation rate of a gas from chemical reaction. Similarly, the heat balance equation can be expressed as

$$\frac{\mathrm{d}U}{\mathrm{d}\tau}=\sum_{l=1}^{L}Q_{in,l}^{k}(\tau)-\sum_{n=1}^{N}Q_{out,n}^{k}(\tau)+Q_{s}^{k}(\tau). \tag{6.128}$$

The first term on the left-hand side is the change rate of the internal energy stored in element k in a zone. The third term on the right-hand side is the energy generation rate resulting from the chemical reaction. The first and second term on this side respectively represent the energy flux flowing into and out of the element, in which heat conduction, convection and radiation to or from the environment may be involved besides the energy flux accompanied with the flow stream. Furthermore, the material balance equations are needed for each component of the air and the exhaust gas which depends upon the chemical reaction process. The numerical results reveal that the transient feature determines the tunnel kiln behaviour, such as fuel consumption, cooling air flow, exhaust flow, heat storage and heat loss. The reader may refer to Yu's work (Yu, 1994) to find the details.

Spang (1972) proposed a dynamic model of a cement kiln which basically consisted of a drum approximately 120–180 m long and 3–6 m in diameter. To make the model as simple as possible while maintaining the essential dynamics, he introduced the following assumptions:

(1) The inside and outside diameters of the kiln are constant.

(2) The solids transported in the gas are not explicitly included in the model.

(3) The clinking rate expressions do not take into account the effect of trace compound concentrations.

(4) Reaction rates are determined by Arrhenius' law.

(5) Specific heats, latent heats, and heats of reactions are independent of temperature and position.

(6) Coefficients of convection and emissivity are independent

of temperature and position.

(7) Conduction in the solid and gas is negligible. The conduction in the axial direction of the wall is also neglected.

(8) The velocity and mass of the solid and gas are constant throughout the length of the kiln.

(9) The heat transfer due to the release of water vapour and CO_2 into the gas can be treated as convective heat transfer.

Based on these assumptions, a set of energy and material balance equations was derived. The energy equations of solid, gas and kiln wall were simplified by averaging out any variation of the kiln and time. According to the practical operation, the energy balance equations for gas, solid and wall yield:

gas:

$$A_g c_{pg} \rho_g u_g \frac{\partial \theta_g}{\partial l} = \beta_1(\theta_w - \theta_g) + \beta_2(\theta_s - \theta_g) + q_f \tag{6.129}$$

solid:

$$A_s c_{ps} \rho_s u_s \frac{\partial \theta_s}{\partial l} = \beta_1(\theta_g - \theta_s) + \beta_3(\theta_w - \theta_s) - c_{ps} \rho_s A_s \frac{\partial \theta_s}{\partial \tau} + A_s q_c \tag{6.130}$$

wall:

$$A_w c_{pw} \rho_w \frac{\partial \theta_w}{\partial \tau} = \beta_1(\theta_g - \theta_w) + \beta_3(\theta_s - \theta_w) + \beta_4(\theta_a - \theta_w) \tag{6.131}$$

where the heat transfer parameters β_i $(i = 1, 2, 3, 4)$ are related to the geometrical dimensions, conduction, convection and radiation heat transfer and gas flow; q_f and q_c depend upon the chemical reaction in the solid and the flame in the gas. Spang (1972) gave the expressions to calculate these parameters.

The other part of the transient model is a set of material balance equations which involve the chemical reaction and material transport. In a cement kiln, the materials may include water, CO_2, $CaCO_3$, $2CaO{\cdot}SiO_2$, $3CaO{\cdot}SiO_2$, $3CaO{\cdot}Al_2O_3$, $3CaO{\cdot}Al_2O_3{\cdot}Fe_2O_3$, SiO_2, Fe_2O_3, Al_2O_3, CaO and fuel. In his work, Spang approximately derived the material balance equation for each component.

The computational approach was implemented on the computer by using a first order finite-difference scheme. According to the numerical results, Spang pointed out that the model appeared to match qualitatively the behaviour of actual kilns fairly well. This was especially true at the burning zone end of the kiln with the all important flame–solid interactions. The result revealed two areas to be refined. The first was indicated by the peak temperatures in the burning zone which seemed somewhat high. This would indicate that emissivity coefficients for each zone were too high. Since there are large physical changes in the material in each zone, changes in emissivity are physically reasonable. The second refinement would be needed because the temperature of the gas in the preheat zone of the model was less sensitive to changes in the preheat zone than had been observed. This was primarily caused by the wall dynamics in the preheat zone.

6.8.2 Dynamics of heating furnaces

Operational processes of heating furnaces are somewhat different from that of kilns. In general, the dynamic models of furnaces pay little attention to chemical reactions occurring in units and the material balance equations are not involved. Heat exchange among the gas, the furnace wall and parts to be heated (the load) is the main part of a transient model. Radiation heat exchange plays a vital role in the transient performance of heating furnaces because they usually operate at temperatures over 1000 K.

The transient batch heating furnace model developed by Chapman *et al.* (1990) simulates the heating process taking place in reheat and heat treatment furnaces. As defined by them, the radiative heat transfer from the flame and combustion products, and from the wall to the load, is usually the dominant mode of heat transfer, with convection heat transfer from the combustion gases making a much smaller contribution. The radiative transfer within the furnace is complicated by the nongrey behaviour of the combustion gases. The presence of selectively absorbing and emitting gases such as CO_2, H_2O, CO, CH_4 and soot and dust particles, requires the model to have some way of describing nongrey effects. The model proposed by Chapman *et al*. (1990) was further validated by Sullivan *et al*. (1992).

To make dynamic simulation possible, a furnace is generally divided into a few zones. In each zone a set of differential energy equations is established for the gas, the wall and the parts to be heated. Sullivan *et al*. (1992) described the main assumptions and features of the dynamic model as follows:

(1) The furnace consists of a single gas zone, five surface (wall) zones and one load zone.

(2) The gas zone is considered to be well-stirred. That is, the temperature and gas composition are assumed to be uniform within the gas volume, with a composition equal to that of the combustion products without dissociation.

(3) The flue gas temperature is equal to the temperature of the furnace gas volume of the furnace gas volume calculated from an energy balance on the gas volume.

(4) The wall and load surface temperature are calculated from one-dimensional transient heat conduction equations for each slab.

(5) Radiation heat exchange is calculated using Hottel's zone method (Hottel and Sarofim, 1964). The surface zones are each treated as being grey and isothermal. The nongrey characteristics of the combustion gases are modelled with a sum-of-grey gases model developed by Smith *et al*. (1982).

The work of Sullivan *et al*. (1992) is described in the following. Here $T_j = \theta_j + 273.15$ is used. By neglecting the energy stored in the gas, the gas energy balance is written as

$$\Delta H_c + H_a\left(T_a\right) = Q_{loss} + H_{cp}\left(T_{cp}\right) \tag{6.132}$$

where ΔH_c represents rate of heat release by combustion, $H_a\left(T_a\right)$ and $H_{cp}\left(T_{cp}\right)$ are enthalpy flow rate of the air and the combustion products, respectively. Q_{loss} is the energy transferred from the combustion products to the load and furnace walls which refer to six surfaces. Q_{loss} can be expressed as

$$Q_{loss} = \sum_{i=1}^{6}\left[A_i h_i (T_g - T_{s,i}) + \sigma(\overrightarrow{GS_i}\, T_g^4 - \overrightarrow{S_i G}\, T_{s,i}^4)\right] \tag{6.133}$$

where the first term on the right-hand side represents the convective heat transfer from the combustion gases to surface zone *i* and the second term stands for the radiative heat exchange between the gas zone and surface zone *i*. $\overrightarrow{GS_i}$ and $\overrightarrow{S_i G}$ are directed-flux exchange areas required by Hottel's zone method. The enthalpy of the combustion products was calculated from

$$H_{cp} = \sum_{l}\left(\int_{T_{ref}}^{T_g} c_{p,l}\,\mathrm{d}T + \Delta h_{f,l}^{o}\right)N_l \tag{6.134}$$

where $\Delta h_{f,l}^{o}$ represents standard heat of formation of combustion product l. The temperature-dependent molar specific heats were approximated by the functions cited in van Wylen and Sonntag (1978) and valid over the temperature range from 300 to 3500 K. The molar flow rates, N_l, were calculated by conducting an atom balance on the generalized combustion equation for a hydrocarbon fuel.

The convective heat transfer coefficient for each surface exposed to the combustion gases was calculated according to an empirical correlation based on experimental measurements (Lebedev and Sokolov, 1976). The direct exchange areas were calculated using the Monte Carlo statistical method (Siegel and Howell, 1981). The total exchange areas ($\overline{GS_i}$ and $\overline{S_i G}$) were then calculated using the procedure suggested by Hottel and Sarofim (1964). The directed-flux exchange areas were calculated from the total exchange areas by using a four grey-gas model proposed by Smith *et al.* (1982):

$$\overrightarrow{S_i S_j} = \sum_{n=1}^{4} A_{n,T_i}(\overline{S_i S_j})_n, \qquad \overrightarrow{GS_j} = \sum_{n=1}^{4} A_{n,T_i}(\overline{GS_j})_n. \tag{6.135}$$

For transient heat conduction in the load and each furnace wall, the energy equation is as

$$\rho c \frac{\partial T}{\partial \tau} = \frac{\partial}{\partial n}\left(k \frac{\partial T}{\partial n}\right) \qquad (6.136)$$

where n is the unit normal into each surface. This equation is sequentially applied to each wall and to the load. The boundary condition on each interior surface is

$$-k_i \left.\frac{\partial T_i}{\partial n}\right|_{\substack{inner \\ surface}} = q_i = \frac{Q_i}{A_i} \qquad (6.137)$$

where the total (convective plus radiative) heat transfer rate is given by

$$Q_i = A_i h_i (T_g - T_{s,i}) + \sigma(\overrightarrow{GS_i}\, T_g^4 - \overrightarrow{S_i G}\, T_{s,i}^4)$$

$$+ \sum_{j=1}^{6} \sigma(\overrightarrow{S_j S_i}\, T_{s,j}^4 - \overrightarrow{S_i S_j}\, T_{s,i}^4). \qquad (6.138)$$

Heat transfer from the outside of the furnace wall to the ambient is by convection and radiation and is expressed as

$$-k_{s,i} \left.\frac{\partial T_i}{\partial x}\right|_{\substack{outer \\ surface}} = h_i\left(T_{s,i} - T_{amb}\right) + \varepsilon_R \sigma\left(T_{s,i}^4 - T_{amb}^4\right). \qquad (6.139)$$

The above-derived differential equations are integrated into the model of furnaces. The numerical method was used to find the transient solution. At each forward time step, eqn (6.133) was solved for the gas zone temperature with a Newton–Raphson procedure. The algorithm developed by Lemonnier *et al.* (1994) can also be used to solve the nonlinear thermal model. Sullivan *et al.* (1992) made a comparison between the theoretical and experimental results for the batch furnaces. For times greater than a threshold which may vary from furnace to furnace, the model overpredicts temperatures to different extents, but faithfully follows the experimental data trends. In addition to the model itself, a number of parameters contribute to the discrepancies between the model predictions and experimental measurements. The assumption of constant refractory properties can be expected to cause overprediction of the load and gas zone temperatures. Treatment of the adiabatic back surface of the load as the floor of the furnace has effectively reduced the mass of the refractory in the furnace and increased the surface area of the load. Both these effects can be expected to contribute to temperature overpredictions within the furnace. Additionally, modelling the load at the floor of the furnace substantially increases the mean beam length of the gas zone and increases the refractory surface area "visible" from the load.

Although no attempt was made to calculate the details of the combustion process or the flowfield, the aforementioned model is able to predict the temperature history of the load within certain errors. This may be a useful prediction for the initial design phase of many batch furnaces. However, the model needs to be improved to bring about better agreement between the model predictions and the experimental measurements. The possible improvements may include a submodel for the combustion process and the flowfield, application of two- or three-dimensional heat conduction equations to the load zone and the furnace walls (or further division of each of these zones into some smaller zones). But such modification costs more computation time. Thus, the model improvement involves trade-offs between the model accuracy and the applicability for simulation, real-time control and design of furnaces.

For some other types of batch furnaces, the load may move during the heating process, and then the energy equation should be modified as

$$\rho c \frac{\partial T_s}{\partial \tau} \pm u_x \frac{\partial T_s}{\partial x} = \frac{\partial}{\partial n}\left(k \frac{\partial T_s}{\partial n}\right) \tag{6.140}$$

where u_x represents the velocity of the load. According to the moving direction, the second term on the left-hand side assumes a negative or positive sign.

6.9 References

Aulds, S. S. and Barron, R. F. (1967) Three-Fluid Heat Exchanger Effectiveness, *International Journal of Heat and Mass Transfer*, **10**, 1457–1462.

Bowman, W. J. (1991) Numerical Modeling of Heat-Pipe Transients, *Journal of Thermophysics and Heat Transfer*, **5**, 374–379.

Bowman, J. and Sweeten, R. W. (1989) Numerical Heat-Pipe Modeling, *AIAA Paper 89-1705*, June, 1989.

Chang, M. Jr, Chow, L. C., Chang, W. S. and Morgan, M. J. (1992) Transient Behaviour of Axially Grooved Heat Pipes with Thermal Energy Storage, *Journal of Thermophysics and Heat Transfer*, **6**, 364–370.

Chapman, K. S., Ramadhyani, S., and Viskanta, R. (1990) Modeling and Parametric Studies of Heat Transfer in a Direct-Fired Batch Reheating Furnaces, *Journal of Heat Treating*, **8**, 137–146.

Chi, S. W. (1976) *Heat Pipe Theory and Practice*, McGraw-Hill, New York.

Danckwerts, P. V. (1953) Continuous Flow Systems — Distribution of Residence Times, *Chemical Engineering Science*, **2**, 1–13.

Daszkowski, T. (1991) *Strömung, Stoff- und Wärmetransport in schüttungsgefüllten Rohrreaktoren*, Dissertation, Universität Stuttgart.

Drew, D. A., Cheng, L. Y. and Lahey, R. T. Jr. (1979) The Analysis of Virtual Mass Transfer in Two-Phase Flow, *International Journal of Multiphase Flow*, **5**, 233–242.

Elkanzi, E. M. and Aly, S. (1992) A Finite Element Solution of the Fixed-Bed Catalytic Reactor Heterogeneous Models, *Computers and Chemical Engineering*, **17**, s485–s492.

Finlayson, A. (1972) *The Method of Weighted Residuals and Variationals Principles*, Academic Press, New York.

Froment, G. F. and Bischoff, K. B. (1979) *Chemical Reactor Analysis and Design*, John Wiley, New York.

Gawdzik, A. and Rakowski, L. (1989) The Methods of Analysis of the Dynamic Properties of the Adiabatic Tubular Reactor with Switch Flow, *Computers Chemical Engineering*, **13**, 1165–1173.

Gupta, J. P. (1986) *Fundamentals of Heat Exchanger and Pressure Vessel Technology*, Hemisphere Publishing Corporation, New York.

Hausen, H. (1929) Über die Theorie des Wärmeaustausches in Regenetatoren, *Zeitschrift für angewandte Mathematik und Mechanik*, **9**, 173–200

Hausen, H. (1942) Vervollständigte Berechnung des Wärmeaustausches in Regeneratoren, *VDI-Beiheft Verfahrenstechnik*, **2**, 31–43

Hausen, H. (1976) *Wärmeübertragung im Gegenstrom, Gleichstrom und Kreuzstrom*, Springer-Verlag, Berlin.

Heggs, P. J. (1986) Heat Exchangers for Heat Recovery, in *High Temperature Heat Exchangers* (ed. By Mori, Y., Sheindlin, A. E. and Afgan, N.), Hemisphere Publishing Corporation, New York.

Hill, A. and Willmott, A. J. (1989) Accurate and Rapid Thermal Regenerator Calculations, *International Journal of Heat and Mass Transfer*, **32**, 465–476.

Hottel, H. C. and Sarofim, A. F. (1964) *Radiative Transfer*,

McGraw-Hill, New York.

Huang, B. J. and Tsuei, J. T. (1985) A Method of Analysis for Heat Pipe Heat Exchangers, *International Journal of Heat and Mass Transfer*, **28**, 553–562.

Issacci, F., Ghoniem, N. M. and Catton, I. (1990), Vapor Flow Patterns during a Start-Up Transient in Heat Pipes, in *AIAA/ASME Fifth Thermophysics and Heat Transfer Conference,* Seattle, WA, June 18–20.

Jefferson, C. P. (1968) Dynamics of Packed Beds with Intraphase Heat or Mass Transfer, *Chemical Engineering Science*, **23**, 509–523.

Kao, S. (1965) Design Analysis of Multistream Hampson Exchangers with Paired Tubes, *Journal of Heat Transfer*, **87**, 202–208.

Kaviany, M. (1991) *Principles of Heat Transfer in Porous Media,* Springer-Verlag, Berlin.

Kays, W. M. and London, A. L. (1984) *Compact Heat Exchangers*, McGraw-Hill Book Company, New York.

Kuo, S. M, Phelan, P. E. and Tien, C. L. (1990) Non-Darcian Analysis of Cryogenic Packed-Bed Regenerators, in *Compact Heat Exchanger* (ed. by Shah, R. K, Kraus, A. D. and Metzger, D.) Hemisphere Publishing Corporation, New York.

Lahey, R. T. Jr, Cheng, L. Y., Drew, D. A. and Flaherty, J. E. (1980) The Effect of Virtual Mass on the Numerical Stability of Accelerating Two-Phase Flows, *International Journal of Multiphase Flow*, **6**, 281–294.

Lebedev, V. I. and Sokolov, V. A. (1976) Study of the Convective Component of Complex Heat Exchange in a Model of a Direct-

Heating Furnace, *Glass and Ceramics*, **33**, 352–354.

Lemonnier, D., Sadat, H. and Saulnier, J. B. (1994) A New Reduction Technique for Non Linear Thermal Models with Conductive and Radiative Couplings, in *The Proceeedings of 10th International Heat Transfer Conference*, **2**, 3-NT-16A.

Li, L. H. (1983) A Numerical Finite Difference Method for Performance Evaluation of a Periodic-Flow Heat Exchanger, *Journal of Heat Transfer*, **105**, 611–617.

Liao, N. S., Wang, C. C. and Tien, C. T. (1988) Analysis of Transient Flow Surge Phenomena in a Single-Tube Condenser, *International Communications in Heat and Mass Transfer*, **15**, 257–268.

Liao, N. S. and Wang, C. C. (1990) Transient Response Characteristics of Two-Phase Condensing Flows, *International Journal of Multiphase Flow*, **16**, 139–151.

Matulla, H. and Orlicek, A. F. (1971) Bestimmung der Wärmeübergangskoeffizienten in einem Doppelrohrwärme-austauscher durch Frequenzganganalyse, *Chemie-Ingenieur-Technik*, 43, 1127–1130.

Mayinger, F. and Schult, M. (1981) Dynamic Behaviour of Double-Phase-Change Heat-Exchangers, in *Heat Exchangers Thermal-Hydraulic Fundaments and Design* (ed. by Kakac, S., Berlges, A. E. and Mayinger, F.), Hemisphere Publishing Corporation, Washington, D.C.

Nahavandi, A. N. and Weinstein, A. S. (1961) A Solution to the Periodic-Flow Regenerative Heat Exchanger Problem, *Applied Scientific Research*, Section A, **10**, 335–348

Neto, S. and Cotta, R. M. (1993) Dynamic Analysis of Double-Pipe Heat Exchangers subject to Periodic Inlet Temperature

Disturbances, *Wärme- und Stoffübertrgung*, **28**, 497–503.

No, C. H. and Kazimi, M. S. (1985) Effects of Virtual Mass on the Mathematical Characteristics and Numerical Stability of the Two-Fluid Model, *Nuclear Science Engineering*, **89**, 197–206.

Patankar, S. V. and Spalding, D. B. (1972) A Calculation Procedure for Heat, Mass and Momentum Transfer in Three-Dimensional Parabolic Flows, *International Journal of Heat and Mass Transfer*, **15**, 1787–1806.

Razelos, P. (1979) An Analytic Solution to the Electric Analog Simulation of the Regenerative Heat Exchanger with Time-Varying Fluid Inlet Temperatures, *Wärme- und Stoffübertragung*, **12**, 59–71.

Riaz, M. (1977) Analytical Solutions for Single- and Two-Phase Models of Packed-Bed Thermal Storage System, *Journal of Heat Transfer*, **99**, 489–492.

Schoenberg, A. A. (1966) *Mathematical Model with Experimental Verification for the Dynamic Behaviour of a Single-Tube Condenser*, NASA TN D-3453.

Schumann, T. E. W. (1929) Heat Transfer: A Liquid flowing through a Porous Prism, *Journal of Franklin Institute*, **208**, 405–416.

Sekulic, D. P., Dzolev, M. and Kmecko, I. (1991) Dynamic Behaviour of a Three Fluid Heat Exchanger: The Experimental Study, in *Experimental Heat Transfer, Fluid Mechanics, and Thermodynamics 1991*, (ed. by Keffer, J. F., Shah, R. K. and Ganic, E. N.), Elsevier, New York, pp. 1338–1343.

Sekulic, D. P. and Herman, C. V. (1987) Transient Temperature Fields in a Three Fluid Heat Exchanger, in *Proceedings of the XVIIth International Congress of Refrigeration*, ***B***, IIF, Vienna,

833–837.

Sekulic, D. P. and Shah, R. K. (1995) Thermal Design Theory of Three-Fluid Heat Exchangers, in *Advances in Heat Transfer*, **26**, 219–328.

Sha, W. T. (1984) Porous-Media Formulation for Multiphase Flow with Heat Transfer, *Nuclear Engineering and Design*, **82**, 93–106.

Shah, R. K. (1981) Thermal Design Theory for Regenerators, in *Heat Exchangers — Thermo-Hydraulic Fundamentals and Design*, (ed. by Kakac, S., Shah, R. K and Bergles, A. E.), Hemisphere Publishing Corporation, New York.

Siegel, R. and Howell, J. R. (1981) *Thermal Radiation Heat Transfer*, Second Edition, McGraw-Hill, New York.

Smith, T. F., Shen, Z. F. and Friedman, J. N. (1982) Evaluation of Coefficients for the Weighted Sum of Gray Gases Model, *Journal of Heat Transfer*, **104**, 602–608.

Sneddon, J. N. (1957) *Elements of Partial Differential Equations*, McGraw-Hill, New York.

Spang, H. A. (1972) A Dynamic Model of a Cement Kiln, *Automatica*, **8**, 309–323.

Stephan, K. (1988) *Wärmeübergang beim Kondensieren und beim Sieden*, Springer-Verlag, Berlin.

Sullivan, P. F., Ramadhyani, S. and Viskanta, R. (1992) A Validation Study of Heat Transfer in a Direct-Fired Batch Reheating Furnace, *HTD-Vol.* **223**, 45–53.

Tilton, D. E., Chow, L. C. and Mahefkey, E. T. (1988) Transient Response of a Liquid Metal Heat Pipe, *Journal of Thermophysics and Heat Transfer*, **2**, 25–30.

Trystram, G., Fahloul, D., Duquenoy, A. and Allache, M. (1992) Dynamic Modelling and Simulation of the Biscuit Baking Oven Process, *Computers in Chemical Engineering*, **17**, s203–s208.

Van Wylen, F. J. and Sonntag, R. E. (1978) *Fundamentals of Classical Thermodynamics*, Second Edition, John Wiley and Sons, New York.

Vortmeyer, D. and Mong, S. L. (1976) Anwendung des Aequivalenzprizipes zwischen Ein- und Zwei-phasenmodellen auf Lösung der Regeneratorgleichungen, *Wärme- und Stoffübertragung*, **9**, 29–37.

Vortmeyer, D. and Schaefer, R. J. (1974) Equivalence of One- and Two-Phase Models for Heat Transfer Processes in Packed Beds: One Dimensional Theory, *Chemical Engineering Science*, **29**, 485–491.

Wadowski, T., Akbarzadeh, A. and Johnson, P. (1991) Characteristics of a Gravity-Assisted Heat Pipe-Based Heat Exchanger, *Heat Recovery Systems and CHP*, **11**, 69–77.

Wakao, N. and Kaguei, S. (1982) *Heat and Mass Transfer in Packed Beds*, Gordon, Breach Science Publication, New York.

Wang, H. and Touber, S. (1991) Distributed and Non-Steady-State Modelling of An Air Cooler, *International Journal of Refrigeration*, **14**, 98–111.

Whalley, P. B. and Hewitt, G. F. (1977) *The Correlation of Liquid Entrainment Fraction and Entrainment Rate in Annular Two Phase Flow*, Report AERE-R9187.

Xuan, Y. and Han, Y. (1996) Dynamic Analysis of Thermal Regenerators, in *Energy and Environment*, in *Proceedings of the International Conference on Energy and Environment*, (ed. by Chen, Z. H., Veziroglu, T. N. and Reay, D. A.), Begell House, New

York.

Xuan, Y. and Roetzel, W. (1993a), Dynamics of Shell-and-Tube Heat Exchangers to Arbitrary Temperature and Step Flow Variations, *AIChE Journal*, **39**, 413–421.

Xuan, Y, and Roetzel, W. (1993b) Stationary and Dynamic Simulation of Multipass Shell and Tube Heat Exchangers with the Dispersion Model for Both Fluids, *International Journal of Heat and Mass Transfer*, **36**, 4221–4231.

Xuan, Y. and Roetzel, W. (1995) Thermal Estimation of Matrix Heat Exchangers, *Journal of Enhanced Heat Transfer*, **2**, 33–41.

Yu, B. (1994) Dynamic Modeling of a Tunnel Kiln, *Heat Transfer Engineering*, **15**, 39–53.

Chapter 7

Transient experimental techniques in heat exchangers

Research on thermal and flow frictional performance of heat exchangers includes both the theoretical and experimental approaches. Although the invention of the computer strengthens the theoretical (numerical) research, it is always limited by the understanding of the complicated processes occurring in exchangers. The experimental research cannot be replaced. For exchangers, the experimental work mainly includes the determination of thermal and frictional data as well as the flow visualization. Here transient experimental techniques of thermal performance are involved. The flow visualisation can be found in the literature (Webb, 1993).

Thermal experiments with an exchanger provide the effectiveness and convective heat transfer coefficient h or the j-factor, that can be achieved by the steady-state or the transient experiment. In general, the steady-state techniques for finding heat transfer coefficients are based on the well-known Wilson plot technique (Wilson, 1915) and the modified technique (Briggs and Young, 1969; Khartabil *et al.*, 1988). Such techniques provide acceptable heat transfer data but there are strict requirements on the experimental conditions. Shah (1990) reviewed the chronological development, advantages and drawbacks of these techniques. The alternative to determining heat transfer data is the transient technique. The application of computers opens the way to the finding of thermal data according to the solution to a so-called inverse heat transfer problem (Beck *et al.*, 1985; Mehta and Jayachandran, 1990). A more direct method is the single-blow test technique which is often adopted for compact surfaces or the packed bed (Pucci *et al.*, 1967; Kays and London, 1984). Hausen (1976) proposed an experimental method to obtain heat transfer data by measuring the periodically oscillating temperature variation in thermal regenerators and heat exchangers. Transient techniques may reduce experimental time and provide heat transfer data with acceptable precision.

7.1 Single-blow transient heat exchanger test

The single-blow test technique is widely used for obtaining heat exchanger performance data of compact heat exchanger surfaces or the packed beds. This technique may have originated from the pioneering work of Anzelius (1926), Nusselt (1927), Schumann (1929) and Hausen (1929) in which the simplest mathematical representation of transient heat transfer in compact exchanger cores or packed beds is involved and the assumed sole mechanism is forced convection between the fluid and the heat transfer surface. Since then it has been applied to various types of compact surfaces.

7.1.1 Fundamental principle of the single-blow test

The single-blow test technique can be used to determine heat transfer data in the way that one matches some mathematical description of the process with the measured temperature response at the exit of compact exchanger cores. Here a suitable mathematical model is necessary. Such a model is traditionally based on the following assumptions:

(1) Thermal and physical properties of the fluid and test core matrix are independent of temperature.

(2) Flow is of the plug-flow pattern.

(3) The tested compact heat transfer surfaces or the packed

beds are homogeneous.

(4) Heat conductivity of both fluid and solid is infinite perpendicular to the flow direction.

(5) Heat conductivity of the fluid is zero and no axial dispersion takes place in the flow direction.

(6) Thermal capacity of the fluid in the test core is negligible compared with the thermal capacity of the core matrix.

According to the application, some of these assumptions may be relaxed and some additional idealizations may be involved, which are related to the complexity of the model and reduction method of the experimental data.

The single-blow technique is based on transient convective heat transfer between the solid surface and the fluid. Thus, the energy balance between the fluid and the solid yields:

$$\frac{C_w}{L}\frac{\partial\theta_w}{\partial\tau} = k_w A_{cw}\frac{\partial^2\theta_w}{\partial l^2} + hP\left(\theta_f - \theta_w\right) \tag{7.1}$$

$$\dot{W}\frac{\partial\theta_f}{\partial l} = hP\left(\theta_w - \theta_f\right) \tag{7.2}$$

where A_{cw} represents the solid plate or matrix cross-sectional area available for heat conduction. P is the flow passage perimeter. If the tested exchanger core or packed bed has compactness β (m^2/m^3) and the frontal area A_{fr}, $P = \beta A_{fr}$. By defining the following parameters and variables

$$u_f = \frac{\dot{m}_f}{\rho_f A_{fr}},\ z = \frac{\tau}{L/u_f},\ x = \frac{l}{L},\ R_{wf} = \frac{C_w}{\rho_f c_{pf} A_{fr} L},$$

$$\lambda_{wf} = \frac{k_w A_{cw}}{\dot{W}L},\ NTU = \frac{hPL}{\dot{W}}$$

one can invert both these equations into the dimensionless forms

$$R_{wf}\frac{\partial t_w}{\partial z} = \lambda_{wf}\frac{\partial^2 t_w}{\partial x^2} + NTU\left(t_f - t_w\right) \tag{7.3}$$

$$\frac{\partial t_f}{\partial x} = NTU\left(t_w - t_f\right) \tag{7.4}$$

where the conductance parameter λ_{wf} accounts for the effect of axial conduction in the solid component. If the solid axial conduction is negligible, $\lambda_{wf} = 0$ and eqn (7.3) becomes

$$R_{wf}\frac{\partial t_w}{\partial z} = NTU\left(t_f - t_w\right). \tag{7.3a}$$

Thus, eqns (7.3a) and (7.4) constitute the simplest and most frequently used mathematical model of transient heat transfer in compact heat exchanger cores or packed beds.

The initial and boundary conditions may be given as

$$t_f(x, 0) = t_w(x, 0) = 0 \tag{7.5}$$

$$t_f(0, z) = f(z). \tag{7.6}$$

The boundary conditions for the exchanger core at both the inlet and outlet may account for heat transfer at these ends. The adiabatic conditions are the simplest ones. For the single-blow test technique, the ideal case corresponds to a step change of the fluid temperature at the inlet. But it is usually difficult to maintain such a step change during an experiment. To achieve it, one should carefully design the experimental apparatus and arrange the experiments.

7.1.2 Transient solutions

The aforementioned model is the basis of the single-blow technique. If the longitudinal conduction of the core is neglected and a step change of the fluid temperature takes place at the inlet, one can obtain the well-known Schumann's solution which has been used by many researchers to determine heat transfer coefficients of compact surfaces and packed beds (Pucci *et al.*, 1967). The analytical temperature response solution for the exponential inlet fluid temperature change $t_f(0,\tau)=1-e^{-\tau/\tau_c}$ was given by Liang and Yang (1975), in which τ_c is the time constant in the fluid inlet temperature function. This solution is very useful because in most cases the inlet fluid temperature variation in experiments can be regarded as an exponential function. The responses to arbitrary inlet temperature variations is easy to obtain according to the theorem of Duhamel (Luo, 1998). If $\lambda_{sf}>0$ in eqn (7.3), no analytical solution is available and one should resort to numerical approaches. Based on the numerical simulation Cai *et al* (1984) extended Liang and Yang's analysis with an empirical formula to include the effect of longitudinal conduction in wall material. Similar cases have been discussed by Mullisen and Loehrke (1986), Heggs and Burns (1988), and Loehrke (1990). Obviously, the Laplace transform can also be used to find the solution to eqns (7.3) and (7.4). According to the given uniform initial conditions, application of the Laplace transform yields

$$R_{wf}sT_w=\lambda_{wf}\frac{\mathrm{d}^2T_w}{\mathrm{d}x^2}+NTU\left(T_f-T_w\right) \tag{7.7}$$

$$\frac{\mathrm{d}T_f}{\mathrm{d}x}=NTU\left(T_w-T_f\right). \tag{7.8}$$

The transformed temperature T_w in the image-domain is obtained as

$$T_w(x,s)=c_1\,\mathrm{e}^{m_1x}+c_2\,\mathrm{e}^{m_2x}+c_3\,\mathrm{e}^{m_3x} \tag{7.9}$$

where the exponents m_i $\left(i=1,\ 2,\ 3\right)$ are the roots of the following polynomial

$$\lambda_{wf} m^3 + \lambda_{wf} NTU m^2 - (R_{wf} s + NTU) m - R_{wf} s NTU = 0 \tag{7.10}$$

Similarly, the transformed temperature T_f is obtained as

$$T_f(x,s) = \alpha_1 c_1 \,\mathrm{e}^{m_1 x} + \alpha_2 c_2 \,\mathrm{e}^{m_2 x} + \alpha_3 c_3 \,\mathrm{e}^{m_3 x} \tag{7.11}$$

where

$$\alpha_1 = \frac{R_{wf} s - \lambda_{wf} m_1^2 + NTU}{NTU}, \quad \alpha_2 = \frac{R_{wf} s - \lambda_{wf} m_2^2 + NTU}{NTU},$$

$$\alpha_3 = \frac{R_{wf} s - \lambda_{wf} m_3^2 + NTU}{NTU}.$$

According to the given boundary conditions, one can determine the unknown coefficients c_1, c_2, and c_3. If the exchanger core at both ends is adiabatic and the inlet temperature of the fluid is $f(z)$ (its transformed form is $F(s)$), these coefficients are obtained as

$$c_1 = \frac{m_3 F(s)}{m_2(\alpha_1 m_3 - m_1)(\mathrm{e}^{m_3} - \mathrm{e}^{m_2}) - m_1(\alpha_2 m_3 - m_2)(\mathrm{e}^{m_3} - \mathrm{e}^{m_1})}$$

$$c_2 = -\frac{m_1(\mathrm{e}^{m_3} - \mathrm{e}^{m_1})}{m_2(\mathrm{e}^{m_3} - \mathrm{e}^{m_2})} c_1, \quad c_3 = \frac{m_1(\mathrm{e}^{m_2} - \mathrm{e}^{m_1})}{m_3(\mathrm{e}^{m_3} - \mathrm{e}^{m_2})} c_1$$

Thus, the temperature variations of the exchanger core and the fluid with an arbitrary incoming value $f(z)$ have been found in the transformed domain. For the specific case of the step change of the fluid at the inlet, $F(s) = 1/s$, which is the ideal inlet forcing function for the single-blow technique. Application of the numerical inverse transform algorithm to these transformed solutions results in temperature profiles of both the exchanger core and the fluid along the flow-direction in the real-time domain. Obviously, under the condition that all other parameters are given, the temperatures of the tested core and the fluid depend only upon the dimensionless number of transfer units *NTU* in which the heat transfer coefficient h appears. The value of this parameter can be easily obtained by matching experimental measurements and theoretical predictions. Generally, the experimental exit fluid temperature $\theta_{f,\exp}(L,\tau)$ is compared with theoretical exit fluid temperature $\theta_{f,theor}(L,\tau)$. During the matching process, the heat transfer coefficient is iteratively altered until the theoretical values approximate the experimental with an acceptable accuracy.

7.1.3 Data reduction techniques

Although it is possible to determine the heat transfer coefficient by matching the measured response with some mathematical description, there exist several data reduction techniques. Each of them has been developed with respect to some characteristics of the theoretically developed breakthrough curves (exit fluid temperature variation) to estimate the heat transfer coefficient. Some of the commonly used data reduction techniques are the direct curve matching, the maximum slope, shape factor, and differential fluid enthalpy methods (Heggs and Burns, 1988).

In the early phase of single-blow technique application, the direct curve matching was used to compare the measured value with the theoretical one. Furnas (1932) first used this technique employing a graphical procedure. He constructed a test using Schumann's assumptions (Schumann, 1929) and matched his experimental coefficient for his test core. This direct curve matching technique required considerable computational effort. With the aid of computers, researchers and engineers have used this trial-and-error approach to minimize some functional relationships between the experimental and theoretical curves. Bradshaw *et al.* (1970) arbitrarily chose to minimize the following expression

$$\sum_{i=1}^{N}\left(\ln\frac{T_{f,th,i}}{T_{f,\exp,i}}\right)^{2} \tag{7.12}$$

where $T_{f,th,i}$ is the absolute (thermodynamic) theoretical fluid outlet temperature (K) at some point in time, $T_{f,\exp,i}$ is the experimental temperature (K) at the same point in time, and N is the number of time intervals on the experimental curve. Selection of such points plays a role to some extent in obtaining the acceptable value of the heat transfer coefficient. Several selection methods have been used by some researchers (Liang and Yang, 1975; Cai *et al.*, 1984; Mullisen and Loehrke, 1986a), which accounted for nonideal step inputs to the experimental test section.

In order to reduce the magnitude of the data reduction effort required by the direct curve matching method, Locke (1950) proposed the maximum slope technique which used the unique relationship between the breakthrough curve and the number of transfer units (*NTU*). Owing to the uniqueness of the maximum slope of the exit fluid temperature curve within a given *NTU* range, Locke developed an explicit relationship between *NTU* and the maximum slope. His analysis is restricted, however, to the core with zero axial conduction. This method is unsuitable for use with systems in which *NTU* is less than 3 (Kohlmayr, 1966).

To allow experimentation in the low *NTU* range, Kohlmayr (1968a, b) proposed another indirect curve matching technique. This involved the evaluation of the centroid of the area between the inlet and outlet fluid temperatures, but there is no evidence that it has been used to analyse experimental data. Mondt and Siegla (1974) suggested using the unique relationship between the theoretical instantaneous step change in the breakthrough curve and *NTU* in the low *NTU* range. This initial rise is valid only for ideal step inputs. The results generated by this method must be viewed with suspicion (Heggs and Burns, 1988).

Baclic *et al.* (1986a) proposed an explicit multipoint technique called the differential fluid enthalpy method. In this method the experimentalist must measure the time constant of the inlet forcing function, and then the temperature difference between the inlet and outlet of the test section up to a specified time. A new procedure for predicting heat transfer coefficients from single-blow data of compact heat exchanger surfaces has been developed by Baclic *et al.* (1986b). They introduced four parameters for describing the characteristics of a testing system and assessing the accuracy of the measured heat transfer coefficient. These parameters are as follows:

(1) the time constant of the exponential inlet temperature signal which accounts for the deviation of the real inlet input from the ideal step change;

(2) maximum fluid temperature rise at various mass flow rates for constant heater power supply;

(3) number of fluid-to-wall heat transfer units;

(4) time constants of the two thermopiles used to measure the temperatures up- and downstream of the tested core.

Along with these apparatus parameters, the differential fluid enthalpy method is used to analyse experimental data.

The uncertainty of single-blow tests shows that it is probably not safe to rely on only one data analysis method (Kohlmayr, 1971). It may be more reasonable to adopt two or more different methods to analyse the same experimental data and accept only results that agree. This approach was used by Rapley and Webb (1983), who used a direct curve-matching technique and then checked their results using either the initial rise or maximum slop method. By means of the same mathematical model, Heggs and Burns (1988) adopted four different data reduction methods for predicting heat transfer coefficients from the same single-blow experimental data about fluid flow through randomly packed beds of spheres over the range $290 < Re_m < 1140$. They found good consistency between the results obtained with least squares and differential fluid enthalpy methods, with acceptable errors. On the other hand, the maximum slope and shape factor methods always predict values of the heat transfer coefficients which are too low.

Although the single-blow test technique is often used to determine fundamental convective heat transfer coefficient correlations, one should pay great attention to some aspects involved in experimental data reduction procedure. The mathematical model should accurately describe the heat transfer process in the compact exchanger core, otherwise the accuracy of the heat transfer coefficient data may be low. Deviations of the inlet fluid temperature history from an ideal step change always exists and it is impossible to execute an experiment with a perfect step change in inlet fluid temperature. The inlet input forcing function of the model has to account for the nonideal step change. Longitudinal conduction in solid components affects the evaluation of the heat transfer coefficients, which may vary with the Reynolds number. For an accurate model, such effects should be taken into account. A more common problem is that the effective core mass which participates in the heat transfer process is not clearly defined. Wind tunnel walls and other support structures may be involved to an unknown degree. Loehrke (1990) pointed out that extraneous wall participation may be most important when small-scale models are tested. The influence of heat transfer to the surrounding walls increases as the sample scale diminishes. The minimum scale that maintains an acceptable level of wall participation depends upon the design of the facility and the sample characteristics. Careful attention to detail can result in a facility that introduces negligible error into the data (Mullisen and Loehrke, 1986). In addition, the traditional single-blow technique fails in the case of variable heat transfer coefficients and deviation from the plug flow as well as longitudinal mixing of the fluid. Recently, Luo (1998) obtained the analytical solutions which take the axial dispersion into account. Figure 7.1 shows the outlet fluid temperature responses to the step and exponential inlet fluid temperature changes for different values of axial dispersive Péclet number *Pe* and the time constant of the exponential inlet change τ_c. Luo pointed out that the transient test techniques using the step or exponential inlet condition would give wrong results if there are flow maldistributions in the exchanger test core and the corresponding axial dispersion coefficient is unknown.

7.1.4 Experimental apparatus

The design principle of an experimental facility is to meet the restrictions imposed by the idealizations of the theoretical analysis. The most important idealization to be met is the simulation of a step input temperature variation of the fluid entering the tested core. Several experimental arrangements have been employed to obtain such a step change in the fluid temperature. Using two fluid sources

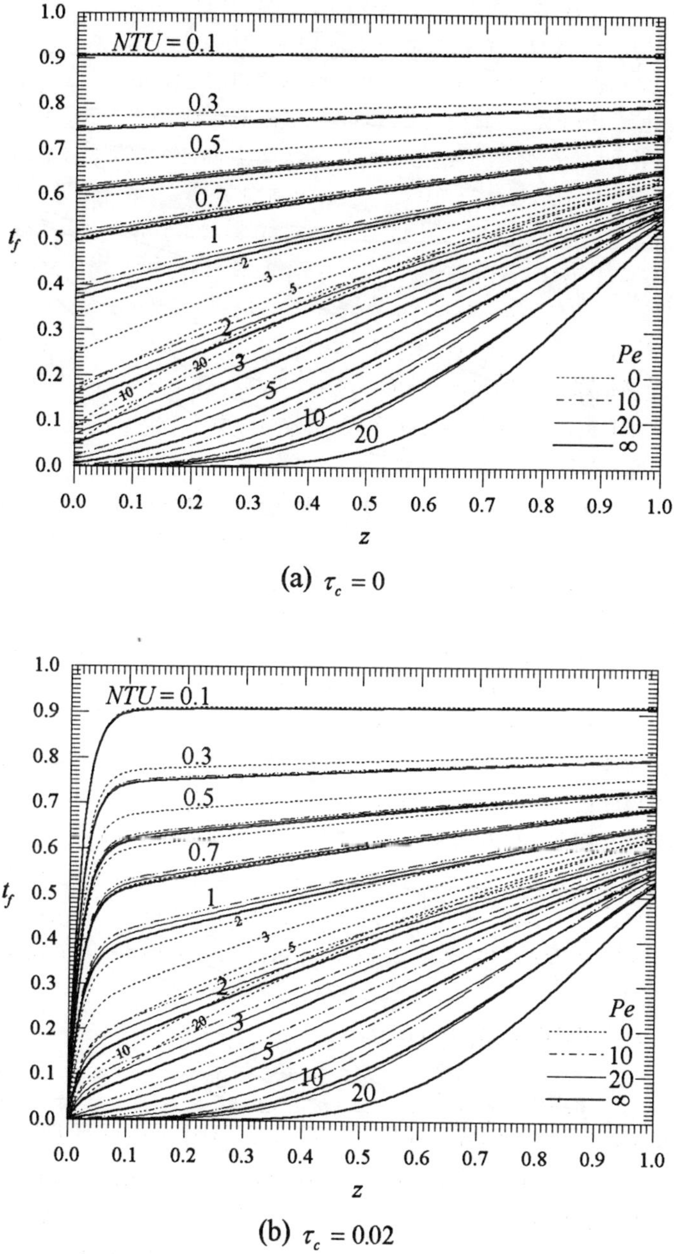

Figure 7.1: Effect of *Pe* on outlet fluid temperature response

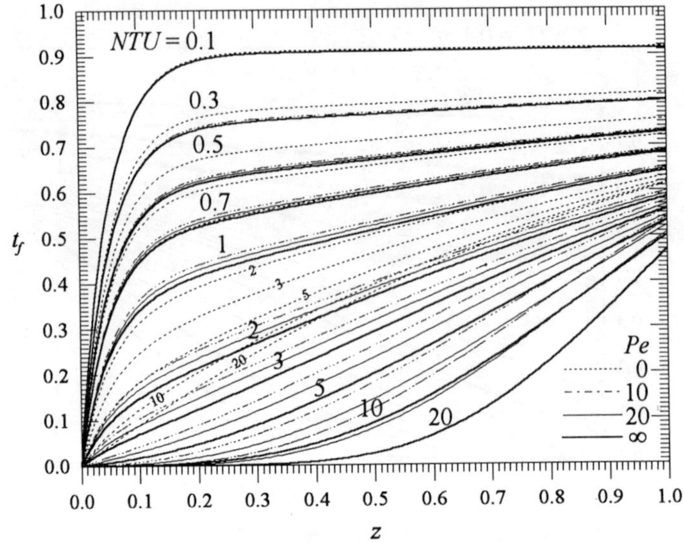

(c) $\tau_c = 0.05$

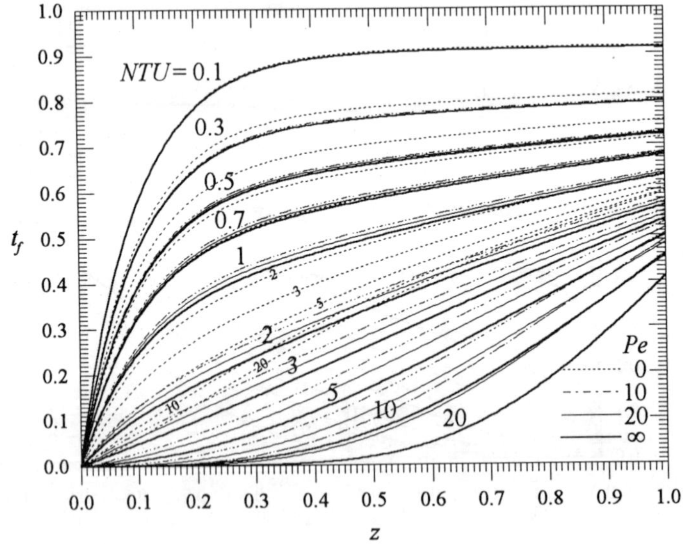

(d) $\tau_c = 0.1$

Figure 7.1: (Cont'd.)

of different temperatures, a valve is installed to switch from one fluid flowing through the core to the other fluid; or the core is mechanically moved from one fluid stream to the other. An alternative concept which is widely used is to employ a single fluid stream with an electrical heater installed upstream of the core. Using a low thermal capacitance heater results in a fast response in fluid temperature to the turning on or off of the electric current to the heater. Pucci *et al.* (1967) described both the valve-switch and the electrical heating types of experimental facilities. Whether the valve-switch type or the electrical heating type is used, the temperature of the fluid stream into a test section is generally 10–20°C above the ambient temperature.

Figure 7.2 schematically shows the entire system of an experimental apparatus with the electrical heating. Generally, the single-blow test facility consists of the wind tunnel, instrumentation, data acquisition and reduction systems and the test core. In regard to the fast response of the experimental system, the electrical heater may be constructed of fine wire mesh screens with a relatively large surface area. The temperatures of the fluid stream entering and leaving the test core are measured by thermopiles consisting of more than ten pairs of thermocouples. A detailed description of such a single-blow test system has been made by Mullisen and Loehrke (1986). Special precaution should be taken in the design of the test section, heater support structure and the core assembly to ensure that the thermal capacity of the tested core dominates during the test period. In Mullisen and Loehrke's apparatus, the interior surface of the test section is lined with foam and the support structure for the plate arrays is made from balsa wood. In addition, the heat capacity of the tested core should be high enough to result in a remarkable temperature variation of the fluid stream between the inlet and the outlet of the core. This temperature variation ought to be sensed and recorded by instruments installed on the experimental unit. To amplify the temperature change signal between up- and downstream of the core, two thermopiles can be installed to respectively measure both the temperature profiles. The analogue voltage output from the thermopiles is further amplified, digitized and fed into a computer. The mass flow rate through the

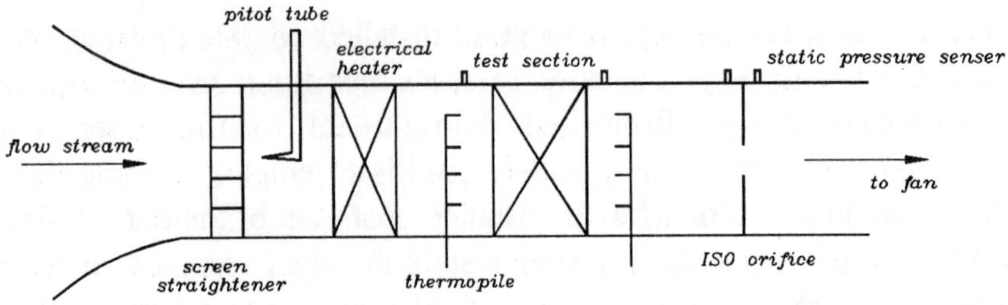

Figure 7.2: Schematic description of a single-blow test apparatus with an electrical heater

core may be measured either by a standard ISO orifice or a pitot static tube. Mullisen and Loehrke (1986) used the orifice at low flow rates and the pitot static tube at high flow rates.

For the given test unit, there are two different test-run procedures. One is that the fluid stream is heated by an electrical heater to a temperature about 15°C over the ambient flows through the core which has the initial temperature being equal to the ambient temperature. During this process the core is heated by the hot fluid stream and the temperature of the fluid stream leaving the core decreases. The other is that after the tested core is first heated to a steady-state temperature level, the electrical heater is shut down and the cold fluid stream flows through the core. During this process the core is cooled and the fluid stream is heated by the core. Both the transient heating and cooling procedures should essentially give the same result on heat transfer coefficients. But the former is often adopted because the cold initial state is more stable than the hot one. Generally, the time needed for a single-blow test run is quite short, but may vary with the value of *NTU*. Mullisen and Loehrke (1986) suggested that typical data collection times last 5 to 60s for *NTU* ranging from 0.5 to 5.0. The optimal dimensionless matching time was given by Cai *et al.* (1984).

As pointed out previously, there are a number of error sources affecting the accuracy of the data. Some of these errors can be easily estimated, but others cannot. For example, the thermal interaction between the facility and a small-scale core may be difficult to estimate accurately. Loehrke (1990) recommended that a calibration should be used to verify either that this interaction has been properly accounted for or that it has a negligible effect on the results. The calibration can consist of a test on a core of known performance. A special core with a simple fluid-passage geometry is preferable because its performance can be calculated directly from the governing equations of fluid mechanics. Loehrke (1990) explained the calibration purpose as follows: it can serve as a benchmark for the entire data generation sequence including measurements and reduction procedures, although this calibration does not ensure the accuracy of tests on cores of other geometry. Millisen and Loehrke (1986) and Cai *et al.* (1984) did the calibration for their own experimental systems.

7.2 The double-blow test technique

As discussed in Section 7.1, the single-blow test technique is widely used to determine heat transfer data for compact exchanger surfaces where heat transfer and friction factor coefficients cannot be directly calculated from theory. During the single-blow test, as the name implies, only one fluid stream flows through the tested core. Thus, the mathematical model of the technique consists of two differential equations: one is for the fluid stream and the other for the core. According to the application experience of the single-blow technique, Gvozdenac (1994) tried to develop a new so-called double-blow test technique. His original purpose may be to obtain the heat transfer coefficients for both sides of an exchanger in one experimental run. Because of the complexity of the experimental system and the mathematical model, this test technique has not been widely used.

7.2.1 Mathematical model of the double-blow technique

In fact, the double-blow test technique is a further application example of dynamic simulation of cross-flow heat exchangers. In the light of his previous work (Gvozdenac, 1986), Gvozdenac (1994) proposed the idea of the double-blow technique. In that paper, he presented a set of differential equations describing transient temperature fields in cross-flow heat exchangers with large wall capacitance, neglecting the thermal flow rates of both the fluid streams. These equations are as follows:

$$-\dot{W}_1 L_1 \left(\frac{\partial \theta_1}{\partial l} + \frac{1}{u_1} \frac{\partial \theta_1}{\partial \tau} \right) = \eta_1 h_1 A_1 (\theta_1 - \theta_w) \tag{7.13}$$

$$\dot{W}_2 L_2 \left(\frac{\partial \theta_2}{\partial l} + \frac{1}{u_2} \frac{\partial \theta_2}{\partial \tau} \right) = \eta_2 h_2 A_2 (\theta_w - \theta_2) \tag{7.14}$$

$$C_w \frac{\partial \theta_w}{\partial \tau} = \eta_1 h_1 A_1 (\theta_1 - \theta_w) - \eta_2 h_2 A_2 (\theta_w - \theta_2) \tag{7.15}$$

where both the fluid streams are unmixed. According to the given initial and inlet conditions, these equations can be solved by the finite-difference method or the Laplace transform method. For the uniform initial conditions and the exponential form of the inlet temperature of fluid 1, Gvozdenac (1986) found the analytic solution of the transient response of gas-to-gas cross-flow exchangers with neither fluid mixed. Such a solution is the basis of the double-blow technique.

7.2.2 Test procedure and data reduction

The double-blow test implies that each of both fluid streams flow over either of two side surfaces of an exchanger core, instead of only one fluid stream flowing through the core as in the application of the single-blow technique. The main part of the double-blow apparatus is a compact exchanger core in which transient heat transfer takes place between the two fluid streams. To some extent, a double-blow apparatus is an experimental platform for transient performance of a compact exchanger. Gvozdenac (1994) described his test apparatus where fluid 1 is electrically heated and flows over one side of the exchanger and fluid 2 is induced into the core from the ambient and flows over the other side. The measured signals of temperatures and flow rates were converted, digitized and fed into a computer for further processing. The procedure of data acquisition is similar to that used in the single-blow test.

In order to analyse the experimental data obtained from the double-blow test run, the data reduction methods (direct curve matching of experiment with theory, trial-and-error approach to minimize some functional relationship between the experimental and theory curves) used for the single-blow technique can be used here without any restrictions. By means of the direct curve matching between theoretically evaluated breakthrough curves and the experimental response curves at discrete times, Gvozdenac (1994) determined the following two parameters

$$NTU = \frac{(\eta hA)_1 (\eta hA)_2}{(\eta hA)_1 + (\eta hA)_2} \frac{1}{\dot{W}_{min}}, \qquad K = \frac{(\eta hA)_1}{(\eta hA)_1 + (\eta hA)_2} \tag{7.16}$$

He concluded that the advantage of the double-blow technique, in comparison with the well-known single-blow method, is the possibility of getting the heat transfer coefficients for both sides of a heat exchanger matrix in one experimental run. But he gave no examples of finding h_1 and h_2 in one test run. In fact, it is only true when heat transfer surfaces of both sides are identical. If the two surfaces are different, it is difficult to uniquely separate h_1 and h_2 from NTU and K, although one does obtain both NTU and K in one experimental run. These two parameters comprehensively represent the heat transfer performance of an exchanger. One cannot expect to determine the accurate values of h_1 and h_2 only by simple algebraic calculation. Additional known conditions or constraints should be involved to find h_1 and h_2 in one experimental run.

7.3 Temperature oscillation technique

As mentioned in the previous section, the transient test technique requires a time-dependent temperature of the incoming stream. In common cases, there are two possible variations, i.e., a step input and a periodic function input. The first input corresponds to the single-blow technique and the second to the so-called temperature oscillation technique. The periodic method consists of imposing a known periodic fluctuation (for example, a sinusoidal wave) on the incoming fluid stream into the test core and measuring the temperature responses of the fluid flowing in and out of the core. In some applications the wall temperature may be measured. The earliest periodic method was proposed by Hausen (1976) more than fifty years ago to determine heat transfer coefficients through analysing the fundamental linear oscillations of the gas temperature in thermal regenerators and was further developed by his scholars Glaser (1938), Langhans (1962) and Yazicizade (1966). Kast (1965) and Matulla and Orlicek (1971) determined heat transfer coefficients in fixed beds and heat exchangers by analyzing the sinusoidal temperature oscillations of the fluid at the inlet and outlet of the apparatus. Stang and Bush (1974) used either the amplitude attenuation or the phase shift of the fluid temperature according to the range of *NTU* to obtain mean heat transfer coefficients. In their model the axial heat conduction in the wall was considered, but the solution of their model cannot satisfy the boundary conditions at the inlet and outlet of the test core, so it may fail if the effect of heat conduction in the wall is significant. Roetzel (1989) and Roetzel and Prinzen (1991) developed alternative methods for the determination of local heat transfer coefficients in tubes. In their investigations the governing equation was derived from the energy balance of the wall so that the effect of the axial dispersion did not arise in the final results, but the wall temperature variation was required which is in many cases difficult to measure. The periodic method was further extended by Roetzel *et al.* (1993a) to include the axial heat conduction in the wall and the axial dispersion in the fluid.

For application of the periodic method to compact surfaces, the differential equations describing temperature variations of the fluid and the core are similar to those for the single-blow technique. Roetzel *et al.* (1993a) proposed a set of differential equations as follows

$$B\frac{\partial t_f}{\partial z}+\frac{\partial t_f}{\partial x}-\frac{1}{Pe}\frac{\partial^2 t_f}{\partial x^2}=NTU\left(t_w-t_f\right) \tag{7.17}$$

$$\frac{\partial t_w}{\partial z} - \lambda \frac{\partial^2 t_w}{\partial x^2} = NTU\left(t_f - t_w\right) \tag{7.18}$$

$$x = 0: \quad t_f - \frac{1}{Pe}\frac{\partial t_f}{\partial x} = f(z), \qquad \frac{\partial t_w}{\partial x} = 0 \tag{7.19}$$

$$x = 1: \quad \frac{\partial t_f}{\partial x} = 0, \qquad \frac{\partial t_w}{\partial x} = 0 \tag{7.20}$$

where $x = \dfrac{l}{L}$, $z = \dfrac{\tau}{C_w / \dot{W}}$, $z_P = \dfrac{\tau_P}{C_w / \dot{W}}$, $B = \dfrac{C_f}{C_w}$, $\lambda = \dfrac{k_w A_{cw}}{\dot{W}L}$.

The inlet temperature variation $f(z)$ is a periodic function

$$f\left(z + z_P\right) = f(z). \tag{7.21}$$

The partial differential equation system (7.17)–(7.21) is solved by means of complex Fourier transform which is defined as

$$F_n = \frac{1}{z_P}\int_0^{z_P} t_f(z)\,\mathrm{e}^{-in\omega z}\,\mathrm{d}z \tag{7.22}$$

and its inverse formula is given by

$$t_f(z) = \sum_{n=-\infty}^{\infty} F_n\,\mathrm{e}^{\mathrm{i}n\omega z} \tag{7.23}$$

where $\mathrm{i} = \sqrt{-1}$, $\omega = 2\pi / z_P$ and z_P the dimensionless period of the temperature oscillation.

Applying eqn (7.22) in eqns (7.17)–(7.20) and eliminating the Fourier transform of the wall temperature, one obtains an ordinary differential equation of 4th order,

$$\sum_{k=0}^{4} a_k \frac{\mathrm{d}^k F_n(x)}{\mathrm{d}x^k} = 0 \tag{7.24}$$

where

$$a_0 = -n^2\omega^2 B + \mathrm{i}n\omega\, NTU(1+B), \tag{7.25}$$

$$a_1 = NTU + \mathrm{i}n\omega, \tag{7.26}$$

$$a_2 = -NTU(\lambda + 1/Pe) - \mathrm{i}n\omega\,(B\lambda + 1/Pe), \tag{7.27}$$

$$a_3 = -\lambda, \tag{7.28}$$

$$a_4 = \lambda/Pe. \tag{7.29}$$

The solution of eqn (7.24) can be written as

$$F_n(x) = \sum_{k=1}^{m} D_k \exp(r_k x) \tag{7.30}$$

in which r_k $(k = 1, 2, \ldots, m)$ are m complex eigenvalues of the complex eigenfunction

$$a_0 + a_1 r + a_2 r^2 + a_3 r^3 + a_4 r^4 = 0 \tag{7.31}$$

and the unknown coefficients D_k $(k = 1, 2, \ldots, m)$ are determined according to the boundary conditions which give

$$\sum_{k=1}^{m} (1 - r_k/Pe) D_k = F_{nin} \tag{7.32}$$

$$\sum_{k=1}^{m} r_k\, \mathrm{e}^{r_k}\, D_k = 0 \tag{7.33}$$

$$\sum_{k=1}^{m} \left[r_k^3/Pe - r_k^2 - (NTU + \mathrm{i}n\omega\, B) r_k \right] D_k = 0 \tag{7.34}$$

$$\sum_{k=1}^{m}\left[r_k^3/Pe - r_k^2 - (NTU + \mathrm{i}n\omega B)r_k\right]\mathrm{e}^{r_k} D_k = 0 \tag{7.35}$$

in which $F_{n_{in}}$ is the Fourier transform of the inlet temperature oscillation.

In general cases $m = 4$. If $\lambda = 0$, there is no longitudinal heat conduction in the wall. In this case, $a_3 = a_4 = 0$, $m = 2$ and only eqns (7.32)–(7.33) are needed. For $Pe = \infty$ but $\lambda \neq 0$, we have $m = 3$ and eqn (7.33) should be eliminated. The simplest case is $1/Pe = \lambda = 0$, in which $a_2 = a_3 = a_4 = 0$, $m = 1$ and only eqn (7.32) is needed.

The inverse transform of eqn (7.30) using eqn (7.23) can be rewritten as

$$t_f(x,z) = t_{fm} + \sum_{n=1}^{\infty} A_n(x)\sin\left[n\omega\, z + \Delta\phi_n(x)\right], \tag{7.36}$$

where

$$A_n(x) = \sqrt{\left\{\mathrm{Re}\left[F_n(x) + F_{-n}(x)\right]\right\}^2 + \left\{\mathrm{Im}\left[F_n(x) - F_{-n}(x)\right]\right\}^2}, \tag{7.37}$$

$$\Delta\phi_n(x) = \arctan\left\{\frac{\mathrm{Re}\left[F_{-n}(x) + F_n(x)\right]}{\mathrm{Im}\left[F_{-n}(x) - F_n(x)\right]}\right\} \tag{7.38}$$

$$t_{f_m} = \frac{1}{z_P}\int_0^{z_P} f(z)\,\mathrm{d}z. \tag{7.39}$$

Since the partial differential equation system is linear, for a sinus-like periodic inlet fluid temperature, only the fundamental frequency of the temperature oscillation, $n = \pm 1$, needs to be considered to evaluate the values of NTU and Pe. Higher harmonics can also be extracted, but their phase shift and amplitude attenuation may yield unreliable values of NTU and Pe, because due to temperature dependent fluid properties additional higher harmonics may be generated between inlet and outlet.

If there is neither axial dispersion in the fluid nor axial heat conduction in solid wall, i.e., $1/Pe = \lambda = 0$, and the fluid temperature at the heat exchanger entrance is a sinus function of time, eqn (7.36) yields

$$t_f(x,z)_{Pe=\infty,\ \lambda=0} = t_{f\,m} + A_{f\,in} \exp\left(-\frac{\omega^2 NTU}{\omega^2 + NTU^2} x\right)$$

$$\times \sin\left[\omega z - \omega\left(\frac{NTU^2}{\omega^2 + NTU^2} + B\right)x\right]. \tag{7.40}$$

Thus, for propagation of the fluid temperature oscillation from the inlet to the outlet of the core one obtains the amplitude attenuation

$$\eta_{f,\,theor} = \exp\left(-\frac{\omega^2 NTU}{\omega^2 + NTU^2}\right) \tag{7.41}$$

and the phase shift

$$\Delta\phi_{f,\,theor} = \omega\left(\frac{NTU^2}{\omega^2 + NTU^2} + B\right). \tag{7.42}$$

The theoretical values of amplitude attenuation and phase shift of the fluid are needed for data processing. But a solution for t_w may not be required for this purpose. As examples, two diagrams of amplitude attenuation and phase shift for $B = 5$, $z_P = 20$ and $B = 0$, $z_P = 1.5$, respectively, are shown in Fig. 7.3.

According to the above-mentioned analysis, only the inlet and outlet temperature variations of the fluid are measured. They are processed with a Fast Fourier Transform algorithm (FFT) to obtain the measured fundamental amplitude attenuation $\eta_{f,\,exp}$ and phase shift $\Delta\phi_{f,\,exp}$ for each experimental run. *NTU* and *Pe* are then determined by solving the nonlinear equation system

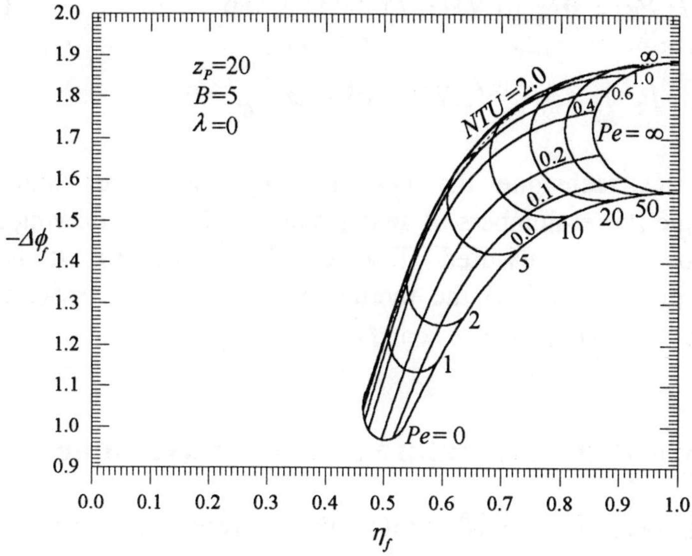

(a) $B = 5,\ z_p = 20,\ \lambda = 0$

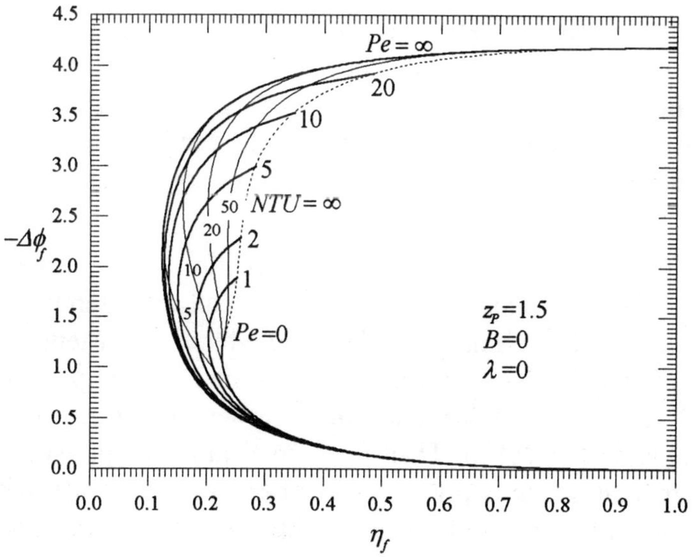

(b) $B = 0,\ z_p = 1.5,\ \lambda = 0$

Figure 7.3: Amplitude attenuation – phase shift diagram

$$f(NTU,\, Pe) = \eta_{f,\,theor}(NTU,\, Pe) - \eta_{f,\,exp} = 0 \tag{7.43}$$

$$g(NTU,\, Pe) = \Delta\phi_{f,\,theor}(NTU,\, Pe) - \Delta\phi_{f,\,exp} = 0 \tag{7.44}$$

Using this technique the heat transfer coefficients and axial dispersive Péclet numbers of dented tubes with water flowing inside the tube were measured (Roetzel *et al.*, 1993a). It is also successfully applied to the thermal performance measurement of plate heat exchangers (Roetzel *et al.*, 1994).

7.4 Extended temperature oscillation technique

As has been mentioned above, the conventional transient test techniques are based on the ideal plug-flow assumption. Although the analytical solution of the fluid temperature responses which takes the axial dispersion in fluid into account is available, one cannot use it to estimate the heat transfer coefficient and axial dispersion coefficient simultaneously (Luo, 1998). One of them should be known if the axial dispersion cannot be neglected. If the axial dispersion coefficient is unknown, the temperature oscillation technique is valid only in certain ranges of parameters, e.g., $C_f/C_w > 1$. Under the consideration that the single-blow method and periodic method have different sensitivities to the heat transfer coefficient and the axial dispersion coefficient, an extended testing technique which combines the single-blow transient testing technique and the temperature oscillation method was proposed by Roetzel and Luo (1998) and then further developed by them (Roetzel and Luo, 1998). Their mathematical model given below is similar to that of the steady-state temperature oscillation with the exception that the axial heat conduction in the wall is now neglected.

$$B\frac{\partial t_f}{\partial z}+\frac{\partial t_f}{\partial x}-\frac{1}{Pe}\frac{\partial^2 t_f}{\partial x^2}=NTU\left(t_w-t_f\right) \tag{7.45}$$

$$\frac{\partial t_w}{\partial z}=NTU\left(t_f-t_w\right). \tag{7.46}$$

In the extended technique, however, a sudden harmonic inlet change in fluid temperature is applied to the tested exchanger instead of a steady oscillation:

$$z=0: \quad t_f=t_w=0 \tag{7.47}$$

$$x=0: \quad t_f-\frac{1}{Pe}\frac{\partial t_f}{\partial x}=1-\cos\omega z \tag{7.48}$$

$$x=1: \quad \frac{\partial t_f}{\partial x}=0 \tag{7.49}$$

where $x=\dfrac{l}{L}$, $z=\dfrac{\tau}{C_w/\left(\dot{m}_f c_{pf}\right)}$, $\omega=\dfrac{2\pi C_w}{\dot{m}_f c_{pf}\tau_P}$, $B=\dfrac{C_f}{C_w}$.

Equations (7.45) and (7.46) are solved by means of Laplace transform and residue theorem. The complete solutions are given by Roetzel and Luo (1998) as follows:

$$t_f(x,z)=1-\cos\omega z-\sum_{j=1}^{\infty}\frac{2\omega^2\beta_j\left[\beta_j\cos\left(\beta_j x\right)+a\sin\left(\beta_j x\right)\right]}{a^2+\beta_j^2+2a}\mathrm{e}^{ax}$$

$$\times\left\{\frac{\cos\omega z+\dfrac{s_j}{\omega}\sin\omega z-\mathrm{e}^{s_j z}}{s_j\left(s_j^2+\omega^2\right)\left[B+NTU^2/\left(s_j+NTU\right)^2\right]}\right.$$

$$+\frac{\cos\omega z+\dfrac{s_{-j}}{\omega}\sin\omega z-\mathrm{e}^{s_{-j}z}}{s_{-j}\left(s_{-j}^{2}+\omega^{2}\right)\left[B+NTU^{2}/\left(s_{-j}+NTU\right)^{2}\right]}\Bigg\}\tag{7.50}$$

$$t_f(x,z)_{B=0}=1-\cos\omega z-\sum_{j=1}^{\infty}\frac{8\omega^{2}a^{2}NTU\beta_j\left(\beta_j^{2}+a^{2}+2aNTU\right)}{\left(\beta_j^{2}+a^{2}\right)\left(\beta_j^{2}+a^{2}+2a\right)}$$

$$\times\left[\frac{\beta_j\cos\left(\beta_j x\right)+a\sin\left(\beta_j x\right)}{NTU^{2}\left(\beta_j^{2}+a^{2}\right)+\omega^{2}\left(\beta_j^{2}+a^{2}+2aNTU\right)^{2}}\mathrm{e}^{ax}\right]$$

$$\times\left\{\exp\left[-\frac{NTU\left(\beta_j^{2}+a^{2}\right)}{\beta_j^{2}+a^{2}+2aNTU}z\right]-\cos\omega z\right.$$

$$\left.+\frac{NTU\left(\beta_j^{2}+a^{2}\right)}{\omega\left(\beta_j^{2}+a^{2}+2aNTU\right)}\sin\omega z\right\}\tag{7.51}$$

$$t_f(x,z)_{Pe=\infty}=\begin{cases}0, & z<Bx\\[2ex] \omega\,\mathrm{e}^{-NTUx}\displaystyle\sum_{n=0}^{\infty}\frac{NTU^{n}}{n!} & \\ \quad\times\left\{\left[C_n(z-Bx)-C_n(0)\right]\sin\left[\omega(z-Bx)\right]\right. & \\ \quad\left.-\left[S_n(z-Bx)-S_n(0)\right]\cos\left[\omega(z-Bx)\right]\right\} & \\ \quad\times\displaystyle\sum_{k=0}^{n}\frac{NTU^{k}x^{k}}{k!}, & z\geq Bx\end{cases}\tag{7.52}$$

$$t_f(x,z)_{Pe=0} = 1 - \cos\omega z$$

$$- \frac{\omega^2}{B(a_1 - a_2)}\left[\frac{1+Ba_1}{a_2^2+\omega^2}\left(\mathrm{e}^{a_2 z} - \cos\omega z - \frac{a_2}{\omega}\sin\omega z\right)\right.$$

$$\left. - \frac{1+Ba_2}{a_1^2+\omega^2}\left(\mathrm{e}^{a_1 z} - \cos\omega z - \frac{a_1}{\omega}\sin\omega z\right)\right] \tag{7.53}$$

$$t_f(x,z)_{Pe=0,\,B=0} = 1 - \cos\omega z$$

$$- \frac{\omega^2 NTU(NTU+1)}{NTU^2 + \omega^2(NTU+1)^2}\left[\mathrm{e}^{-\frac{NTU}{NTU+1}z} - \cos\omega z + \frac{NTU}{\omega\,(NTU+1)}\sin\omega z\right] \tag{7.54}$$

where $a = Pe/2$, β_j is the jth positive root of the eigenfunction

$$\tan\beta = \frac{2a\beta}{\beta^2 - a^2} \tag{7.55}$$

$$s_{\pm j} = -\frac{1}{2B}\left[NTU(1+B) + \frac{a^2+\beta_j^2}{2a}\right]$$

$$\pm \frac{1}{2B}\sqrt{NTU^2(1+2B) + 2NTU\frac{a^2+\beta_j^2}{2a} + \left(NTUB - \frac{a^2+\beta_j^2}{2a}\right)^2}$$

$$(j = \pm 1,\ \pm 2,\ \cdots,\ \pm\infty) \tag{7.56}$$

$$a_{1,2} = -\frac{1}{2B}\left[NTU(1+B) + 1\right]$$

$$\pm \frac{1}{2B}\sqrt{NTU^2(1+2B) + 2NTU + (1-NTUB)^2}\,. \tag{7.57}$$

$C_n(z)$ and $S_n(z)$ are calculated from the recurrence formulas

$$C_0(z) = \frac{e^{-NTUz}}{NTU^2 + \omega^2}(\omega \sin \omega z - NTU \cos \omega z) \tag{7.58}$$

$$C_n(z) = \int z^n e^{-NTUz} \cos \omega z \; dz$$

$$= \frac{z^n e^{-NTUz}}{NTU^2 + \omega^2}(\omega \sin \omega z - NTU \cos \omega z) + n \frac{NTU C_{n-1}(z) - \omega S_{n-1}(z)}{NTU^2 + \omega^2} \tag{7.59}$$

$$S_0(z) = -\frac{e^{-NTUz}}{NTU^2 + \omega^2}(\omega \cos \omega z + NTU \sin \omega z) \tag{7.60}$$

$$S_n(z) = \int z^n e^{-NTUz} \sin \omega z \, dz$$

$$= -\frac{z^n e^{-NTUz}}{NTU^2 + \omega^2}(\omega \cos \omega z + NTU \sin \omega z)$$

$$+ n \frac{NTU S_{n-1}(z) + \omega C_{n-1}(z)}{NTU^2 + \omega^2}. \tag{7.61}$$

It is obvious that the above solutions contain two parts: one is the initial stage component, and the other is the steady oscillation component. The initial stage component has an exponential term in time and declines rapidly. If the value of the initial stage component approaches zero, steady-state oscillations are reached and the solution given by Roetzel *et al.* (1993a) can be used.

During the experiment a sudden harmonic inlet temperature variation of the air flow is generated using a computer-controlled compact electric heater. Because of its thermal capacity a sinusoidal power supply voltage does not yield a sinusoidal inlet temperature variation in the initial stage. The heating power is modified according to the Fourier analysis of the measured inlet fluid temperature to achieve a satisfactory sinusoidal inlet temperature variation. By testing, the electric power is set to an initial value and thereafter the inlet and outlet temperatures are the same and keep steady, and the precalculated power control voltage is applied on the electric power source. The inlet and outlet temperatures as well as the mass flow rate are measured continuously and saved for evaluating the heat transfer coefficient and dispersion coefficient. The average time needed for one test run is reported to be about 20 minutes.

The whole curve-matching technique is used in data reduction so that the influence of the temperature fluctuation can be dramatically reduced. After a test run, the inlet temperature variation is analysed by means of fast Fourier transform to obtain the inlet boundary condition, i.e., the amplitude and the phase shift. The theoretical outlet temperature response to this harmonic inlet temperature variation is calculated and the measured outlet temperature variation is then compared with the theoretical curve. The heat transfer coefficient and axial dispersion coefficient are evaluated with the criterion that the standard error of the dimensionless temperature deviation between measured values and theoretical results becomes a minimum.

Typical inlet and outlet temperature variations from Roetzel and Luo (1998) are shown in Fig. 7.4. The experiment was conducted on a "standard" test core consisting of 48 copper plates with dimensions of 1 mm (thickness) × 300 mm (width) × 500 mm (length in flow direction) to form 49 uniformly spaced parallel channels. Figure 7.4 shows that the theoretical matching curve (smooth curve) and the measured outlet temperatures (points) are almost identical.

The extended temperature oscillation technique combines the advantages of both single-blow technique and oscillation technique. If the value of heat capacity ratio B is small, the outlet temperature response to the oscillation component is more sensitive to the axial dispersion and/or to large values of NTU; on the other hand, the outlet temperature response to the step change component is more sensitive to small values of NTU. This additional component enlarges the parameter validity region of the oscillation technique.

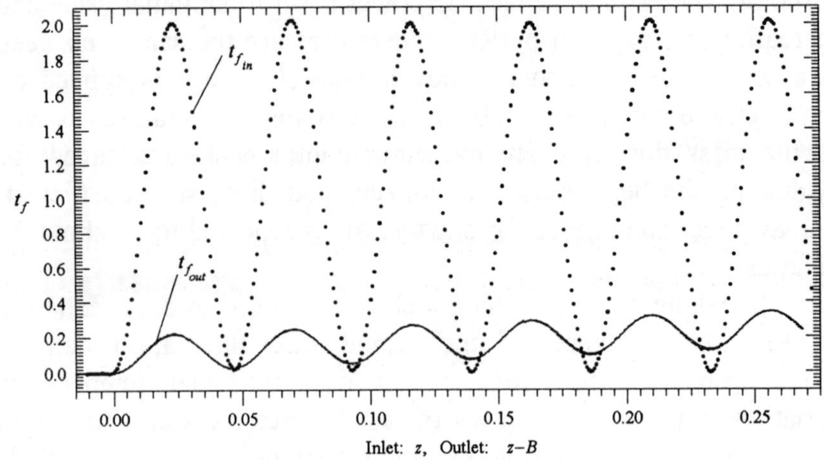

Figure 7.4: Measured inlet and outlet temperatures (Re = 900.6; $\dot{m}_f c_f$ = 117.6 W/K; C_w = 25.26 kJ/K; B = 0.00185; z_P = 0.04655; NTU = 2.536; Pe = 20.80; ···· measured, — calculated)

7.5 Oscillation technique for local heat transfer coefficient

Another application of the periodic method is to determine heat transfer performance within ducts and shell-and-tube heat exchangers through which a fluid flows. The fundamental principle is that the heat transfer coefficient is found by means of the phase lag or amplitude attenuation between the fluid temperature oscillation and the solid wall temperature oscillation. For the experimental fluid temperature wave $t_f(\tau)$, the following coefficients of the first harmonic are obtained with the Fourier analysis

$$u_s = \frac{1}{\pi}\int_0^{2\pi} t_f(\tau)\sin\omega\tau \, \mathrm{d}(\omega\tau) \text{ and } u_c = \frac{1}{\pi}\int_0^{2\pi} t_f(\tau)\cos\omega\tau \, \mathrm{d}(\omega\tau) \tag{7.62}$$

Thus, the first harmonic is represented as

$$t_{f,1}(\tau) = u_{f,1} \sin(\omega\tau + \varphi_f) \tag{7.63}$$

where amplitude $u_{f,1} = \sqrt{u_s^2 + u_c^2}$, phase angle $\varphi_{f,1} = \arctan(u_c/u_s)$, period τ_P and angle velocity $\omega = 2\pi/\tau_P$.

Whether conduction resistance of the solid wall is taken into account or not may result two different mathematical models (Roetzel, 1989): (1) The model A (zero conduction resistance) and (2) The model B (finite conduction resistance). In the following both these models are briefly introduced

(1) The model A

If the conduction resistance of the solid wall is negligible, the lumped-parameter model for the wall of a tube is given as

$$C_w \frac{\mathrm{d}t_w}{\mathrm{d}\tau} = h_i A_i (t_f - t_w) - h_o A_o (t_w - t_a). \tag{7.64}$$

If the first harmonic of the oscillatory temperature wave of the tube wall is known as $u_{w,1} = u_{o,1} \sin\omega\tau$, the phase shift $\varphi_{f,1}$ and amplitude $u_{f,1}$ of the fluid temperature wave are respectively expressed as

$$\varphi_{f,1} = \arctan\left(\frac{C_w\omega}{h_i A_i + h_o A_o}\right) \text{ and } u_{f,1} = u_{o,1}\sqrt{\left(\frac{C_w\omega}{h_i A_i}\right)^2 + \left(1 + \frac{h_o A_o}{h_i A_i}\right)^2}. \tag{7.65}$$

From the phase shift $\varphi_{f,1}$ and $u_{f,1}/u_{o,1}$, both heat transfer coefficients h_1 and h_o can be determined from the above formulas. If $h_o << h_i$ (insulated tube or free convection outside and turbulent flow inside), the evaluation of phase or of amplitude is sufficient and both methods should give the same coefficient.

(2) The model B

By introducing the dimensionless variable $x = r\sqrt{\omega/(2a)}$ and $z = \omega\tau$, the energy equation accounting for radial conduction is represented as

$$2\frac{\partial t_w}{\partial z} = \frac{1}{x}\frac{\partial t_w}{\partial x} + \frac{\partial^2 t_w}{\partial x^2} \tag{7.66}$$

For the case that the third type of boundary condition is imposed on both the inner and outer side, Roetzel (1989) derived an approximate solution for the phase shift as well as the amplitude attenuation. In fact, insertion of $t_{w,1}(x,z) = G_1(x)\mathrm{e}^{\mathrm{i}z}$ (if it is necessary to account for all harmonic waves, $t_w(x,z) = \sum_{n=-\infty}^{\infty} G_n(x)\mathrm{e}^{\mathrm{i}nz}$ into eqn (7.66) yields

$$\frac{\mathrm{d}^2 G_1(x)}{\mathrm{d}x^2} + \frac{1}{x}\frac{\mathrm{d}G_1(x)}{\mathrm{d}x} - 2\mathrm{i}G_1(x) = 0 \tag{7.67}$$

This expression is the image-variable Bessel equation with zero order. According to the given boundary conditions, one can easily obtain the solution to eqn (7.67) as well as the related phase shift and amplitude attenuation.

In the light of the above-mentioned principle of the temperature oscillation technique, Roetzel (1989), and Xuan and Roetzel (1994) carried out experiments in a tube. The original experimental unit (Roetzel, 1989) is illustrated in Fig. 7.5. The test section consisted of a thermally insulated ϕ 17 mm × 1 mm copper tube, 60 cm in length. A rotary valve driven by a small motor made the hot and cold fluid streams alternately flow through the test tube and the rotational speed could be controlled to change the period of the incoming temperature oscillation. The temperatures were measured by thermocouples and the flow rate was sensed by a turbo-flowmeter. The temperature fluctuation of the hot water source and the cold water source was controlled to an accuracy below 0.1°C to maintain a stable temperature amplitude during the whole experiment. The cold stream and the hot stream from the stable cold and hot source alternately flowed through the tube and a flow with a periodic temperature oscillation was formed within the tube. Since the pump worked continuously, the switch-time of the rotary valve was far smaller than the period of the fluid temperature variation and the stable cold and hot sources dampened the possible fine pressure fluctuation with the tube, the flow rate fluctuation induced by valve switching was negligible.

In order to examine this periodic method, some experiments were carried out in the testing system shown in Fig. 7.5 (Xuan and Roetzel, 1994). The temperature of the fluid stream through the test tube changes periodically between 30–40°C and the experimental Reynolds number lies in the range $4\times10^3 - 32\times10^3$. The readings were recorded corresponding to three different periods ($\tau_P = 7.7$, 12.3 and 15.1 s) to analyse the effect of the temperature oscillation period on the determination of heat transfer coefficient. Hausen's

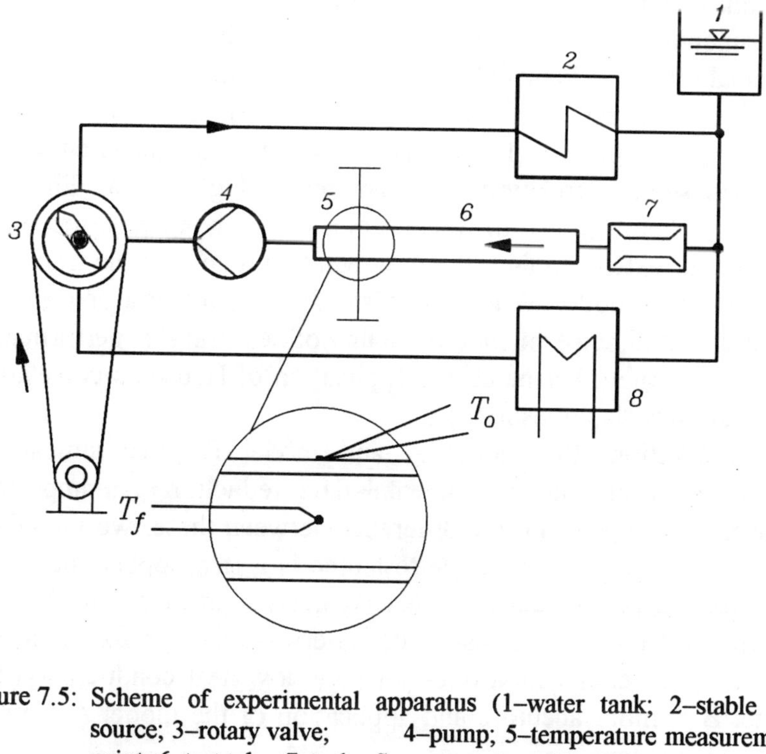

Figure 7.5: Scheme of experimental apparatus (1–water tank; 2–stable hot source; 3–rotary valve; 4–pump; 5–temperature measurement point; 6–test tube; 7–turbo-flowmeter; 8–stable cold source)

correlation (Hausen, 1976) for heat transfer coefficients in a tube is used to check the accuracy of the transient measurement method. By differentiation of Hausen's equation for the mean heat transfer coefficient with respect to the flow length one arrives at the following equation for the local Nusselt number at the flow length *L*:

$$Nu = 0.0235\left(Re^{0.8} - 230\right)\left(1.8Pr^{0.3} - 0.8\right)\left[1 + \frac{1}{3}\left(\frac{d}{L}\right)^{2/3}\right]. \tag{7.68}$$

In this equation *d* is the inside tube diameter and *L* the flow length at which the outside wall temperature and the inside heat transfer coefficient are measured.

As mentioned before, there are two possible approaches to determine heat transfer coefficients: one can use either the experimental phase difference or the amplitude attenuation between the fluid temperature wave and the outside wall wave. The experimental data reduction has shown that both the approaches principally lead to identical results, especially for cases $Re > 6000$. The heat transfer coefficients obtained from experimental data reduction (by the model A or model B) and the calculated values from Hausen's correlation are illustrated in Fig. 7.6 and Fig. 7.7. For $Re > 6000$, both of them coincide with each other to a satisfactory extent. In fact, the flow within a tube is in the transition zone for $Re < 6000$. Whether the steady-state method or transient method is used, experimental feasibility of heat transfer performance is poor. On the other hand, the application of Hausen's correlation in this zone is also questionable.

In addition, the model A and model B gave almost no difference during the experimental data reduction. An important parameter of estimating the difference between these two models is the Biot number *Bi*. Since the test tube is a thin copper tube with very great heat conductivity, *Bi* is in general smaller than 0.1 and the effect of conduction resistance is very small, even for higher h_i. For a tube or duct with a thick wall and low heat conductivity, the model B is more accurate and application of the model A may be limited.

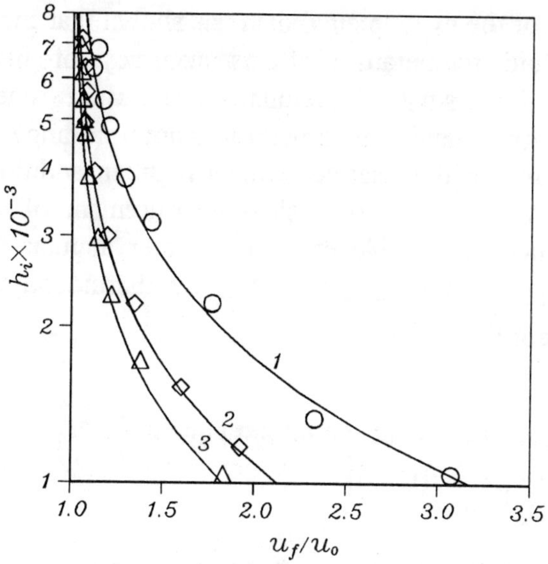

Figure 7.6: Heat transfer coefficients obtained by the amplitude attenuation (τ_p(s): 1–7.7; 2–12.3; 3–15.1)

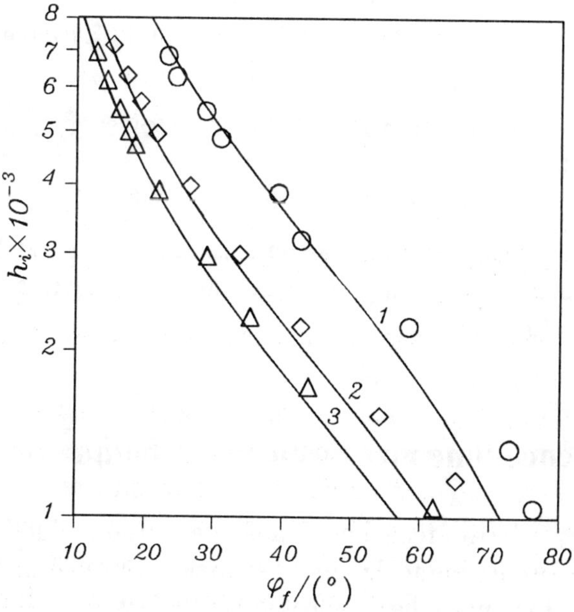

Figure 7.7: Heat transfer coefficients obtained by the phase shift (τ_p(s): 1–7.7; 2–12.3; 3–15.1)

Analysis of the experimental data has shown that the oscillatory period of fluid temperature affects accuracy of heat transfer coefficients. For smaller periods, temperature oscillation is considerably strong and the temperature variation range of the solid tube is narrow and the relative error is high. In addition, the heat capacity of the thermocouple delays measurement of temperature and may induce remarkable error for greater angular frequencies. Thus, the measured temperature $t_{f,c}$ by thermocouples may be corrected according to the formula

$$t_f - t_{f,c} = \frac{u_f}{\sqrt{1+\left[h_c A_c/(C_c\omega)\right]^2}} \sin\left(\omega\tau + \arctan\frac{h_c A_c}{C_c\omega}\right) \tag{7.69}$$

where the subscript c represents the thermocouple.

In cases when the phase difference $\varphi_f > 45°$, the heat transfer coefficients obtained by the experimental phase shift are of poor accuracy. The reason is that the curve slope of the function $y = \tan\varphi_f$ sharply increases and a fine variation of φ_f may induce a great change of y, a small experimental error of φ_f may result in a large error in the heat transfer coefficient. However, application of the amplitude attenuation can still yield satisfying data. Thus, if the phase shift $\varphi_f > 45°$ and the period is smaller ($\tau_P < 12\,\text{s}$), it is recommended that experimental data reduction should be carried out by means of the amplitude attenuation which satisfies the condition $u_f/u_o > 1.1$.

7.6 Residence time measurement technique

In the single blow techniques and the original and extended temperature oscillation techniques two unknowns can or have to be determined: the mean heat transfer coefficient or the number of transfer units *NTU* as well as the axial dispersion coefficient or the Péclet number *Pe*.

This is an advantage because both parameters are necessary for the calculation of transient processes in heat exchangers and one single experiment would yield two consistent values for the dispersion model. On the other hand the simultaneous estimation of two unknowns has disadvantages because the accuracy could be too low in some regions of the parameters. It would be very useful, if one of both unknowns could also be measured in a separate experiment.

One way of doing this could be the measurement of the local heat transfer coefficient using the oscillation method described in the above section. However, the method is inconvenient for heat exchangers because the local fluid and wall temperatures inside the exchanger would have to be measured at numerous points to find a suitable mean heat transfer coefficient.

A very attractive alternative is the determination of the axial dispersion coefficient or the dispersive Péclet number Pe using residence time measurements. Roetzel and Balzereit (1997) developed a suitable transient measurement technique for the fast and convenient determination of the axial dispersion coefficient in a flow channel of a heat exchanger. In front of the inlet a salt solution pulse is injected into the steady water flow stream which propagates with dispersion through the exchanger. The concentration profiles are measured with a newly developed device (Köllmann and Balzereit, 1996) at the inlet and outlet of the flow channel under consideration. Comparison of the calculated and measured salt concentration profiles of the inlet and outlet yields the mass dispersion coefficient, which can be assumed to be equal to the thermal dispersion coefficient (or Péclet number Pe).

The method corresponds to the limiting thermal case of an adiabatic flow channel, i.e., $NTU = 0$. The same differential equations are valid as those for the temperature oscillation technique of Section 7.3. Replacing the fluid temperature t_f in eqn (7.17) with the salt concentration (mass fraction) c and redefining the dimensionless time according to $z = \tau \dot{V}/V$ yields with $NTU = 0$

$$\frac{\partial c}{\partial z} + \frac{\partial c}{\partial x} - \frac{1}{Pe}\frac{\partial^2 c}{\partial x^2} = 0. \tag{7.70}$$

The second equation (7.18) is no longer necessary, which demonstrates the simplicity of the method.

The initial and boundary conditions are, according to Danckwerts (1953):

$$z = 0: \quad c = 0 \tag{7.71}$$

$$x = 0: \quad c - \frac{1}{Pe}\frac{\partial c}{\partial x} = f(z) \tag{7.72}$$

$$x = 1: \quad \frac{\partial c}{\partial x} = 0\,. \tag{7.73}$$

Applying the Laplace transform with respect to z and the reverse transform with direct integration by the method of residues finally yields for the Dirac pulse $f(z) = \delta(z)$, the following concentration profile at the outlet:

$$c(x,z) = \sum_{j=1}^{\infty} \frac{2\beta_j\left[\beta_j \cos\left(\beta_j x\right) + a \sin\left(\beta_j x\right)\right]}{a^2 + \beta_j^2 + 2a} \mathrm{e}^{ax - \frac{a^2+\beta_j^2}{2a} z} \tag{7.74}$$

where $a = Pe/2$, β_j is the jth positive root of the eigenfunction

$$\tan\beta = \frac{2a\beta}{\beta^2 - a^2}\,. \tag{7.55}$$

Figures 7.8–7.10 show the calculated and measured outlet concentration profiles of plate heat exchangers. Calculations reveal that a Dirac impulse can be assumed instead of the real impulse at the inlet without changing the result.

The salt pulse method yields excellent results for high Péclet numbers, which cannot be determined with steady-state methods. This method has also been applied to shell-and-tube heat exchangers and the results show the influence of different geometric parameters, such as number of baffles and gap size between baffles and shell

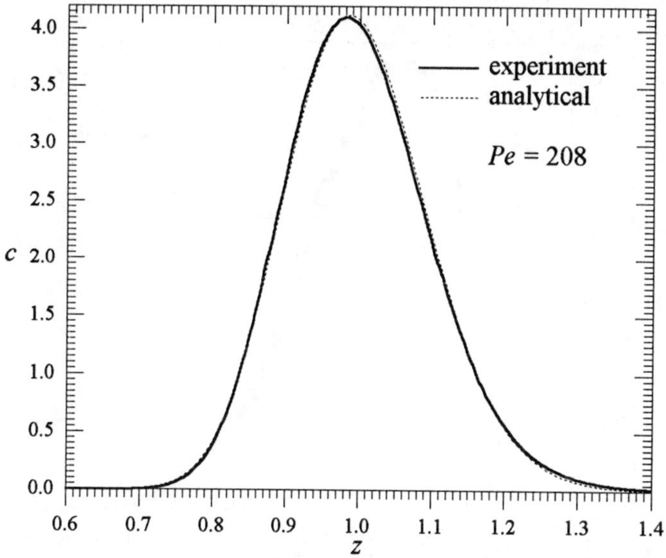

Figure 7.8: Measured and calculated system responses to the experimental inlet signal for the plate heat exchanger in a four-pass configuration, $Pe = 208$, $Re = 730$

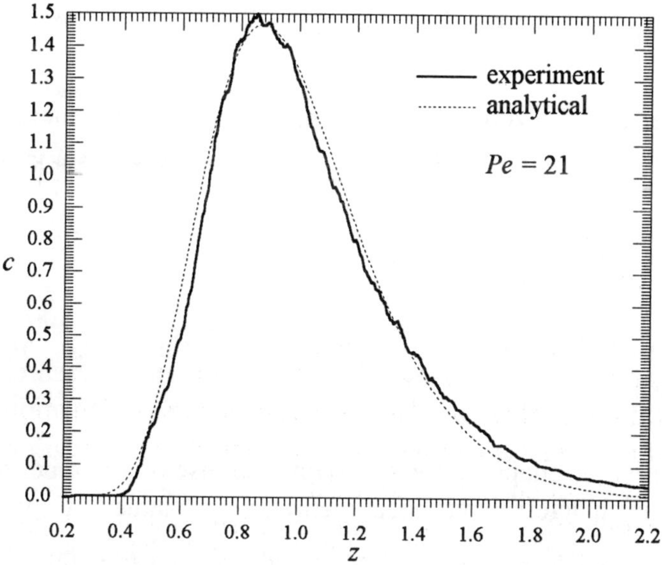

Figure 7.9: Measured and calculated system responses to the experimental inlet signal for the plate heat exchanger in Z configuration, $Pe = 21$, $Re = 668$

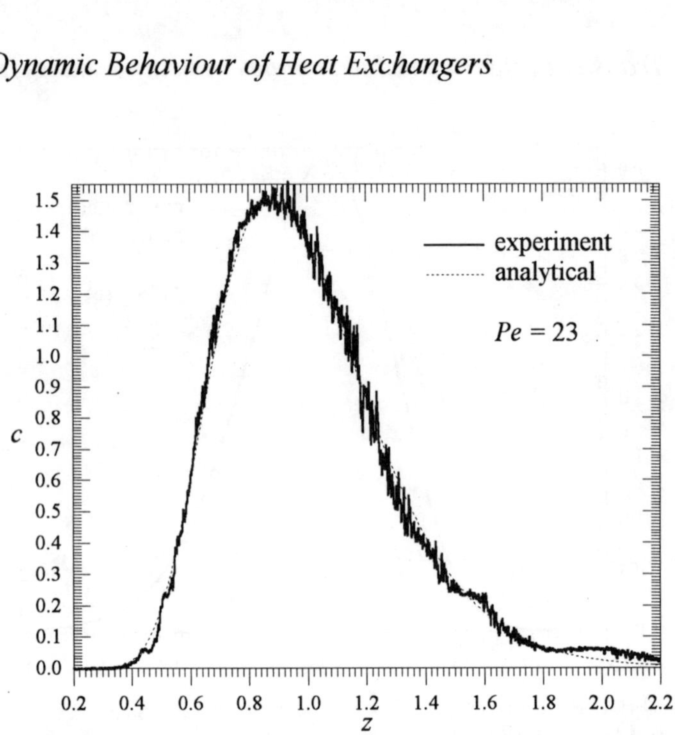

Figure 7.10: Measured and calculated system responses to the experimental inlet signal for the plate heat exchanger in U configuration, $Pe = 23$, $Re = 65$

(Roetzel and Balzereit, 1998). The residence time measurement technique together with the aforementioned transient test techniques is a useful tool for the determination of *Pe* and *NTU* of heat exchangers.

7.7 Experimental research on dynamics of heat exchangers

In the previous chapters, the transient response of a number of types of heat exchangers have been theoretically handled. Experimental research is another approach to the analysis of the dynamics of heat exchangers, which can be used to check the theoretical model. On the other hand, it is a method that compensates for the drawbacks of the theoretical model and deals with the problems which may be difficult for the theoretical model, such as arbitrary simultaneous variations of the incoming flow rates and temperatures.

7.7.1 Experiments on shell-and-tube heat exchangers

Stermole and Larson (1963) studied the dynamic response of a double-pipe steam–water heat exchanger both experimentally and theoretically. Their experimental data showed that a single partial differential equation model adequately represents the dynamics of a steam–water heat exchange system. This model was based on the assumption that the steam temperature was uniform over the length of the exchanger at any time, and only a single energy balance equation was needed for the water phase.

In their papers, Roetzel and Xuan (1993b) and Xuan and Roetzel (1993) conducted experimental and theoretical study on the dynamics of a shell-and-tube heat exchanger with baffles. They examined transient response to arbitrary incoming temperature and abrupt flow rate disturbances as well as the effect of maldistribution of a fluid stream on the dynamic behaviour of exchangers. A diagram of the experimental system is shown in Fig. 7.11. The working fluid was water and the temperature of the warm water was controlled by a platinum-resistance thermometer (Pt 100). During the experiment, the inlet temperature changed on the shell side whcre warm water was heated by an electric heater. Cold water flowed on the tube side. The flow rate was suddenly altered by adjusting the rotational speed of a pump motor. Five magnetic valves controlled by a computer and HP44725 were installed in the experimental unit in order to achieve uniform initial conditions in the test apparatus and a sudden inlet temperature change. The electromagnetic valves *a*, *b* and *e* were open but *c* and *d* were closed while preparing for measurement and heating up the water. During this phase, cold water flowed on both sides. At the instant of measurement, these valves were switched, i.e., *a*, *b* and *e* were closed while *c* and *d* were opened, so that an abrupt temperature change occurred at the shellside inlet. Temperatures and flow rates were measured by thermocouples and turbo-flowmeters, respectively. The inlet and outlet temperatures as well as the flow rates were read and recorded continuously by a computer and HP3852A system which consists of an HP44708A relay multiplexer with thermocouple compensation and HP44701A digital integrating voltmeter.

The test heat exchanger was made of V2A-steel and had 19 baffles. The inner shell diameter was 106.8 mm and the tube bundle consisted of 40 plain tubes, 819 mm long. The configuration of the exchanger was described in detail in a previous work (Xuan, 1991). To reduce the heat loss, the exterior of the exchanger used was thermally insulated. Since the clearance between the shell and baffle may be the main cause of shellside flow maldistribution, during the experiment, this clearance was altered by polyamide rings of different thickness which could be embedded on the outer edges of the baffles to examine its influence on the transient behaviour of the equipment. The experiment was performed in the following Reynolds number ranges: $Re_1 > 3000$ and $Re_2 > 3000$. By exchanging the locations of the shellside or tube-side inlet and outlet, parallel and countercurrent arrangements were achieved.

Each measurement took about 200 s for recording the profiles of the shellside and tubeside temperatures. The incoming temperature variation was not confined to a step change, which

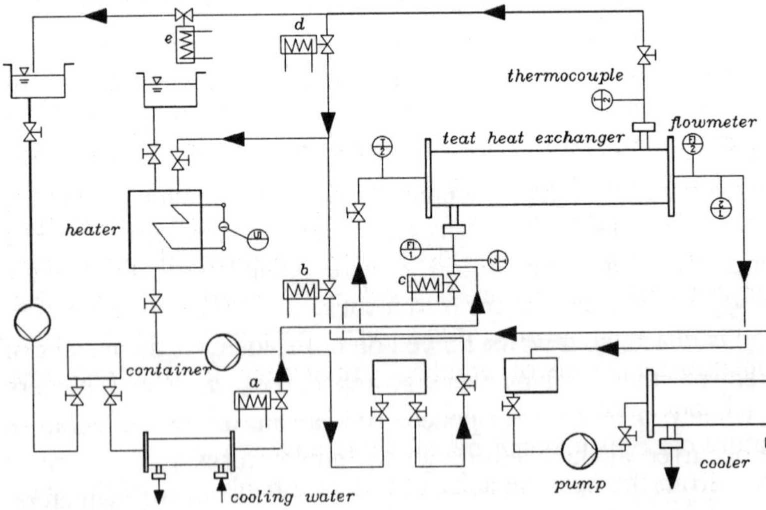

Figure 7.11: Scheme of the transient experimental system

caused no mathematical difficulty for comparison between theoretical and experimental results since the previously derived theory is suitable for describing the transient response to arbitrary inlet temperature changes. For the purpose of comparison, the correlations obtained from the previous steady-state experiments (Xuan, 1991) on the same exchanger were used. The steady-state experiment showed that, with different constant geometrical parameters, the Péclet number *Pe* depends mainly upon the clearance between the shell and baffle and is practically independent of flow rates.

The transient measurements were performed for clearances of $\delta = 2.45$, 1.0, and 0.25 mm. Along with the theoretical results from the dispersed plug-flow model, some experimental inlet temperatures and outlet responses are illustrated in the following figures in which the triangle stands for the experimental shellside exit temperature, the circle for the experimental tubeside exit temperature, the solid line for the theoretical results, and the dashed for the experimental inlet temperature. Figures 7.12 and 7.13 are transient responses to the inlet temperatures in a countercurrent exchanger and a cocurrent exchanger, respectively. The relevant inlet temperature profiles are also plotted. Comparison between theoretical and experimental results in the diagrams shows a satisfactory agreement. Furthermore, this agreement was reproduced for different experimental runs at the same Reynolds number and the same clearance. Because of heat exchange in the equipment as well as the finite power of heater and the finite cooling capacity of cooler, the shellside inlet temperature slowly decreased with time and the tubeside inlet temperature slowly increased throughout the experiment. Accurate profiles of inlet temperature changes can ensure the success of the theoretical prediction, although arithmetic means are a good approximation if the curve slope of the inlet temperatures are not very steep. Such approximations may be of practical importance for the real-time control of industrial process since the precise history of the inlet temperature changes may not be known in advance.

Apart from the heat capacity of the tube bank, that of the shell can affect the transient behaviour of the exchanger, which has been identified by predicting outlet responses to the measured inlet

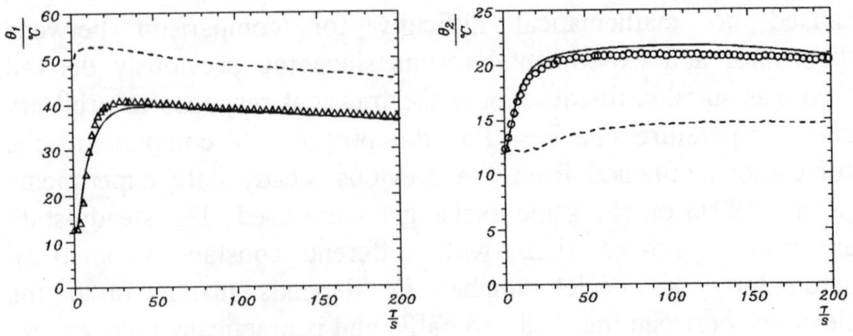

(a) shellside fluid (b) tubeside fluid

Figure 7.12: Transient temperatures of two fluids at the clearance $\delta = 2.45\,\text{mm}$ ($Pe = 3.6$) with countercurrent arrangement

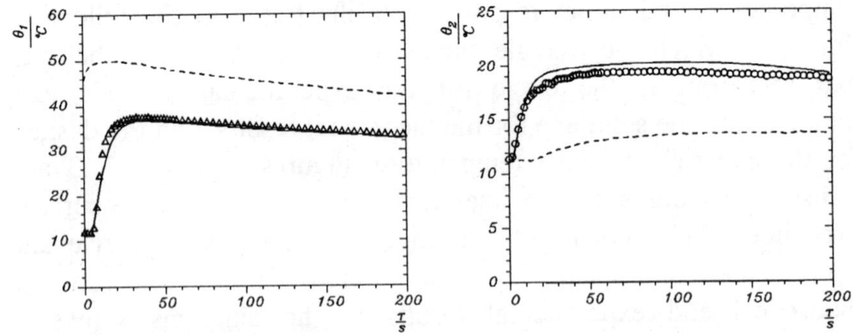

(a) shellside fluid (b) tubeside fluid

Figure 7.13: Transient temperatures of two fluids at clearance $\delta = 1.0\,\text{mm}$ with cocurrent arrangement

temperature changes. Thus, the thermal capacitance of the shell should not be neglected but should be included in the model. However, its effect depends not only upon the actual capacity but also upon the heat exchange process between the shellside fluid and the shell. The smaller the clearance between the tube bank and the shell, the smaller is the leakage stream through this clearance; and poorer heat exchange between the shellside fluid and the shell means that the shell influence on the transient behaviour decreases. The experiment has shown that the shell influence is negligible for the clearance $\delta < 0.25$ mm.

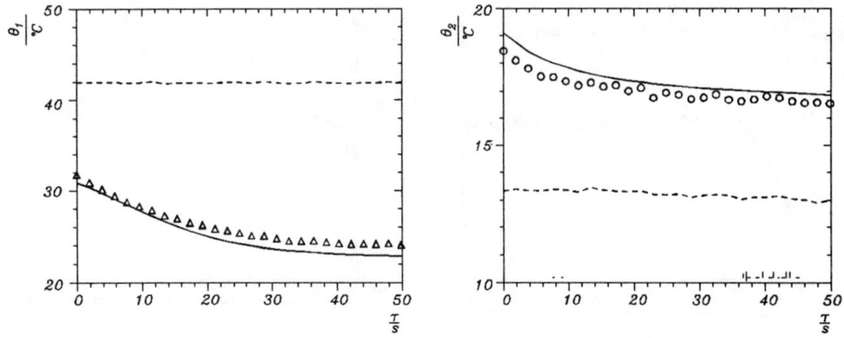

(a) shellside fluid (b) tubeside fluid

Figure 7.14: Responses of exit temperatures to a sudden decrease of the shellside flow rate ($\delta = 0.25$ mm and countercurrent arrangement)

Flow maldistribution causes a deterioration of the heat transfer performance and reduces the temperature changes in a heat exchanger. However, in the conventional plug-flow model, a larger apparent temperature difference rather than the actual value is assumed and a lower apparent heat transfer coefficient must be used to maintain agreement between theoretical and experimental results. In contrast, the dispersed plug-flow model provides the possibility of accounting for the effect of maldistribution by the dimensionless parameter *Pe*. It is expected that the effect of flow maldistribution becomes stronger for a high values of *NTU* and laminar flow. In other words, application of the dispersed plug-flow model will be of greater importance for high values of *NTU* and low values of *Pe*.

Figure 7.14 shows the exit temperature responses to an abrupt disturbance of the shellside flow rate, i.e., the shellside Reynolds number decreased from $Re_1 = 7700$ to $Re_1 = 2770$ with a relatively constant tubeside Reynolds number $Re_2 = 3477$ and countercurrent arrangement. The clearance between the tube bank and shell is $\delta = 0.25$ mm, in which the leakage stream is very small and this type of maldistribution has little effect ($Pe = 42$ from Xuan, 1991). In this case, the effect of shell heat capacitance is negligible.

Figure 7.15 corresponds to the clearance $\delta = 2.45$ mm and countercurrent arrangement, and a sudden increase of shell-side flow rate with the Reynolds number from $Re_1 = 3213$ to

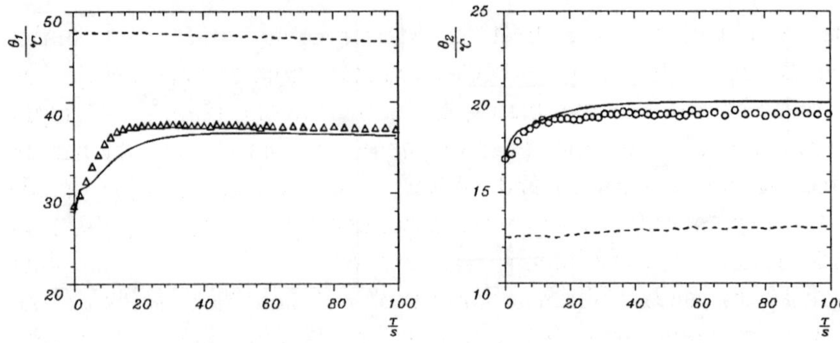

(a) shellside fluid (b) tubeside fluid

Figure 7.15: Responses of exit temperatures to an abrupt increase of the shellside flow rate ($\delta = 2.45$ mm and countercurrent arrangement)

$Re_1 = 9774$ and a constant $Re_2 = 3213$. Because of the wider gap, the leakage stream increased and the intensity of flow maldistribution became stronger. If this effect were neglected and a greater value of the Péclet number was inserted, there would be a greater deviation between the calculated and experimental values. In this case, the effect of shell heat capacity was noticeable, especially in the strongly transient phase ($\tau < 20$ s). Comparison has shown a satisfactory agreement between the calculated and experimental results, which demonstrates that the dispersed plug-flow model is valid for predicting the transient behaviour of exchangers. Furthermore, quite a good reproduction of such agreement for different runs at the same Reynolds number, clearance and flow arrangement was achieved. During the experiment, the shell- and tubeside inlet temperature did not remain constant. In other words, both temperature variations and flow rate disturbances occur, which are allowed for in the theoretical calculation.

7.7.2 Experiments on plate heat exchangers

To validate the theoretical model, Das *et al.* (1995) carried out experiments to measure the transient behaviour of two welded-plate heat exchangers with identical construction but different numbers of plates under different operating conditions. The tested exchangers are two welded-plate heat exchangers with inlet and outlet connections at the same end (U-connection port arrangement). The first one has 14 effective plates, excluding the end plates, with eight channels on one side and seven on the other. The second one has 28 effective plates, with 15 and 14 channels, respectively. The details of the experimental set-up were described by Das *et al.* (1995). A wide range of experiments was carried out for both the plate exchangers using different combinations of fluid flow rates on the two sides ranging from 0.1 to 0.8 l/s. The experimental results were compared with those obtained from the conventional plug-flow model and the dispersed plug-flow model. It was found that the dispersion model gave a better simulation than the conventional plug-flow model. A typical example was that computation by means of the dispersed plug-flow model with $Pe = 3.5$ gives a response that yielded a steady-state temperature being equal to the experimental value within the limits of experimental error. However, the traditional plug-flow model corresponding to infinite Pe failed.

Furthermore, the effect of some parameters like NTU and thermal flow rate ratio R was studied experimentally. As expected, an increase in NTU_1 increases the effectiveness by decreasing the hot side and increasing the cold side exit temperatures. The role of NTU_1 is not significant on the initial delay period. The rise period of the temperatures was found to increase with NTU_1 for the cold side and decrease with it for the hot side. With an increase in the number of plates, the delay period increases considerably and the rise period marginally.

Using two experimental procedures such as the well-known frequency response approach and the pulse technique, Khan *et al.* (1988) studied the dynamic characteristics of a countercurrent plate heat exchanger consisting of 21 plates. During the frequency response testing, a sinusoidal disturbance was imposed upon the hot stream entering the plate exchanger by means of a signal generator connected to a pneumatic control valve. The response of the temperature of the cold stream leaving the exchanger was recorded at 4 s intervals and the experimental programme was designed to cover a frequency range between 0.0063 and 0.44 rad/s. This response to the sinusoidal disturbance in the hot stream may be expressed as

$$\theta_{c,exit}(\tau) = \theta_c + A\sin(\omega\tau + \varphi) \tag{7.75}$$

where A is the amplitude of the output disturbance, ω the angular frequency of the applied disturbance, φ the phase shift between input and output, and θ_c the steady-state cold stream exit temperature. All these can be obtained by fitting the experimental response.

During the pulse test, rectangular pulses of different amplitudes were introduced into the inlet flow of the hot stream via the pneumatic valve. The outlet temperature of the cold stream was recorded at 4 s intervals. Comparing the experimental data with those obtained from the lumped parameter model, Khan *et al.* (1988) found that the transfer function relating the outlet temperature of the cold stream to the flow rate of the hot stream is best described by an overdamped second-order lag coupled with a first-order lead with dead time.

7.8 References

Anzelius, A. (1926) Über Erwärmung vermittels stromender Medien, *Zeitschrift für Angewandte Mathematik und Mechanik*, **6**, 291–294.

Baclic, B. S., Heggs, P. J., and Ziyan, H. A. (1986a) The Differential Fluid Enthalpy Method, in *Proceeding of the 8th International Heat Transfer Conference*, San Francisco, **5**, 2616–2622.

Baclic, B. S., Gvozdenac, D. D., and Badlja, D. (1986b) Performance Identification of the Apparatus for Testing Compact Heat Exchanger Surfaces, in *Advances in Heat Exchanger Design*, (ed. by Shah, R. K. and Pearson, J. R.), HTD Vol. 66, pp. 13–20, ASME, New York.

Beck, J. V., Blackwell, B., and Clair, S. R. St. (1985) *Inverse Heat*

Conduction, Ill-Posed Problems, Wiley, New York.

Bradshaw, A. V., Johnson, A., McLachlan, N. H., and Chin, Y. T. (1970) Heat Transfer between Air and Nitrogen and Packed Beds of Nonreacting Solids, *Transactions of Institution of Chemical Engineers*, **48**, 77–84.

Briggs, D. E. and Young, E. H. (1969) Modified Wilson Techniques for Obtaining Heat Transfer Correlations for Shell and Tube Heat Exchangers, *Chemical Engineering Progress Symposium Series*, **65**, 35–45.

Cai, Z. H., Li, M. L., Wu, Y. W., and Ren, H. S. (1984) A Modified Selected Point Matching Technique for Testing Compact Heat Exchanger Surfaces, *International Journal of Heat and Mass Transfer*, **27**, 971–978.

Danckwerts, P. V. (1953) Continuous Flow Systems. Distribution of Residence Time, *Chemical Engineering Science*, **2**, 1–13.

Das, S. K., Spang, B., and Roetzel, W. (1995) Dynamic Behaviour of Plate Heat Exchangers — Experiments and Modelling, *Journal of Heat Transfer*, **117**, 859–864.

Furnas, C. C. (1932) Heat Transfer from Gas Stream to Bed of Broken Solids, *U.S. Bureau of Mines, Bulletin No.* 361.

Glaser, H. (1938) Der Wärmeübergang in Regeneratore, *Z. VDI, Beiheft Verfahrenstechnik*, **4**, 112–125.

Gvozdenac, D. D. (1986) Analytical Solution of the Transient Response of Gas-to-Gas Crossflow Heat Exchangers with Both Fluids Unmixed, *Journal of Heat Transfer*, **108**, 722–727.

Gvozdenac, D. D. (1994) Experimental Prediction of Heat Transfer Coefficients by Use of a Double-Blow Method, *Wärme- und Stoffübertragung*, **29**, 361–365.

Hausen, H. (1929) Über die Theorie des Wärmeaustausches in Regeneratoren, *Zeitschrift für Angewandte Mathematik und Mechanik*, **9**(3), 173–200.

Hausen, H. (1976) *Wärmeübertragung im Gegenstrom, Gleichstrom und Kreuzstrom,* Springer-Verlag, Berlin.

Heggs, P. J. and Burns, D. (1988) Single-Blow Experimental Prediction of Heat Transfer Coefficients — A Comparison of Four Commonly Used Techniques, *Experimental Thermal and Fluid Science*, **1**, 243–251.

Kast, W. (1965) Messung des Wärmeüberganges in Haufwerken mit Hilfe einer temperatur-modulierten Strömung, *Allgemeine Wärmetechnik*, **12** (6/7), 119–125.

Kays, W. M. and London, A. L. (1984) *Compact Heat Exchangers,* McGraw-Hill, New York.

Khan, A. R., Baker, N. S., and Wardle, A. P. (1988) The Dynamic Characteristics of a Countercurrent Plate Heat Exchanger, *International Journal of Heat and Mass Transfer*, **31**, 1269–1278.

Khartabil, H. F., Christensen, R. N., and Richards, D. E. (1988) A Modified Wilson Plot Technique for Determining Heat Transfer Coefficients, in *The Proceedings of 2nd UK National Conference on Heat Transfer*, **2**, 1331–1357.

Kohlmayr, G. F. (1966) Exact Maximum Slopes for Transient Matrix Heat Transfer Testing, *International Journal of Heat and Mass Transfer*, **9**, 671–680.

Kohlmayr, G. F. (1968a) Extension of the Maximum Slope Method to Arbitrary Upstream Fluid Temperature Changes, *Journal of Heat Transfer*, **90**,130–134.

Kohlmayr, G. F. (1968b) An Indirect Curve Matching Method for

Transient Matrix Heat-Transfer Testing in the Low NTU-Range, *International Journal of Heat and Mass Transfer*, **11**, 567–581.

Kohlmayr, G. F. (1971) Implementation of Direct Curve Methods for Transient Matrix Heat Transfer, *Applied Scientific Research*, **24**, 127–148.

Köllmann, K. and Balzereit, F. (1996) Meßverfahren zur zeitaufgelösten Konzentrationsbestimmung in strömenden elektrolytischen Flüssigkeiten, *Technisches Messen*, **63**, 458–464.

Langhans, W. U. (1962) Wärmeübergang und Druckverlust in Regeneratoren mit rostgitterartiger Speichermasse, *Archiv für Eisenhüttenwesen*, **33**, 347–353 and 441–451.

Liang, C. Y. and Yang, W. J. (1975) Heat Transfer and Friction Loss Performance of Perforated Heat Exchanger Surfaces, *Journal of Heat Transfer*, **97**, 9–15.

Locke, G. L. (1950) Heat Transfer and Flow Friction Characteristics of Porous Solids, Tech. Report, No. 10, Department of Mechanical Engineering., Stanford University.

Loehrke, R. I. (1990) Evaluating thc Results of the Single-Blow Transient Heat Exchanger Test, *Experimental Thermal and Fluid Science*, **3**, 574–580.

Luo, X. (1998) *Das axiale Dispersionsmodell für Kreuzstromwärmeübertrager*, Fortschritt-Berichte VDI, Reihe 19, Wärmetechnik/Kältetechnik, Nr. 109, VDI-Verlag GmbH, Düsseldorf.

Matulla, H. and Orlicek, A. F. (1971) Bestimmung der Wärmeübergangskoeffizienten in einem Doppelrohrwärme-austauscher durch Frequenzganganalyse, *Chemie-Ingenieur-Technik*, **43**, 1127–1130.

Mehta, R. C. and Jayachandran, T. (1990) Determination of Heat Transfer Coefficient Using Transient Temperature Response Chart, *Wärme- und Stoffübertragung*, **26**, 1–5.

Mondt, J. R. and Siegla, D. C. (1974) Performance of Perforated Plate Heat Exchanger Surfaces, *Journal of Engineering for Power*, **96**, 81–87.

Mullisen, R. S. and Loehrke, R. I. (1986a) A Transient Heat Exchanger Evaluation Test for Arbitrary Fluid Inlet Temperature Variation and Longitudinal Core Conduction, *Journal of Heat Transfer*, **108**, 370–376.

Nusselt, W. (1927) Die Theorie des Winderhitzers, *Zeitschrift des Vereines Deutscher Ingenieure*, **71**(3), 85–91.

Pucci, P. F., Howard, C. P., and Piersall, C. H. Jr. (1967) The Single-Blow Transient Testing Technique for Compact Heat Exchanger Surfaces, *Journal of Engineering for Power*, **89**, 29–40.

Rapley, C. W. and Webb, A. L. (1983) Heat Transfer Performance of Ceramic Regenerator Matrices with Sine-Duct Shaped Passages, *International Journal of Heat and Mass Transfer*, **26**, 805–816.

Roetzel, W. (1989) Measurement of Heat Transfer Coefficients in Tubes by Temperature Oscillation Analysis, *Chemical Engineering Technology*, **12**, 379–387.

Roetzel, W. and Balzereit, F. (1997) Determination of Axial Dispersion Coefficients in Plate Heat Exchangers Using Residence Time Measurements, *Revue Générale de Thermique*, **36**, 635–648.

Roetzel, W. and Balzereit, F. (1998) Axial Dispersion in Shell-and-Tube Heat Exchangers, in *Proceedings of the International Conference on Heat Exchangers for Sustainable Development*, 15–18 June 1998, Lisbon, Portugal, pp. 569–581.

Roetzel, W. and Luo, X. (1997) Extended Temperature Oscillation Measurement Technique for Heat Transfer and Axial Dispersion Coefficients, in *Compact Heat Exchangers for the Process Industries*, (ed. by Shah, R. K.), Begell House, New York, pp. 381–388.

Roetzel, W. and Luo, X. (1998) Extended Temperature Oscillation Measurement Technique for Heat Transfer and Axial Dispersion Coefficients, *Revue Générale de Thermique*, **37**, 277–283.

Roetzel, W., Luo, X. and Xuan, Y. (1993a) Measurement of Heat Transfer Coefficient and Axial Dispersion Coefficient Using Temperature Oscillations, *Experimental and Fluid Science*, **7**, 345–353.

Roetzel, W. and Prinzen, S. (1991) Measurement of Local Heat Transfer Coefficients Using Temperature Oscillations, in *Experimental Heat Transfer, Fluid Mechanics, and Thermodynamics 1991*, (ed. by Keffer, J. F., Shah, R. K. and Ganic, E. N.), pp. 497–504, Elsevier, New York.

Roetzel, W. and Xuan, Y. (1993b) Transient Behaviour of Shell and Tube Heat Exchangers, *Chemical Engineering Technology*, **16**, 296–302.

Roetzel, W., Das, S. K. and Luo, X. (1994) Measurement of the Heat Transfer Coefficient in Plate Heat Exchangers Using a Temperature Oscillation Technique, *International Journal of Heat and Mass Transfer*, **37**, Suppl. 1, 325–331.

Schumann, T. E. W. (1929) Heat Transfer: A Liquid Flowing through a Porous Prism, *Journal of the Franklin Trust Institute*, **208,** 405–416.

Shah, R. K. (1990) Assessment of Modified Wilson Plot Techniques for Obtaining Heat Exchanger Design Data, in *The Proceedings of the 9th International Heat Transfer Conference*, **5**, 14-HX-9.

Stang, J. H. and Bush, J. E. (1974) The Periodic Method for Testing Compact Heat Exchanger Surfaces, *Journal of Engineering for Power*, **96**, 87–94.

Stermole, F. J. and Larson, M. A. (1963) Dynamic Response of Heat Exchangers to Flow Rate Changes, *I&EC Fundamentals*, **2**, 62–67.

Webb, R. L. (1994) *Principles of Enhanced Heat Transfer*, A Wiley-Interscience Publication, New York.

Wilson, E. E. (1915) A Basis for Rational Design of Heat Transfer Apparatus, *Transactions of American Society of Mechanical Engineers*, **37**, 47–70.

Xuan, Y. (1991) *Thermische Modellierung mehrgängier Rohrbündelwärmeübertrager mit Umlenkblechen und geteiltem Mantelstrom*, Fortschritt-Berichte VDI, Reihe 19, Wärmetechnik/Kältetechnik, Nr. 52, VDI-Verlag GmbH, Düsseldorf.

Xuan, Y. and Roetzel, W. (1993) Dynamics of Shell-and-Tube Heat Exchangers to Arbitrary Temperature and Step Flow Variations, *AIChE Journal*, **39**, 413–421.

Xuan, Y. and Roetzel, W. (1994) Transient Measurement of Heat Transfer Coefficients inside a Tube, *Journal of Chemical Industry and Engineering* (China), **45**, 756–759.

Yazicizade, A. Y. (1966) Untersuchung der Wärmeübertragung und des Druckabfalls in Regeneratoren mit körniger oder schachtartig aufgebauter Speichermasse, *Glastechnische Berichte*, **39**, 203–217.

Chapter 8

Miscellaneous problems

In this chapter attention is paid to some problems which represent new developments of dynamic models of heat exchangers, applications of the dynamic theory as well as control and regulation. All these problems are related to the transient behaviour of heat exchangers.

8.1 Hyperbolic axial dispersion model

In all of the previous chapters, both the traditional plug-flow model and the dispersed plug-flow model have been applied to modelling the transient behaviour of a variety of heat exchangers. It is well known that the plug-flow model is based on the assumption that the flow velocity of a fluid stream remains constant in both transverse and longitudinal directions and the fluid stream flows as a plug. The conventional plug-flow model is widely used because of its simplicity, but fails to take various types of flow maldistribution into account.

The dispersed plug-flow model is an improvement on the traditional plug-flow model. It introduces an apparent axial dispersion term into the energy or mass balance equation. Such a dispersion model originated from a description of longitudinal mixing. The most notable is the work of Danckwerts (1953) and

Taylor (1954) who, in pioneering papers, introduced the concept of longitudinal dispersion superimposed on plug flow to describe the flow pattern in which not all fluid elements travel at equal speed through a system. As mentioned in the previous chapters, the dispersion model has been applied to analysing the steady-state and transient performance of various heat exchangers. The axial dispersion as well as other factors causing deviation of the real flow pattern from the ideal plug flow is characterized by the Péclet number *Pe* based on the macro-dispersion coefficient, which may be considered as an apparent thermal conductivity. It is inherently implied that the dispersion or disturbance resulting from all possible types of flow maldistribution has an infinite propagation velocity and the dispersion effect arrives simultaneously everywhere within a system. This is like the propagation of a thermal signal that is described by the Fourier heat conduction law. It is well known that the Fourier law is empirical and based on the assumption of an infinite propagation velocity of thermal disturbances. This law is valid only for slow temperature changes. In some cases such as thermal wave propagation at cryogenic temperatures, extremely quick temperature changes and very high heat flux (for example, during laser heated process), the traditional Fourier law may not be suitable any more.

Chester (1963) proposed a "non-Fourier" (also known as "hyperbolic" or "wave") model of conduction which was found to be of practical importance near absolute zero temperature. This formulation assumes a heat flux in the form

$$\dot{\mathbf{q}} + \frac{a}{C^2}\frac{\partial \dot{\mathbf{q}}}{\partial \tau} = -\lambda \nabla \theta \tag{8.1}$$

The resulting hyperbolic heat conduction or telegraph equation has the following form (Morse and Feshbach, 1953; Eckert and Drake, 1972):

$$\frac{a}{C^2}\frac{\partial^2 \theta}{\partial \tau^2} + \frac{\partial \theta}{\partial \tau} = a\nabla^2 \theta \tag{8.2}$$

where C represents the conduction wave propagation velocity. If $C\to\infty$, this equation turns to the conventional parabolic Fourier conduction equation. However, it seems to be logical to distinguish axial dispersion from molecular conduction, when considering the problem from the point of view of the propagation velocity. The axial dispersion is a macroscopic disturbance which propagates through the fluid and hence cannot propagate at infinite speed.

An extension of the conventional axial dispersion model for heat exchangers in which the finite propagation velocity of the axial dispersion is taken into account was first proposed by Roetzel and Spang (1994, 1995). The wave model was further developed and applied to the transient analysis of heat exchangers (Luo and Roetzel, 1995; Roetzel and Das, 1995). Shortly after Roetzel and Spang, Westerterp *et al.* (1995a, 1995b) published a similar model for longitudinal mass dispersion in chemical reactors. Compared with Chester's equation (8.1) their constitutive equation for the dispersion flux (mass dispersion in their case) contained in addition the derivation of the source term (chemical reactions) with respect to the driving force (concentration) and a third parameter for flow maldistribution. Westerterp and co-workers subsequently applied their wave model to tubular reactors (Kronberg *et al.*, 1996; Benneker *et al.*, 1997).

8.1.1 General hyperbolic energy equations

The starting-point of the derivation is the energy equation for a fluid moving with uniform and constant velocity u in the l-direction in a channel of constant cross-section A_c:

$$\rho_f c_{pf}\left(\frac{\partial\theta_f}{\partial\tau}+u\frac{\partial\theta_f}{\partial l}\right)=-\frac{\partial\dot{q}_l}{\partial l}+\sum_{i=1}^{N}\frac{P_i}{A_c}\dot{q}_{si} \tag{8.3}$$

The axial heat flux $\dot{q}_l$ is caused by conduction or dispersion and $\dot{q}_{si}$ are the convective heat fluxes from N wetted walls of perimeter P_i to the fluid in the channel. In a shell-and-tube heat exchanger for instance, the shellside fluid would wet the tube walls and the shell.

The dispersive axial heat flux is predominant compared with the conduction and can be expressed according to Chester's conduction law using the hyperbolic operator (Luo and Roetzel, 1995)

$$H = 1 + \frac{a}{C^2}\left(\frac{\partial}{\partial \tau} + u\frac{\partial}{\partial l}\right) \tag{8.4}$$

yielding

$$H\dot{q}_l = -\mathcal{D}\frac{\partial \theta_f}{\partial l} \tag{8.5}$$

in which the apparent thermal conductivity $\mathcal{D}$ is called the dispersion coefficient. The convective heat fluxes from the walls to the fluid could be expressed in a similar way:

$$H_{si} = 1 + \frac{a_{si}}{C_{si}^2}\left(\frac{\partial}{\partial \tau} + u_{si}\frac{\partial}{\partial l}\right) \tag{8.6}$$

and

$$H_{si}\dot{q}_{si} = h_i\left(\theta_{wi} - \theta_f\right) \tag{8.7}$$

where the velocity u_{si} is the mean boundary layer velocity or film velocity, $0 \le u_{si} \le u$, and the ratio a_{si}/C_{si}^2 the relaxation time for hyperbolic heat conduction in the boundary layer "i" (film) near the wall. Substituting in eqn (8.3) the heat fluxes $\dot{q}_l$ and $\dot{q}_{si}$ according to eqns (8.5) and (8.7), respectively, yields for constant parameters

$$\rho_f c_{pf}\left(\frac{\partial \theta_f}{\partial \tau} + u\frac{\partial \theta_f}{\partial l}\right) = \frac{\mathcal{D}}{H}\frac{\partial^2 \theta_f}{\partial l^2} + H\sum_{i=1}^{N}\frac{P_i}{A_c H_{si}}h_i\left(\theta_{wi} - \theta_f\right). \tag{8.8}$$

For further evaluation this equation has to be multiplied by the factor $H\prod_{i=1}^{N} H_{si}$.

This energy equation for a fluid in a thermal regenerator or heat exchanger takes dispersive and convective relaxation times into account and is an extension of the known telegraph equation or hyperbolic heat conduction equation.

The energy equation for the wall can be derived in a similar way. For the stationary wall $u_w = 0$ and the energy equation (8.3) can be expressed as

$$\rho_w c_{p_w} \frac{\partial \theta_w}{\partial \tau} = -\frac{\partial \dot{q}_{wl}}{\partial l} - \sum_{j=1}^{M} \frac{P_j}{A_{cw}} \dot{q}_{sj} \,. \tag{8.9}$$

The summation indicates that the wall may be wetted by M fluids, e.g., in a shell-and-tube heat exchanger the tube wall is wetted by the tubeside and the shellside fluids. The convective heat fluxes can again be expressed using eqns (8.6) and (8.7)

$$H_{sj}\dot{q}_{sj} = h_j\left(\theta_w - \theta_{fj}\right) \tag{8.10}$$

with individual values of a_{sj}/C_{sj}^2, u_{sj} and consequently different operators H_{sj}. The hyperbolic axial wall heat conduction can be taken into account using the operator

$$H_w = 1 + \frac{a_w}{C_w^2}\frac{\partial}{\partial \tau} \tag{8.11}$$

and

$$H_w \dot{q}_{wl} = -k_w \frac{\partial \theta_w}{\partial l} \,. \tag{8.12}$$

The elimination of the heat fluxes $\dot{q}_{wl}$ and $\dot{q}_{sj}$ in eqn (8.9) is done in the same way as with the fluid energy equation. Substituting eqns (8.12) and (8.10) into eqn (8.9) yields the equation

$$\rho_w c_{pw} \frac{\partial \theta_w}{\partial \tau} = \frac{k_w}{H_w} \frac{\partial^2 \theta_w}{\partial l^2} + \sum_{j=1}^{M} \frac{P_j h_j}{A_{cw} H_{sj}} \left(\theta_{fj} - \theta_w \right) \tag{8.13}$$

which has to be multiplied by the factor $H_w \prod_{j=1}^{M} H_{sj}$.

Equations (8.8) and (8.13) represent the complete energy equations for fluids and walls in heat exchangers or regenerators taking hyperbolic heat conduction, convection and dispersion into account. However, the above equations are too complicated for general application. Fortunately they can be simplified under realistic assumptions.

8.1.2 Heat exchanger model

Depending on the simplifications made, different models can be derived from the general equations (8.8) and (8.13). The most realistic one is the heat exchanger model (Roetzel and Spang, 1995). In this model the relaxation time for convective heat transfer and that for axial heat conduction in the wall is neglected, i.e., $H_s = 1$ and $H_w = 1$. Introducing in eqn (8.8) a third sound velocity Mach number (Roetzel *et al.*, 1998) $M = u/C$ together with the known dimensionless groups and variables $NTU = hA/\dot{W}$, $Pe = uL/a_f^*$, $x = l/L$ and $z = u\tau/L$ yields the dimensionless hyperbolic energy equation for the fluid

$$\frac{\partial \theta_f}{\partial z} + \frac{\partial \theta_f}{\partial x} + \frac{M^2}{Pe} \left(\frac{\partial^2 \theta_f}{\partial z^2} + 2 \frac{\partial^2 \theta_f}{\partial z \partial x} + \frac{\partial^2 \theta_f}{\partial x^2} \right)$$

$$- \frac{M^2 NTU}{Pe} \left[\frac{\partial}{\partial z} \left(\theta_w - \theta_f \right) + \frac{\partial}{\partial x} \left(\theta_w - \theta_f \right) \right]$$

$$= \frac{1}{Pe} \frac{\partial^2 \theta_f}{\partial x^2} + \sum_{i=1}^{N} NTU_i \left(\theta_{wi} - \theta_f \right) \tag{8.14}$$

which turns to the traditional parabolic axial dispersion equation if $M \to 0$ and to the usual nondispersive plug-flow energy equation if $Pe \to \infty$.

The energy equation for the wall is derived from the general equation (8.13) with $H_w = H_s = 1$. Introducing a dimensionless wall heat conductivity

$$\lambda = \frac{a_w}{uL} \tag{8.15}$$

and a heat capacity ratio of fluid and wall inside the channel

$$B = \frac{C_f}{C_w} \tag{8.16}$$

yields the wall energy equation of the hyperbolic heat exchanger model

$$\frac{\partial \theta_w}{\partial \tau} = \lambda \frac{\partial^2 \theta_w}{\partial x^2} + \sum_{i=1}^{N} BNTU_i \left(\theta_f - \theta_w\right). \tag{8.17}$$

8.1.3 The boundary conditions

As in all previous derivations it is assumed that the axial dispersion takes place only inside the exchanger between the inlet and outlet cross-section. In the channels before and behind the exchanger the axial dispersion coefficient (apparent axial heat conductivity of the fluid) is zero. The dispersive system is a closed system. The degree of axial dispersion is expressed by the Péclet number $0 \le Pe \le \infty$. $Pe = 0$ means complete axial mixing and $Pe = \infty$ nondispersive plug-flow. The propagation velocity can be expressed by a Mach number *M*. $M < 1$ is the subsonic dispersive flow and $M > 1$ the supersonic dispersive flow. One has to distinguish between these two cases (Roetzel *et al.*, 1998).

For subsonic flow a temperature step appears at the inlet cross-section when the nondispersive flow enters the dispersive region. If the fluid is cooled in the exchanger, heat is sucked off from the inlet channel by a dispersive heat flow inside and a conductive heat flow (zero conductivity and infinite temperature gradient) outside the exchanger. This causes a temperature drop at the inlet cross-section.

The subsonic situation describes the real axial mixing, also named as backmixing, which is only possible if the mixing effect propagates with a higher velocity than the flow velocity. Otherwise the backmixing effect would never reach the inlet cross-section.

From an energy balance at the inlet cross-section Roetzel and Das (1995) derived the following inlet condition for subsonic dispersive flow:

$$\theta_f^{(-)} - \theta_f^{(+)} = -\frac{1-M^2}{Pe}\frac{\partial \theta_f^{(+)}}{\partial x} + \frac{M^2}{Pe}\frac{\partial \theta_f^{(+)}}{\partial z} \tag{8.18}$$

which becomes the known Danckwerts condition when $M \to 0$. Under steady-state conditions the temperature drop decreases to zero when $M \to 1$.

Unlike the subsonic case, for supersonic flow there is no feedback of information from the exit to the entry. As pointed out by Roetzel *et al.* (1998), the first inlet condition is

$$\theta_f^{(-)} = \theta_f^{(+)} . \tag{8.19}$$

This condition is realistic if no backmixing but pure maldistribution takes place, e.g., different flow velocities in the tubes of a bundle.

The second boundary condition refers also to the inlet. In the inlet cross-section the fluid experiences a sudden temperature gradient in the flow direction causing an axial dispersive heat flow. However, because the axial heat flow is zero in the inlet flow channel, it is still zero and remains zero at the inlet cross-section although a sudden driving force exists.

$$\dot{q}_x + \frac{a}{C^2}\left(\frac{\partial \dot{q}_x}{\partial \tau} + u\frac{\partial \dot{q}_x}{\partial l}\right) = -\mathcal{D}\frac{\partial \theta_f^{(+)}}{\partial l}. \tag{8.20}$$

For $x = 0$: $\quad \dot{q}_x = 0, \qquad \frac{\partial \dot{q}_x}{\partial \tau} = 0. \qquad (8.21)$

From this consideration the second inlet condition can be derived (Roetzel *et al.*, 1998).

$$\frac{\partial \theta_f^{(+)}}{\partial x} = \frac{NTU\left(\theta_w^{(+)} - \theta_f^{(+)}\right) - \frac{\partial \theta_f^{(+)}}{\partial z}}{1 - \frac{1}{M^2}} \tag{8.22}$$

where NTU is defined with the heat transfer coefficient between θ_f and θ_w. If under steady-state conditions the overall heat transfer coefficient is used in *NTU*, the wall temperature θ_w has to be replaced by the other fluid temperature.

For the outlet special considerations are necessary. In the subsonic case, the Danckwerts outlet condition

$$\left.\frac{\partial \theta_f}{\partial x}\right|_{x=1} = 0 \tag{8.23}$$

may be a good approximation although an axial heat flow may still exist at the outlet. In the supersonic case no outlet condition is necessary because one has two inlet conditions.

At the outlet an axial dispersive heat flow exists which must be taken into account in the energy balance. To fulfil the energy balance this axial heat flow should be compensated by a temperature change at the outlet similar to the inlet temperature drop in the subsonic case. Although from the thermodynamic point of view such a temperature rise (for a cooled fluid) could not take place with real axial heat conduction, this temperature rise could be explained by the mixing of different flow streams. The temperature change is the difference between the adiabatic mixing outlet temperature and the local mean temperature for heat transfer (Roetzel *et al.*, 1998).

The hyperbolic dispersion model seems to be able to distinguish between pure maldistribution and real backmixing. With an appropriate value of the Mach number it is possible to determine which of the two effects is predominant. However, the hyperbolic model is still far from being a suitable tool for the calculation of steady-state and transient processes in heat exchangers and further theoretical and experimental research is necessary to develop this model.

8.2 Control application of dynamic model of heat exchangers

In general, P-, PI-, and PID control algorithms are widely used in the process industries (for example, control of heat exchangers) even though many advanced control methods have been developed. Although they were originally used for linear systems, these algorithms can also be applied to nonlinear systems by linearizing the nonlinear problems. Control of a system or an apparatus involves the transfer function which may be considered as the basis of traditional P-, PI- and PID control algorithms. It is defined as the ratio of the input to the output of a system in the frequency domain. If the input is expressed as $X(\tau)$ and its Laplace transform is $\overline{X}(s)$, and the output is $Y(\tau)$ and the transformed form is $\overline{Y}(s)$, the transfer function of this system is given as

$$G(s) = \frac{\overline{Y}(s)}{\overline{X}(s)}. \tag{8.24}$$

The transfer function $G(s)$ is the ratio of the output image-function to the input image-function under the uniform initial condition. The transfer function is a form of mathematical model of a system and reveals the transfer relationship between input and output. It is related to the characteristics of a system itself and is independent of the way in which the input changes. To determine the transfer function of a system, two main mathematical models have been used, i.e., the lumped and the distributed parameter model. The former is quite simple and dependent on only the time variable, but may be of poor accuracy. The latter takes the change characteristics with both time and space variables into account and partial differential equations are involved. It is more complicated than the former, but may be of better accuracy. Both these models find use in the dynamic simulation and control of heat exchangers, although the distributed parameter model is mainly used in this monograph.

A simple controller for heat exchangers was briefly outlined by Shah (1979), in which the deviation between the measured output temperature and the set point value was reduced to zero by adjusting the inlet temperature of the hot stream (Fig. 8.1). This may be the simplest example of the control of heat exchangers, but reflects the fundamental control loop. Preising and Ma (1995) described the principle of a model-based controller of heat exchangers. This section mainly introduces some transfer functions obtained from dynamic models of different heat exchangers rather than the control algorithm which is beyond this monograph. These transfer functions may be necessary for designing controllers.

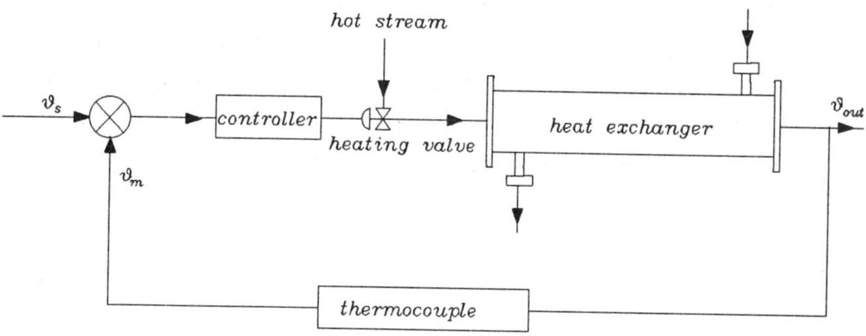

Figure 8.1: Temperature controller for heat exchangers

Case study: Tubular heat exchangers

By assuming that the wall capacitance is negligible and the steam temperature θ_s is uniform over the length of the exchanger at any time, Stermole and Larson (1963) proposed a mathematical description of a distributed parameter steam–water heat exchanger system as follows:

$$\frac{\partial \theta}{\partial \tau} = B(\theta_s - \theta) - u\frac{\partial \theta}{\partial x} \tag{8.25}$$

where $B = \dfrac{KP}{\rho c_p A}$. This equation can be applied to analysing the water temperature variation in a steam heater.

If the flow rates and inlet water temperature remain constant but the steam temperature varies, one can obtain the following transfer function by using the Laplace transform

$$G_s = \frac{\bar{\theta}(L,s)}{\bar{\theta}_s(s)} = \frac{B}{B+s}\left[1 - \mathrm{e}^{-\frac{(B+s)L}{u}}\right]. \tag{8.26}$$

In the light of this transfer function, one is able to design a controller to regulate the outlet water temperature by adjusting the inlet steam temperature. In some cases, one can also adjust the steam flow rate for the same objective. Stainthorp and Axon (1965) studied the response of process fluid temperature to steam flow perturbations and derived the transfer function relating process fluid temperature to steam flow perturbations in a single concentric tube for the constant inlet temperature of the process fluid

$$\frac{\bar{\theta}(L,s)}{\bar{f}_s(s)} = \frac{\beta_1'}{\alpha_1'}\left[1 - \mathrm{e}^{-\frac{\alpha_1' L}{u}}\right] \tag{8.27}$$

where $\bar{f}_s(s)$ is the Laplace transform of a change in steam flow, u is the mean fluid velocity and α_1' as well as β_1' are expressions containing system parameters.

A mathematical model was derived by Schoenberg (1966) to predict the dynamic behaviour of a single-tube condenser. The tube is assumed to contain a mobile liquid–vapour interface and to be cooled by a gas or liquid stream in crossflow. The model consists of transfer functions relating the condensing pressure and the interface position to various disturbance variables. The transfer functions are derived from the linearized form of the fundamental heat transfer and flow equations of the process. They found that some of the values of the coefficients of the transfer functions had little influence on the predicted frequency response. If these values were neglected, a much simpler form of the transfer function was arrived at, which agreed at the lower frequencies with the original more complex model. The detailed calculation procedure is described in their report.

Tayeb *et al.* (1991) constructed the block diagram of the feedback control loop containing a shell-and-tube heat exchanger and calculated the closed loop responses under both servo and regular operations. The transfer function was obtained for a proportional method and process reaction curve method. Thus, they approximately reduced the original complex forms of transfer functions to simpler forms, which may simplify the design of controllers.

In Chapter 3, a series of distributed parameter dynamic models for tubular heat exchangers have been proposed. From those transformed solutions, one can easily obtain the transfer function pertinent to the particular type of an exchanger with the given flow arrangement. By means of the transformed solutions (3.19) and (3.20), one can determine the transfer functions relating the dynamic response of the outlet temperatures to the inlet disturbances for a heat exchanger with cocurrent flow. Here $G_1(s)$ and $G_2(s)$ represent the transfer functions of the outlet responses of fluid 1 and 2, respectively

$$G_1(s) = \frac{T_1(1,s)}{F(s)}, \quad G_2(s) = \frac{T_2(1,s)}{F(s)} \tag{8.28}$$

where $F(s)$ is the inlet temperature disturbance. If it occurs with fluid 1,

$$G_1(s) = \frac{\lambda_2 - a_{11}}{\lambda_2 - \lambda_1} e^{\lambda_1} - \frac{\lambda_1 - a_{11}}{\lambda_2 - \lambda_1} e^{\lambda_2}$$

$$G_2(s) = \frac{(\lambda_1 - a_{11})(\lambda_2 - a_{11})}{a_{12}(\lambda_2 - \lambda_1)} \left(e^{\lambda_1} - e^{\lambda_2}\right). \tag{8.29a}$$

If the inlet temperature disturbance takes place with fluid 2,

$$G_1(s) = -\frac{a_{12}}{\lambda_2 - \lambda_1} \left(e^{\lambda_1} - e^{\lambda_2}\right)$$

$$G_2(s) = -\frac{\lambda_1 - a_{11}}{\lambda_2 - \lambda_1} e^{\lambda_1} + \frac{\lambda_2 - a_{11}}{\lambda_2 - \lambda_1} e^{\lambda_2}. \tag{8.29b}$$

In the similar way, the transfer functions for a heat exchanger with countercurrent flow can be obtained from the distributed parameter solutions (3.25) and (3.26). Also, one can derive the transfer functions for multipass shell-and-tube heat exchangers from the related solutions in the Laplace transform domain. These distributed parameter solutions may correspond to either the plug-flow model or the dispersed plug-flow model. These transfer functions are of complex forms and they are difficult to apply to the design of practical controllers. In some cases, it is found that the outlet responses may be well-approximated by using a first- or second-order system (Das *et al.*, 1995). Therefore, one may use the following transfer functions pertinent to the first- or second-order system to replace the complicated forms of the transfer functions obtained directly from the transformed solutions. For this purpose, the first- and second-order transfer functions are given

$$G_{1st} = \frac{K e^{-\tau_d s}}{1 + \tau_c s} \qquad \text{(the first-order system)} \tag{8.30a}$$

$$G_{2nd} = \frac{K(s+\xi)\omega_n^2\,e^{-\tau_d s}}{\xi(s^2+2\varsigma\omega_n s+\omega_n^2)} \quad \text{(the second-order system).} \qquad (8.30b)$$

The parameters such as delay time τ_d and time constant τ_c for the first-order system, as well as the natural frequency ω_n, damping ratio ζ, lead frequency ξ, and delay time τ_d, are found by curve-fitting. This is done by minimizing the square of the errors between the simplified and the real temperature calculated by the distributed parameter model, i.e.,

$$Error_{1st} = \sum\left(t_{1st} - t_{cal}\right)^2 \qquad (8.27a)$$

$$Error_{2nd} = \sum\left(t_{2nd} - t_{cal}\right)^2. \qquad (8.27b)$$

The optimization is performed by selecting a suitable algorithm.

Obviously, the temperatures and/or flow rates of all incoming streams into an exchanger may simultaneously be disturbed at the inlets and one inlet disturbance causes outlet responses in all of the streams leaving the apparatus. Therefore, an exchanger may be generally considered as a one-input multi-output (OIMO) system or multi-input multi-output (MIMO) system. The control of multivariable systems is far beyond the scope of this monograph. One may refer to the literature for such problems. Roffel and Rijnsdorp (1982), Melo and Friedly (1992), Lin and Wu (1982) as well as many other researchers deal with the multivariables system.

8.3 Effect of fouling on the performance of heat exchangers

Fouling of heat transfer surfaces accompanies the operation of a heat exchanger. Heat exchanger fouling is a dynamic process and continues until either dynamic equilibrium (i.e., the amount of foulant deposited on the surface is equal to that removed by a fluid stream) is reached or until it becomes necessary to clean the exchanger. The time involved may vary from case to case, for example, a few hours, several months or even years. Fouling affects the dynamic behaviour and performance of heat exchangers. In general, fouling reduces the overall heat transfer coefficient and degrades performance of heat transfer units.

Although some research work on fouling has been done, the fouling problem of the heat transfer surface remains a challenge to the technical skills and acumen of engineers and scientists (Bott, 1992). People often face the difficulty how to estimate the effect of fouling on design and rating of heat exchangers. As pointed out by Bott, the design of exchangers often involves the use of sophisticated computer codes based on well-established heat transfer correlations, but with a static allowance made for the effects of fouling. The effect of the fouling layer on the heat transfer process is usually expressed by the fouling resistance R_f. For the design of heat exchangers, fouling resistances are assigned to provide an excess surface area in order to maintain satisfactory performance after some fouling has occurred. It is important to make a reasonable specification of the fouling resistance, but it is often not easy to do so. TEMA (Chenoweth, 1990) recommended a series of fouling values for different applications.

8.3.1 Fouling mechanism

The fouling of a heat transfer surface varies with time and is difficult to be predicted. The mechanism may be related with the dynamic exchange of mass, heat and momentum. It is affected by a variety of factors. Among the various factors which affect the accumulation of deposits on surfaces, the most important ones are the fluid velocity, the concentration and supersaturation levels of the scale forming ions and the temperature (Bott, 1990). Epstein (1978, 1983) proposed the following categories of fouling: (1) crystallization fouling; (2) particulate fouling; (3) chemical reaction fouling; (4) corrosion fouling; (5) biological fouling; (6) solidification fouling. This category methodology has been widely accepted. Strong or weak interactions exist among all these types of fouling, which result in fouling of heat transfer surfaces.

The fouling mechanism is so complicated that at present one cannot propose an accurate description of the formation process. On the basis of some optimizations, different fouling models have been developed. A generally accepted fouling model is the Kern–Seaton equation (Kern and Seaton, 1959). They expressed the net rate of fouling per unit surface area as follows:

$$\frac{\mathrm{d}\, m_{fr}}{\mathrm{d}\,\tau} = \dot{m}_d - \dot{m}_{fr} \tag{8.32}$$

where $\dot{m}_d$ and $\dot{m}_{fr}$ are the deposition and removal rates, respectively. If it is assumed that components and properties of a foulant layer are uniform on the whole surface and along the thickness direction, the fouling resistance is defined as

$$R_{fr} = \frac{m_{fr}}{k_{fr}\,\rho_{fr}}. \tag{8.33}$$

Insertion of this expression into the differential relationship (8.32) yields

$$\frac{\mathrm{d}\, R_{fr}}{\mathrm{d}\,\tau} = \frac{\dot{m}_d - \dot{m}_{fr}}{k_{fr}\rho_{fr}} = \varphi_d - \varphi_{fr}, \tag{8.34}$$

in which the variation rate of the fouling resistance with time depends upon the deposit rate $\dot{m}_d$ and removal rate $\dot{m}_{fr}$. This expression is fundamental and versatile, but the forms of $\dot{m}_d$ and $\dot{m}_{fr}$ are strongly related to the types of foulant forming. For fouling that is forming in a forced convection flow, the transfer rate $\dot{m}$ may be expressed as (Andritsos and Karabelas, 1992):

$$\dot{m} = h_m\left(C_b - C_{eq}\right) \tag{8.35}$$

where C_b is the bulk concentration of the transferable

crystal-building ionic species, molecules or small particles, C_{eq} the equilibrium concentration and h_m a mass transfer coefficient. For the turbulent flow in a smooth pipe with the inner diameter d_i, analogy between mass and heat transfer yields

$$Sh = \frac{h_m d_i}{D} = 0.023\ Sc^{0.4}\ \mathrm{Re}^{0.8} \tag{8.36}$$

where Sh is the Sherwood number and D the molecular diffusion coefficient.

8.3.2 Effect of the fouling resistance on design and rating of heat exchangers

For the steady-state design and rating of heat exchangers, the overall heat transfer coefficient K is defined as

$$\frac{1}{KA} = \left[\frac{1}{\eta_o hA} + \frac{R_{fr}}{A}\right]_1 + \left[\frac{1}{\eta_o hA} + \frac{R_{fr}}{A}\right]_2 + \frac{R_w}{A} \tag{8.37}$$

where η_o is the overall (thermal) effectiveness of the finned surface and $\eta_o = 1$ for a bare surface. In such a way the fouling resistance R_{fr} is incorporated into the design and rating of thermal units. The choice of R_{fr} determines the extra surface area that has to be added to a heat exchanger to maintain the required performance until the level of fouling represented by R_{fr} is reached. A constant fouling resistance is generally assigned for calculation, although the fouling formation is a dynamic process. This constant value may be a mean over the operation period or an estimate of the possible maximum. There exists considerable deviation among the estimated values of the fouling resistance from different sources. The authorized and widely accepted data are given by TEMA (Chenoweth, 1990).

The reason for using the roughly estimated fouling resistance is that it is extremely difficult to propose an accurate and time-dependent formula for the fouling resistance. However, one may give an approximate expression for $R_{fr} = f(\tau)$. There exist some empirical models for predicting the fouling resistance. The falling rate model (Reitzer, 1964), the linear increase model (Watkinson and Epstein, 1970), and the asymptotic model are widely spread. The first assumes that the deposit rate of foulant is related to the temperature difference as:

$$\dot{m}_d = h_r^n B^n (\theta_i - \theta_b)^n . \tag{8.38}$$

This model may be suitable for the fouling on the evaporation surface. The second assumes that the fouling resistance is proportional to the operation time

$$R_{fr} = K^* \tau \tag{8.39}$$

where K^* is the proportionality coefficient. The third assumes that the deposit rate $\dot{m}_d$ is constant and the removal rate $\dot{m}_r$ is proportional to the mass of the deposited foulant. Thus, the expression (8.32) is rewritten as (Kern and Seaton, 1959)

$$\frac{\mathrm{d}\, m_{fr}}{\mathrm{d}\,\tau} = \dot{m}_d - b m_{fr} \tag{8.40}$$

where b is the proportional constant (1/s). Integration of (8.34) yields

$$R_{fr} = R_{fr}^* \left(1 - \mathrm{e}^{-\tau/\tau_c}\right) \tag{8.41}$$

where $R_{frr}^* = \dfrac{\dot{m}_d}{b \rho_{fr} k_{fr}}$ and τ_c may be explained as the mean time that a foulant component is deposited on the heat transfer surface or the time which is needed for the foulant layer to reach an asymptotic threshold at the constant rate $\dot{m}_d$. The asymptotic model (8.41) has been used by many researchers to study the fouling characteristics of heat transfer apparatus. One can also find other theoretical and empirical models for predicting the fouling resistance. In most cases, however, the expression (8.41) is applicable. In general, the fouling resistance is experimentally estimated from the following expression

$$R_{fr} = \left(\frac{1}{K}\right)_{clean} - \left(\frac{1}{K}\right)_{fouling} . \tag{8.42}$$

It means that the fouling resistance can be indirectly measured by measuring the overall heat transfer coefficients at the clean state $(\tau = 0)$ and at the pollutant state (the operation time $\tau > 0$). There are different methods and instruments used to measure the foulant layer and the fouling resistance (Rabas and Chenoweth, 1991; Yang and Xu, 1995).

8.3.3 Effect of the fouling resistance on the dynamics of heat exchangers

Without question, the dynamic process of foulant deposit on a heat transfer surface affects the transient behaviour of heat exchangers. In the previous chapters, an approximate method has been introduced to take the effect of fouling resistance on dynamics of heat exchangers into account. Such a treatment may be suitable for the case that the fouling process is in dynamic equilibrium or the fouling deposit develops slowly. In the following it is assumed that the heat conductivity rather than the heat capacitance of the foulant layer exerts an influence on the dynamics of heat exchangers. For a two-fluid heat exchanger, the application of the lumped parameter model to the wall yields

$$C_w \frac{\partial \theta_w}{\partial \tau} = \dot{q}_1 A_1 + \dot{q}_2 A_2 \tag{8.43}$$

where $\dot{q}_1$ and $\dot{q}_2$ can be approximately derived as

$$\dot{q}_1 = \frac{\theta_1 - \theta_w}{1/h_1 + \delta_{fr}/k_{fr_1}} \quad \text{and} \quad \dot{q}_2 = \frac{\theta_2 - \theta_w}{1/h_2 + \left(\delta_{fr}/k_{fr}\right)_2} . \tag{8.44}$$

Thus, the lumped parameter equation for the wall is rewritten as

$$C_w \frac{\partial \theta_w}{\partial \tau} = h_1^* A_1 (\theta_1 - \theta_w) + h_2^* A_2 (\theta_2 - \theta_w) \tag{8.45}$$

where $h_i^* = 1/h + (\delta_{fr}/k_{fr})_i$. According to the fouling type, the $R_{fr1,2}$ which represent the fouling resistance on both of the surfaces can be replaced by the particular formulae. In most cases, the asymptotic expression (8.41) is applicable.

By substituting h_1^* and h_2^* for h_1 and h_2 in the differential equations for the fluid streams, one derives the transient equations for both of the streams, taking into account the effect of the foulant layers on the heat transfer surface. Obviously, the equations accounting for the fouling effect are formally the same as those described in the previous chapters, but they may be nonlinear since the apparent heat transfer coefficients h_1^* and h_2^* may vary with the temperature. The general solution approach should be the numerical method, although the Laplace transform may still be used in some cases.

Here it is assumed that the heat capacitance of the foulant layer is negligible, compared with the capacitance of the wall. If this idealization is severely violated, one should account for the heat capacitance of the foulant layer, and then the problem becomes complicated. One intuitive approach is to incorporate the foulant capacitance into the wall capacitance, so C_w in eqn (8.45) is replaced by $C_w^* = C_w + C_{fr1} + C_{fr2}$. This treatment physically violates the expression (8.45) since it implies that the wall and the foulant are in thermal equilibrium, but it does take the foulant capacitance into account and is of a simple form. The original idea of introducing C_w^* is to find a simple way of handling the foulant capacitance. The strict method of treating the foulant capacitance is to apply the Fourier's conduction law to the wall and foulant layer, respectively,

$$\left(\rho c_p\right)_i \frac{\partial \theta_i}{\partial \tau} = k_i \left(\frac{\partial^2 \theta}{\partial x^2} + \frac{n}{x} \frac{\partial \theta}{\partial x} \right)_i \quad \left(i = w \text{ or } fr_{1,2}\right) \tag{8.46}$$

where $n = 0$ for a plate and $n = 1$ for a cylinder.

It should be kept in mind that the solution region for eqn (8.46) may vary with time. This equation and the equations for the fluid streams are coupled. From the given initial and boundary conditions, the finite-difference method can be used to find the transient profiles and the effect of fouling on the dynamic behaviour of heat exchangers. On the other hand, the above-mentioned procedure may be adopted to estimate the foulant forming on the surface by means of the experimental data.

8.4 Dynamic design of heat exchangers

Up to now, it is almost without any exception that all types of heat exchangers have been designed according to the steady-state performance, i.e., the steady-state design procedure has been carried out to determine the necessary heat transfer surface area and geometrical structure to satisfy the prescribed heat load under certain constrains. There are two different steady-state approaches used for design and rating of heat exchangers: the mean temperature difference method and the method of effectiveness–number of heat transfer units (Peters and Timmerhaus, 1980; Kays and London, 1984). Whether either of these approaches is used, all design and operating parameters are time-independent in the design and rating. Strictly speaking, it is not true to assume constant design operating parameters.

All of the design parameters in the input block of the exchanger design program are nominal values. All possible disturbance factors in the practical process may make the operating conditions deviate from the nominal design values. Some typical examples are the development of the fouling resistance on the heat transfer surface, variation of heat transfer coefficient, disturbances of temperatures and flow rates of incoming fluid streams. In such cases, the heat exchanger designed according to the given steady-state nominal design parameters may fail to satisfy the prescribed process demand. On the other hand, the most important unsolved problem in process control is how to quantitatively incorporate dynamic controllability into conventional steady-state design. As pointed out by Elliott and Luyben (1995), it has been widely recognized that much can be gained by simultaneously designing the plant with its control structure. All of the aspects mentioned here may be involved in the dynamic design. Reduced capital costs, improved process dynamics and a satisfactory operation state are advantages of the dynamic design. Compared with the steady-state design method, the dynamic design approach reveals the comprehensive effect of all of the time-dependent design parameters on heat exchanger sizing values.

8.4.1 Dynamic design based on simulation models

With respect to the time-dependent variations or disturbances of some design parameters, it is unquestionable that the dynamic and steady-state simulators for heat exchangers can be applied to the design and rating of heat exchangers. For a simulation problem, the heat exchanger and inlet conditions are given; what has to be determined are the heat duty and outlet conditions. Being different from the simulation, however, the design problem is to determine a set of unit sizing parameters in the light of the specified stream parameters and operation conditions which are under some constraints.

As mentioned before, the variable operation conditions may include the inlet temperatures and flow rates, heat transfer coefficients and fouling resistances. With the exception of the fouling resistance, which increases with operation time, the other parameters may either increase or decrease with operation time. All these variations or some of them may simultaneously take place during the operation of heat exchangers. To maintain good operation of heat exchangers and the process specifications, it is necessary and reasonable to account for the effect of these parameters on the size of heat exchangers during the design phase, which is the reason that this monograph tries to propose the dynamic design approach.

To outline the dynamic design procedure, the methodology similar to that in the commercial program DYNSIM (Gani *et al.*, 1992) is used in the following description. The dynamic design and analysis problems are defined as follows: the study of the behaviour of heat exchangers due to changes/disturbances in exchanger sizing parameters (for example, geometrical structure, heat transfer surface area, volume, arrangement) ($\mathbf{d}_1$) and/or manipulative variables (for example, heat duty and inlet parameters) ($\mathbf{d}_2$), determination of the new values of the process stream/unit variables (for example, outlet parameters) (**P**), and the study of the effect of specified values of **P** on $\mathbf{d}_1$ and/or $\mathbf{d}_2$. The simulation and design of heat exchangers are performed under the constraints (**c**). For the dynamic simulation, all variables belonging to $\mathbf{d}_1$ and $\mathbf{d}_2$ are specified and all elements of **P** are calculated. For the design problem of heat exchangers, the variables belonging to **P** and $\mathbf{d}_2$ may be specified and the sizing elements of $\mathbf{d}_1$ are calculated. For the design and analysis of heat exchangers, the energy equation may be implicitly expressed as

$$f\left(\mathbf{y},\ \mathbf{c},\ \mathbf{d}_1,\ \mathbf{d}_2,\ \mathbf{P},\ \tau\right)=0 \tag{8.47}$$

where **y** is related to energy accumulation terms. Vector **y** represents the set of differential variables for the dynamic simulation and the set of algebraic variables for the steady-state simulation. If all parameters remain constant, one obtains the steady-state solution by setting $\tau \to \infty$.

For the dynamic simulation, one can use the explicit expressions developed in the previous chapters. For the dynamic design problem, iterative evaluation may be involved. During the design procedure, some steps in the steady-state design are still used. The first step may be to select the type of heat transfer surface and exchanger configuration. According to the changeable design input parameters ($\mathbf{d}_2$ and **P**), one can obtain the changeable sizing parameters ($\mathbf{d}_1$) by defining the given constraints by means of the expression (8.47). With respect to the desired specification range and the most frequent variations in operation, one can estimate the reliability of the design results and select a group of sizing parameters as the final design. Here only the fundamental principles of the dynamic design of heat exchangers have been described. There is a lot to be done in order to apply the dynamic design to heat exchangers.

8.4.2 Fuzzy design method of heat exchangers

Various disturbance factors appearing in the process may make operation conditions (or parameters) change in certain ranges. Such variations may be ambiguous and not be quantitatively described with precision. It may not be suitable to use the traditional methods (for example, probability) to describe these variations with uncertainty. But one can apply the fuzzy theory to the description of uncertain variations. Kraslawski and Nyström (1993) pointed out that "fuzziness" is quite a different notion from probability: Probability describes the objective uncertainty obtained on the basis of a large number of observations; fuzziness describes the uncertainty which has a subjective meaning and the degrees of possession of a given property. Sincc its appearance in the 1960s, the fuzzy theory has found wide use in a variety of scientific and engineering branches for control, decision-making, simulation, design and data analysis.

Conventional human and computer-aided design methods require highly precise descriptions of design input parameters (for example, inlet temperatures, flow rates, etc.). Heat exchangers manufactured from the "precise" design results according to the precisely given input parameters may fail to maintain the process specifications in cases of vague variations of operation conditions. Therefore, it is very meaningful to account for the effect of the vague and imprecise variations of operation conditions during the preliminary design stages. Application of the fuzzy set theory provides the possibility of describing the vague variations of design parameters of thermal apparatus with a set of fuzzy notions and to develop the fuzzy design method of thermal apparatus. In this way, one can estimate the effect of the imprecise change of design input parameters on the design of thermal apparatus.

Three fuzzy notions such as *fuzzy set, membership function* and *fuzzy number* are involved in application of the fuzzy theory. A fuzzy set is a set whose elements belong to it partially. The membership function is a characteristic function which illustrates the degree of membership of an element of a fuzzy set. A fuzzy number is one of the most important terms in engineering applications. It is a fuzzy set whose elements are numbers. An example of fuzzy number is "about 1200". The fuzzy design of heat exchangers is based on the Vertex method (Dong and Shah, 1987). The fundamental principle is described as follows: for the fuzzy set consisting of the fuzzy variations of all design parameters, the first step is to determine the membership function for each fuzzy variable (it means a map μ from a set X to the interval [0,1]). The membership function may be determined from experiences or theoretical and empirical estimation of parameter variations and the most popular forms are triangular and trapezoidal representations. The next is to construct the α-cut of the fuzzy set pertinent to each design parameter by selecting different values $0 \le \alpha \le 1$. Such a α-cut representation is a subset of the fuzzy variable whose membership values are over α values. Here α-cut is a discretization technique on the membership value domains of variables instead of on variable domains themselves. The final step is to make use of the extension principle of Zadeh (1975) and to insert all fuzzy design variables related to α-cuts into the design model which was originally designed for the "precise" nominal design parameters in order to find the fuzzy objective (i.e., a fuzzy set) — the function of all fuzzy design variables and the corresponding membership function.

For each design variable, there exists a fuzzy set $\tilde{A}_j$ related to the set U_j

$$\tilde{A}_j \bullet U_i \to [0,\ 1] \tag{8.48}$$

The membership function $\mu_{\tilde{A}_j}(u_j)$ is a map from the set U_j to the interval [0,1] and takes the values between 0 and 1. Its value represents the membership degree of element u_j to the fuzzy set $\tilde{A}_j$. The closer $\mu_{\tilde{A}_j}(u_j)$ approaches 1, the greater is the membership degree of u_j to $\tilde{A}_j$. If $\mu_{\tilde{A}_j}(u_j)=1$, for example, element u_j absolutely belongs to the fuzzy set $\tilde{A}_j$; if $\mu_{\tilde{A}_j}(u_j)=0$, u_j does not belong to $\tilde{A}_j$. The value interval of each variable x_j is expressed as

$$x_j=[a_j,\ b_j]=\{x_j|a_j\le x\le b_j\} \tag{8.49}$$

where a_j and b_j are respectively the lower and the upper bound related to the given design variable. The real number pair $[a_j, b_j]$ is called the interval number. Fuzzy set $\tilde{A}$ is defined as:

$$\tilde{A}=\{x,\mu_{\tilde{A}}(x)\}\quad \forall x\in X,\quad \mu_{\tilde{A}}(x)=[0,1] \tag{8.50}$$

where x is an element of $\tilde{A}$. The α-cut of a fuzzy set $\tilde{A}$ is defined as

$$\tilde{A}_\alpha=\{x|\mu_{\tilde{A}}\ge\alpha\} \tag{8.51}$$

where α is the threshold value $(0\le\alpha\le1)$. A fuzzy set is an extension of a conventional set, but an α-cut is a conventional set divided from a fuzzy set. The concept of α-cut links up the fuzzy set and the conventional set.

By means of the extension principle, the fuzzy design method uses the α-cut of a fuzzy set and the interval analysis method (Dong and Shah, 1987) to establish contact between the fuzzy mathematical concepts and non-fuzzy concepts (i.e., the fuzzy set and the conventional set). In this way, the "precise" design program which is originally applied to the "real" parameters is extended to the fuzzy number.

For $\alpha\in[0,1]$, the membership function of a fuzzy variable can be discretized by selecting a series of α values $\alpha=\{\alpha_1,\ \alpha_2,\ \ldots,\ \alpha_m\}$ in order to obtain an α-cut pertinent to the given fuzzy set. Obviously, the α-cut depends upon the form of membership function related to the given fuzzy design variable. For an arbitrary α_i, the believable range of design parameter x_j is

$X_{\sim j}^{(\alpha_i)} = \left[a_j^{(\alpha_i)}, b_j^{(\alpha_i)}\right]$. In the real domain, a "precise" design procedure for thermal apparatus may be expressed as a multivariate function

$$y = f(x_1, x_2, \ldots, x_n) \tag{8.52}$$

where x_i is a real design variable in the set $X = \{x\}_i$ and y may be the heat transfer surface area of a heat exchanger or other design objectives. If x_i is fuzzy variable and $x_i \in X_{\sim i}$, a fuzzy function corresponding to $y = f(x_1, x_2, \ldots, x_n)$ may be described as

$$Y_{\sim} = f(X_{\sim 1}, X_{\sim 2}, \ldots, X_{\sim n}). \tag{8.53}$$

According to the extension principle, the membership function of fuzzy variable $Y_{\sim}$ yields

$$\begin{aligned}\mu_{Y_\sim}(y) &= \bigvee_{\substack{x_1 \in X_{\sim 1}, \ldots, x_n \in X_{\sim n} \\ y = f(x_1, x_2, \ldots, x_n)}} \left[\mu_{X_{\sim 1}}(x_1) \wedge \mu_{X_{\sim 2}}(x_2) \wedge \ldots \wedge \mu_{X_{\sim n}}(x_n)\right] \\ &= \sup_{\substack{x_1 \in X_{\sim 1}, \ldots, x_n \in X_{\sim n} \\ y = f(x_1, x_2, \ldots, x_n)}} \left\{\min\left[\mu_{X_{\sim 1}}(x_1) \wedge \mu_{X_{\sim 2}}(x_2) \wedge \ldots \wedge \mu_{X_{\sim n}}(x_n)\right]\right\}.\end{aligned} \tag{8.54}$$

If $X = [a, b]$ is defined as an interval variable whose value is an interval number $[a, b]$, function $f(x_1, x_2, \ldots, x_n)$ contains n interval variables. All these interval variables form a n-dimensional rectangular $X_1 \times X_2 \times \cdots \times X_n$ with 2^n vertices. The coordinates of these vertices are combinations of n pairs of end points of interval numbers related to n variables, i.e., $C_k = (e_1, e_2, \ldots, e_n)_k$ $(k = 1, 2, 3, \ldots, 2^n)$ in which e_1 is either a_1 or b_1, e_2 is either a_2 or b_2, and so on. Since the range of a fuzzy number varies with the threshold value α, the n-dimensional rectangular and the coordinates of their 2^n vertices vary with α, and then C_k should be expressed as $C_k^{(\alpha_i)} = \left(e_1^{(\alpha_i)},\ e_2^{(\alpha_i)},\ \ldots,\ e_n^{(\alpha_i)}\right)_k$.

Generally, the accuracy of the fuzzy design method increases with the number of selected α_i points. For each $C_k^{(\alpha_i)}$ of 2^n combinations pertinent to n design variables, the design results are obtained as $y_k = f\left(e_1^{(\alpha_i)},\ e_2^{(\alpha_i)},\ \ldots,\ e_n^{(\alpha_i)}\right)_k$. Therefore, the interval of the design objective corresponding to α_i is found by means of the vertex method (Dong and Shah, 1987)

$$Y_D^{(\alpha_i)} = \left[\bigwedge_k f\left(C_k^{(\alpha_i)}\right),\ \bigvee_k f\left(C_k^{(\alpha_i)}\right)\right]$$

$$= \left\{\min\left[f\left(C_1^{(\alpha_i)}\right), f\left(C_2^{(\alpha_i)}\right), \ldots, f\left(C_{2^n}^{(\alpha_i)}\right)\right],\right.$$

$$\left.\max\left[f\left(C_1^{(\alpha_i)}\right), f\left(C_2^{(\alpha_i)}\right), \ldots, f\left(C_{2^n}^{(\alpha_i)}\right)\right]\right\} \tag{8.55}$$

This expression is the basis of fuzzy design. To explain the application of the fuzzy method to the design of thermal apparatus, a heat-pipe heat exchanger is taken as an example. The flow rate of the flue gas $V_{01} = 5000\,\text{m}^3/\text{h}$ (at the standard state) and the inlet temperature $\theta'_{01} = 300°\text{C}$, the flow rate of air to be preheated $V_{02} = 4500\,\text{m}^3/\text{h}$ (at the standard state) and the inlet temperature $\theta'_{02} = 20°\text{C}$. The outer diameter of the carbon-steel tube is $d_o = 25\,\text{mm}$ and its wall thickness $\delta_w = 2\,\text{mm}$. The circular plain fin is attached to the evaporation and condensation parts. The outer fin diameter is $d_f = 50\,\text{mm}$, the thickness $\delta_f = 2\,\text{mm}$ and the span between two adjacent fins $a_f = 6\,\text{mm}$. The carbon-steel–water heat pipes are in a staggered arrangement with the transverse pitch $S_1 = 65\,\text{mm}$ and the longitudinal pitch $S_2 = 56.3\,\text{mm}$. The outlet temperature of the flue gas is given as $\theta''_1 = 180°\text{C}$ to avoid the low temperature sulphate corrosion of flue gas against the heat pipe. All these parameters are nominal values.

Since there exists a variety of factors that influence the operation processes, the flow rates and inlet temperatures of the flue gas and air in practical operations fluctuate within some ranges deviating from the "precisely" given nominal design parameters. The fluctuations may be of uncertainty. It is assumed that the flue gas flow rate, the air flow rate and the inlet gas temperature are the fuzzy variables involved in the design of the preheater and the related membership functions of these variables belonging to the fuzzy sets $\tilde{V}_1$, $\tilde{V}_2$ and $\tilde{\theta}_1'$ are given as

$$\mu_{\tilde{V}_1}(V_1)=\begin{cases}\dfrac{V_1-V_{1a}}{V_{01}-V_{1a}} & V_{1a}\le V_1\le V_{01}\quad \left(V_{1a}=V_{01}-a_1\Delta V_1\right)\\[2ex] \dfrac{V_{1b}-V_1}{V_{1b}-V_{01}} & V_{01}<V_1\le V_{1b}\quad \left(V_{1b}=V_{01}+b_1\Delta V_1\right)\\[2ex] 0 & \text{other } V_1\end{cases} \tag{8.56}$$

$$\mu_{\tilde{V}_2}(V_2)=\begin{cases}\dfrac{V_2-V_{2a}}{V_{02}-V_{2a}} & V_{2a}\le V_2\le V_{02}\quad \left(V_{2a}=V_{02}-a_2\Delta V_2\right)\\[2ex] \dfrac{V_{2b}-V_2}{V_{2b}-V_{02}} & V_{02}<V_2\le V_{2b}\quad \left(V_{2b}=V_{02}+b_2\Delta V_2\right)\\[2ex] 0 & \text{other } V_2\end{cases} \tag{8.57}$$

$$\mu_{\tilde{\theta}_1'}(\theta_1')=\begin{cases}\dfrac{\theta_1'-\theta_{1a}'}{\theta_{01}'-\theta_{1a}'} & \theta_{1a}'\le \theta_1'\le \theta_{01}'\quad \left(\theta_{1a}'=\theta_{01}'-a_3\Delta\theta_1'\right)\\[2ex] \dfrac{\theta_{1b}'-\theta_1'}{\theta_{1b}'-\theta_{01}'} & \theta_{01}'<\theta_1'\le \theta_{1b}'\quad \left(\theta_{1b}'=\theta_{01}'+b_3\Delta\theta_1'\right)\\[2ex] 0 & \text{other } \theta_1'\end{cases} \tag{8.58}$$

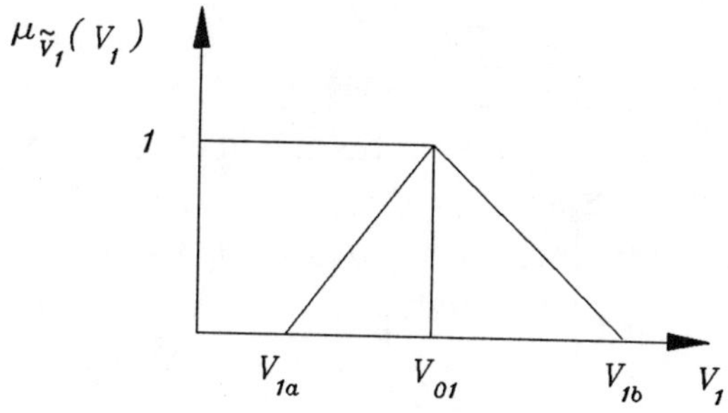

Figure 8.2: The membership function of fuzzy design variable $\tilde{V}_1$

where a_i, b_i and c_i represent uncertain variations of the design parameters.

The membership function of design variable $\tilde{V}_1$ is shown in Fig. 8.2 and the design variables $\tilde{V}_2$ and $\tilde{\theta}_1'$ have similar figures. Obviously, these membership functions are of a triangular form. For each threshold value α_i, the intervals of the fuzzy design variables $\tilde{V}_1$, $\tilde{V}_2$ and $\tilde{\theta}_1'$ are $V_{1,\alpha} = \left[V_{1a,\alpha}, V_{1b,\alpha}\right]$, $V_{2,\alpha} = \left[V_{2a,\alpha}, V_{2b,\alpha}\right]$ and $\tilde{\theta}_{1,\alpha}' = \left[\theta_{1a,\alpha}', \theta_{1b,\alpha}'\right]$. The coordinates of the eight vertices of the three-dimensional rectangle formed from the intervals of these three fuzzy design variables yield as follows: $\left[V_{1a,\alpha}, V_{2a,\alpha}, \theta_{1a,\alpha}'\right]$, $\left[V_{1a,\alpha}, V_{2a,\alpha}, \theta_{1b,\alpha}'\right]$ $\left[V_{1a,\alpha}, V_{2b,\alpha}, \theta_{1a,\alpha}'\right]$, $\left[V_{1a,\alpha}, V_{2a,\alpha}, \theta_{1b,\alpha}'\right]$, $\left[V_{1b,\alpha}, V_{2a,\alpha}, \theta_{1a,\alpha}'\right]$, $\left[V_{1b,\alpha}, V_{2a,\alpha}, \theta_{1b,\alpha}'\right]$, $\left[V_{1b,\alpha}, V_{2b,\alpha}, \theta_{1a,\alpha}'\right]$, $\left[V_{1b,\alpha}, V_{2b,\alpha}, \theta_{1b,\alpha}'\right]$. The number of heat pipes is the design objective of a heat pipe preheater. Originally, the number F_N of heat pipes is expressed as a function of the "precise" design variables V_1, V_2 and θ_1':

$$F_N = f(V_1, V_2, \theta_1'). \tag{8.59}$$

For the fuzzy design variables $\tilde{V}_1$, $\tilde{V}_2$ and $\tilde{\theta}_1'$, the fuzzy design objective yields

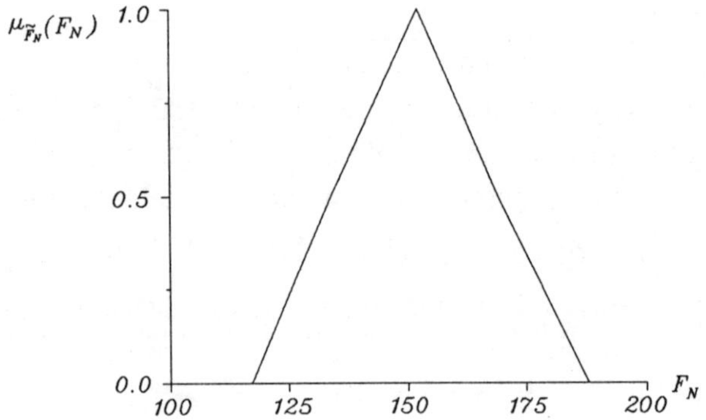

Figure 8.3: The membership function of the design objective (the number of heat pipes) ($a_1 = 1.2$, $b_1 = 1.5$, $a_2 = 1.2$, $b_2 = 1.4$, $a_3 = 1.5$, $b_3 = 1.2$, $\Delta V_1 = 100\ \mathrm{m^3/h}$, $\Delta V_2 = 90\ \mathrm{m^3/h}$, $\Delta\theta_1' = 15°\mathrm{C}$)

$$\widetilde{F}_N = \left(f\widetilde{V}_1, \widetilde{V}_2, \widetilde{\theta}_1'\right). \tag{8.60}$$

According to the expression (8.60), one can obtain a fuzzy set about the design objective (the number of heat pipes). The corresponding membership function is illustrated in Fig. 8.3. The variation interval of the threshold value α reveals the effect of uncertainty of fuzzy design variables on the design result. Fluctuations of the inlet temperature of the hot fluid and the flow rates of hot and cold fluid streams result in variations of heat duty of air preheater and the overall heat transfer coefficient, and then the required surface area (here the number of heat pipes) changes. The design imprecision concerns the uncertainty of the design variables and the interval and membership function of the fuzzy set of design variables confines the interval of the fuzzy set of design results. The membership function of design results depends upon the form of the membership function of the fuzzy design variables, In the example, the membership function of fuzzy design variables is assumed to be triangular, and then the membership function of fuzzy design results is also triangular. In general, the membership function of the fuzzy function $Y_{\sim} = f\left(X_{\sim 1}, X_{\sim 2}, \dots, X_{\sim n}\right)$ corresponding to the multivariable function $y = f(x_1, x_2, \dots, x_n)$ is determined by the expression (8.54).

For the fuzzy design procedure, computation cost remarkably increases rapidly with an increase in the number of fuzzy design variables involved. If the membership function of a fuzzy design variable is complicated, one should select further points $\alpha_i \{\alpha \in [0, 1]\}$. For each α_i $(i = 1, 2, \dots, M)$, one may determine the variation interval of the fuzzy design variable by the given membership function in order to obtain more accurate design results. In this case, it may be difficult to reveal the effect of vague variations of design variables on the design objective by choosing only three points $(\alpha = 0,\ 0.5,\ 1.0)$.

Indeed, probability may also be used to deal with the design uncertainty of thermal apparatus, but it analyses the effect of design variables on the design objective one by one and compensates such effect only by enlarging heat transfer surface areas, which may lead to over-design. On the contrary, the fuzzy design method systematically and comprehensively analyses the transfer process of uncertain variations of design variables in the design procedure and the effect of fuzzy design variables on the design objective. The fuzzy theory may be more suitable for the analysis of the design uncertainty of thermal apparatus (either thermal or frictional computation).

8.5 References

Andritsos, N. and Karabelas, A. J. (1992) Crystallization Fouling: The Effect of Flow Velocity on the Deposition Rate, in *Fouling Mechanisms — Proceedings of the Eurotherm Seminar Nr.* 23, 29–38.

Benneker, A. H., Kronberg, A. E. and Westerterp, K. R. (1997) Longitudinal Mass and Heat Dispersion in Tubular Reactors, *Industrial and Engineering Chemistry Research*, **36**, 2031–2040.

Bott, T. R. (1990) *Fouling Notebook*, Institution of Chemical Engineers, London.

Bott, T. R. (1992) Heat Exchanger Fouling. The Challenge, in *Fouling Mechanisms — Proceedings of the Eurotherm Seminar Nr.* **23**, 3–10.

Chenoweth, J. M. (1990) Final Report of the HTRI/TEMA Joint Committee to Review the Fouling Section of the TEMA Standards, *Heat Transfer Engineering*, **11**, 13–107.

Chester, M. (1963) Second Sound in Solids, *Physics Review*, **131**, 2013–2015.

Das, S. K., Spang, B. and Roetzel, W. (1995) Dynamic Behaviour of Plate Heat Exchangers — Experimental and Modeling, *Journal of Heat Transfer*, **117**, 859–864.

Danckwerts, P. V. (1953) Continuous Flow Systems. Distribution of Residence Time, *Chemical Engngineering Science*, **2**, 1–13.

Dong, Q. and Shah, H. C. (1987) Vertex Method for Computing Functions of Fuzzy Variables, *Fuzzy Sets and Systems*, **24**, 65–78.

Eckert, E. R. G. and Drake Jr. R. M. (1972) *Analysis of Heat and Mass Transfer*, McGraw-Hill, New York.

Elliott, T. R. and Luyben, W. L. (1995) Capacity-Based Economic Approach for the Quantitative Assessment of Process Controllability during the Conceptual Design Stage, *Industrial and Engineering Chemistry Research*, **34**, 3907–3915.

Epstein, N. (1978) Fouling in Heat Exchangers, in *Proceedings of the 6th Int. Heat Transfer Conference*, **6**, 235–253.

Epstein, N. (1983) Thinking about Heat Transfer Fouling: A 5×5 Matrix, *Heat Transfer Engineering*, **4**, 43–56.

Gani, R., SØrensen, E. L., and Perregaard, J. (1992) Design and Analysis of Chemical Processes through DYNSIM, *Industrial and Engineering Chemistry Research*, **31**, 244–254.

Kays, W. M. and London, A. L. (1984) *Compact Heat Exchangers*, McGraw-Hill, New York.

Kern, D. Q. and Seaton, R. E. (1959) A Theoretical Analysis of Thermal Surface Fouling, *Chemical Engineering Progress*, **4**, 258–262.

Kraslawski, A. and Nyström, L. (1993) Fuzzy Sets. Introduction to Theory and Applications in Process Engineering, *Chemical Engineering Technology*, **16**, 291–295.

Kronberg, A. H., Benneker, A. H. and Westerterp, K. R. (1996) Wave Model for Longitudinal Dispersion: Application to the Laminar-Flow Tubular Reactor, *AIChE* Journal, **41**, 3133–3145.

Lin, P. and Wu, Y. C. (1982) Identification of Multi-Input Multi-Output Linear Systems from Frequency Response Data, *Transactions of ASME, Journal of Dynamic Systems, Measurement and Control*, **104**, 58–64.

Luo, X. and Roetzel, W. (1995) Extended Axial Dispersion Model for Transient Analysis of Heat Exchangers, in *4th UK National Conference on Heat Transfer*, IMechE Conference Transactions 1995-2, C510/141/95, Mechanical Engineering Publications Limited, London, pp. 411–416.

Melo, D. L. and Friedly, J. C. (1992) On-Line, Closed-Loop Identification of Multivariable Systems, *Industrial and Engineering Chemistry Research*, **31**, 274–281.

Morse, P. M. and Feshbach, H. (1953) *Methods of Theoretical Physics*, McGraw-Hill, New York.

Peters, M. S. and Timmerhaus, K. D. (1980) *Plant Design and Economics for Chemical Engineers*, McGraw-Hill, New York.

Preising, H. and Ma, X. H. (1995) Modellgetützte Regelung von Wärmeaustauschern, *Chemie-Ingenieur-Technik*, **67**, 1472–1476.

Rabas, T. J. and Chenoweth, J. M. (1991) *Fouling and Enhancement Interactions*, HTD-Vol. 164, ASME, New York.

Reitzer, B. J. (1964) *Int. Engng. Chem.*, **2**, 345–348.

Roetzel, W. and Das, S. K. (1995) Hyperbolic Axial Dispersion Model: Concept and Its Application to a Plate Heat Exchanger, *International Journal of Heat and Mass Transfer*, **38**, 3065–3076.

Roetzel, W. and Spang, B. (1994) Axiales Dispersionsmodell für Regeneratoren mit endlicher Ausbreitungsgeschwindigkeit thermischer Störungen, Mitteilung aus dem Institut für Thermodynamik, Universität der Bundeswehr, Hamburg.

Roetzel, W. and Spang, B. (1995) Axiales Dispersionsmodell für Wärmeübertrager mit endlicher Ausbreitungsgeschwindigkeit thermischer Störungen, UniForschung, Forschungsmagazin der Universität der Bundeswehr Hamburg, no 3, 37–47.

Roetzel, W., Spang, B., Luo, X. and Das, S. K. (1998) Propagation of the Third sound Wave in Fluid: Hypothesis and Theoretical Foundation, *International Journal of Heat and Mass Transfer*, **41**, 2769–2780.

Roffel, B. and Rijnsdorp, J. E. (1982) *Process Dynamics Control and Protection*, Ann Arbor Science Publishers, Ann Arbor, MI.

Schoenberg, A. A. (1966) *Mathematical Model with Experimental Verification for the Dynamic Behaviour of a Single-Tube Condenser*, NASA TN D-3453.

Shah, M. J. (1979) *Engineering Simulation Using Small Scientific Computers*, Prentice-Hall, Inc., Englewood Cliffs, NJ.

Stainthorp, F. P. and Axon, A. G. (1965) The Dynamic Behaviour of a Multipass Steam-Heated Exchanger — I: Response to Steam Temperature and Steam Flow Perturbations, *Chemical Engineering Science*, **20**, 107–119.

Stermole, F. J. and Larson, M. A. (1963) Dynamic Response of Heat Exchangers to Flow Rate Changes, *I&EC Fundamentals*, **2**, 62–67.

Tayeb, A. M., El-Rifai, M. A., El-Tawil, Y. A., and Abdel-Monem, R. M. (1991) Calculation of Optimal Controller Settings and Transient Responses of a Shell-and-Tube Heat Exchanger, *Energy Conversion Management*, **32**, 431–438.

Taylor, G. (1954) The Dispersion of Matter in Turbulent Flow through a Pipe, *Proceeding of the Royal Society of London*, **A223**, 446–468.

Watkinson, A. P. and Epstein, N. (1970) Particulate Fouling of Sensible Heat Exchangers, in *Proceedings of the 4th International Heat Transfer Conference*, **1**, 1–12.

Westerterp, K. R., Dil'man, V. V., and Kronberg, A. E. (1995a) Wave Model for Longitudinal Dispersion: Development of the Model, *AIChE Journal*, **41**, 2013–2028.

Westerterp, K. R., Dil'man, V. V., and Kronberg, A. H. (1995b) Wave Model for Longitudinal Dispersion: Analysis and Applications, *AIChE Journal*, **41**, 2029–2039.

Yang, S. R and Xu, Zh. M. (1995) *Fouling and Countermeasure of Thermal Apparatuses*, Science Press, Beijing.

Zadeh, L. A. (1975) The Concept of a Linguistic Variable and Its

Application to Approximate Reasoning, *Information Science*, **8**, 199–247.

Appendix

Numerical inversion of the Laplace Transform

The Laplace transform of a function $f(\tau)$ (for $\tau \geq 0$) is defined by

$$F(s) = L\{f(\tau)\} = \int_0^{\infty} f(\tau) \mathrm{e}^{-s\tau} \,\mathrm{d}\tau \tag{A.1}$$

and the inverse transform is given by

$$f(\tau) = L^{-1}\{F(s)\} = \frac{1}{2\pi\,\mathrm{i}} \int_{\sigma-\mathrm{i}\infty}^{\sigma+\mathrm{i}\infty} F(s) \mathrm{e}^{s\tau} \,\mathrm{d}s \tag{A.2}$$

where the constant σ is chosen sufficiently large such that

$$\int_0^{\infty} \mathrm{e}^{-\sigma\tau} |f(\tau)| \mathrm{d}\tau < \infty \tag{A.3}$$

Generally speaking, the Laplace transform may be easier than its inverse transform. Although there exist some tables and formulas that can be applied to the analytical inversion of the Laplace transform, it cannot be expected to perform the inverse transform analytically for all cases. In fact, one should resort to the numerical inversion algorithms in most scientific and technical applications.

There are many different numerical inverse Laplace transforms in the literature. Some inversion algorithms may be especially valid for some applications, but fail to deal with the other cases. It is almost impossible to apply an algorithm to all cases. If an algorithm is used, its accuracy should be checked by employing other inversion techniques. Here the Gaver–Stehfest algorithm (Stehfest, 1970) and the inverse transform based on the Fourier series (Ichikawa and Kishima, 1972) are respectively introduced. Both these algorithms are widely adopted for the determination of the inverse Laplace transform. They are also used to find transient responses of heat exchangers throughout this monograph.

1. The Gaver–Stehfest algorithm for approximate inversion

Using the calculation method developed by Gaver (1966), Stehfest (1970) proposed an approximate inversion algorithm of the Laplace transform as follows:

$$f_a(\tau) = \frac{\ln 2}{\tau} \sum_{i=1}^{N} V_i F\left(\frac{\ln 2}{\tau} i\right) \tag{A.4}$$

where N must be an even integer and V_i is given as

$$V_i = (-1)^{N/2+i} \sum_{k=[(i+1)/2]}^{\min(i,N/2)} \frac{k^{N/2}(2k)!}{(N/2-k)!k!(k-1)!(i-k)!(2k-i)!}.$$

Since V_i depends upon N only, the array V is to be evaluated only once for cases of repeated procedure calls with the same N. Theoretically $f_a(\tau)$ becomes more accurate for the greater N. Practically, however, rounding errors worsen the results if N becomes too large because a larger V_i leads to greater values of f_a, which reflects the unboundness of the inverse Laplace transform. Generally, N lies in the range $8 < N < 22$. To reduce truncation errors, double precision arithmetic is recommended. It appears that this algorithm works very well when discontinuous or oscillatory solutions are not anticipated. If there are oscillatory components or discontinuities in $f(\tau)$ in the region $\tau > 0$, it is not expected that accurate values of $f(\tau)$ will be obtained by means of this algorithm. When evaluating an unknown function from its Laplace transform, nevertheless, one should compare the results corresponding with different values of N to estimate whether the function is smooth enough and select the optimum N. In the most cases $N = 18$ will give the best results if the double precision arithmetic is used. In addition, it is preferable to check the results by other inversion algorithms.

2. Numerical inversion algorithm based on Fourier series technique

The Laplace transform is an integral transform and its applicability is restricted to the functions whose integral transforms are convergent. The principle and application of numerical inversion of the Laplace transform by means of Fourier series technique was proposed by Dubner and Abate (1968) as well as Ichikawa and Kishima (1972). Several Fourier transform algorithms were summarized by Davies and Martin (1979). The basis of this algorithm hinges on the fact that in evaluating the inverse Laplace transform integral there is freedom in choosing the integration contour. By introducing a special substitution, the Laplace transform integral can be converted into the Fourier integral. By virtue of selecting the particular contour, the integral is well approximated and the error may be as small as desired.

By setting $s = \sigma + \mathrm{i}\omega$, the Laplace transform pairs are converted into the Fourier transform as follows:

$$F(\sigma + \mathrm{i}\omega) = \int_0^\infty f(\tau)\,\mathrm{e}^{-\sigma\tau}\,\mathrm{e}^{-\mathrm{i}\omega\tau}\,\mathrm{d}\tau \tag{A.5}$$

$$f(\tau) = \frac{\mathrm{e}^{\sigma\tau}}{2\pi}\int_{-\infty}^{\infty} F(\sigma + \mathrm{i}\omega)\,\mathrm{e}^{\mathrm{i}\omega\tau}\,\mathrm{d}\omega\,. \tag{A.6}$$

According to the procedure described in the above-mentioned literature, one constructs the following periodic functions with period $2T$.

$$f_n(\tau) = g(\tau + 2nT), \quad f_n(\tau) = f_n(\tau + 2nT)$$

$$(n, m = 0,\ 1,\ 2,\ \ldots\ ;\ 0 \le \tau \le 2T) \qquad \text{(A.7)}$$

where $g(\tau) = f(\tau)\,\mathrm{e}^{-\sigma\tau}$. The Fourier series representation of each $f_n(\tau)$ is given by

$$f_n(\tau) = \sum_{k=-\infty}^{\infty} C_{n,k}\,\mathrm{e}^{\mathrm{i}k(\pi/T)\tau} \qquad \text{(A.8)}$$

and coefficients $C_{n,k}$ are given by

$$C_{n,k} = \frac{1}{2T}\int_0^{2T} f_n(\tau)\,\mathrm{e}^{-\mathrm{i}k(\pi/T)\tau}\,\mathrm{d}\tau\,. \qquad \text{(A.9)}$$

Furthermore, consider the following infinite series

$$f_p(\tau) = \mathrm{e}^{\sigma\tau}\sum_{n=0}^{\infty} f_n(\tau)\,. \qquad \text{(A.10)}$$

Combining expressions (A.7), (A.8) and (A.9), one has

$$f_p(\tau) = \frac{\mathrm{e}^{\sigma\tau}}{2T}\sum_{k=-\infty}^{\infty}\left\{\int_0^{\infty} f(\tau)\,\mathrm{e}^{-\sigma\tau}\,\mathrm{e}^{-\mathrm{i}k(\pi/T)\tau}\,\mathrm{d}\tau\right\}\mathrm{e}^{\mathrm{i}k(\pi/T)\tau}\,. \qquad \text{(A.11)}$$

The integral in the bracket is reduced to $F(\sigma + \mathrm{i}\,k\pi/T)$ by using (A.5), so that one has

$$f_p(\tau) = \frac{\mathrm{e}^{\sigma\tau}}{2T}\sum_{k=-\infty}^{\infty} F\left(\sigma + \frac{\mathrm{i}\,k\pi}{T}\right)\mathrm{e}^{\mathrm{i}k(\pi/T)\tau}\,. \qquad \text{(A.12)}$$

On the other hand, the integral (A.6) can be approximated as

$$f(\tau)=\frac{e^{\sigma\tau}}{2\pi}\sum_{k=-\infty}^{\infty}F(\sigma+\mathrm{i}k\Delta\omega)\,e^{\mathrm{i}k\Delta\omega\tau}\Delta\omega\,. \tag{A.13}$$

By setting $\Delta\omega=\frac{\pi}{T}$, one has

$$f(\tau)=\frac{e^{\sigma\tau}}{2T}\sum_{k=-\infty}^{\infty}F\left(\sigma+\frac{\mathrm{i}k\pi}{T}\right)e^{\mathrm{i}k(\pi/T)\tau}=f_p(\tau)\,. \tag{A.14}$$

Considering the relation

$$F\left(\sigma+\frac{\mathrm{i}k\pi}{T}\right)e^{\mathrm{i}k(\pi/T)\tau}+F\left(\sigma-\frac{\mathrm{i}k\pi}{T}\right)e^{-\mathrm{i}k(\pi/T)\tau}$$

$$=2\,\mathrm{Re}\left\{F\left(\sigma+\frac{\mathrm{i}k\pi}{T}\right)e^{\mathrm{i}k(\pi/T)\tau}\right\} \tag{A.15}$$

one can rewrite expression (A.14) as

$$f(\tau)=\frac{e^{\sigma\tau}}{2T}\left\{\sum_{k=-\infty}^{-1}F\left(\sigma+\frac{\mathrm{i}k\pi}{T}\right)e^{\mathrm{i}k(\pi/T)\tau}+\sum_{k=1}^{\infty}F\left(\sigma+\frac{\mathrm{i}k\pi}{T}\right)e^{\mathrm{i}k(\pi/T)\tau}\right.$$

$$\left.+F(\sigma)\right\}=\frac{e^{\sigma\tau}}{T}\left\{\frac{1}{2}F(\sigma)+\mathrm{Re}\left[\sum_{k=1}^{\infty}F\left(\sigma+\frac{\mathrm{i}k\pi}{T}\right)e^{\mathrm{i}k(\pi/T)\tau}\right]\right\}. \tag{A.16}$$

In practical applications, the infinite series may be truncated by N terms. By means of this expression, one can get numerical values of $f(\tau)$ at any point τ in the interval $0\le\tau\le 2T$ from the Laplace transform $F(s)$ corresponding to $f(\tau)$. From this expression, some different forms may be derived to perform the numerical inverse transform. By setting $\tau=T$, the expression (A.16) becomes

$$f(T) = \frac{e^{\sigma T}}{T}\left[\frac{1}{2}F(\sigma) + \mathrm{Re}\sum_{k=1}^{N}(-1)^k F\left(\sigma + \frac{\mathrm{i}k\pi}{T}\right)\right]. \tag{A.17}$$

With an similar derivation method, Dubner and Abate (1968) proposed the following inverse Laplace transform for the interval (0, *T*/2)

$$f(\tau) = \frac{2e^{\sigma\tau}}{T}\left[\frac{1}{2}F(\sigma) + \mathrm{Re}\sum_{k=1}^{N}F\left(\sigma + \frac{\mathrm{i}k\pi}{T}\right)\cos\frac{k\pi}{T}\tau\right] \tag{A.18}$$

The accuracy of these numerical inverse algorithms is discussed in the literature cited above. Obviously, the accuracy mainly depends upon the parameter σ and the term N. The round-off and truncation errors are magnified by the factor $e^{\sigma\tau}$ if the selected σ is greater than necessary. For the expression (A.17), σT may be selected as a constant (for example, $\sigma T = 4$ or $\sigma T = 5$). Once $\sigma\tau$ is large, the term N should be accordingly increased in order to obtain a good numerical solution. In some cases, N may approach 1000 or even more. To save computational time, the Fast Fourier Transform (FFT) can be used. By setting

$$\tau_n = \frac{2nT}{N} \qquad (n = 0,\ 1,\ \dots,\ N-1) \tag{A.19}$$

the expression (A.16) becomes

$$f(\tau_n) = \frac{e^{\sigma\tau_n}}{T}\left[\mathrm{Re}\sum_{k=0}^{N-1}F\left(\sigma + \frac{\mathrm{i}k\pi}{T}\right)e^{\mathrm{i}2\pi nk/N} - \frac{1}{2}F(\sigma)\right] \tag{A.20}$$

where $N = 2^{\gamma}$ and γ is an integer. One can adapt the existing FFT programme to determine the value of f_n given by

$$f_n = \sum_{k=0}^{N-1}F\left(\sigma + \frac{\mathrm{i}k\pi}{T}\right)e^{\mathrm{i}2\pi nk/N}. \tag{A.21}$$

Insertion of f_n into (A.20) yields the numerical value of $f(\tau_n)$.

Compared with the Gaver–Stehfest algorithm, the approach based on the Fourier series may find more applications. It can also be applied to functions with oscillatory components and discontinuities, although it may take much computational time.

3. References

Davies, B. and Martin, B. (1979) Numerical Inversion of the Laplace Transform: a Survey and Comparison of Methods, *Journal of Computational Physics*, **33**, 1–32.

Dubner, H. and Abate, J. (1972) Numerical Inversion of Laplace transform by Relating them to the Finite Fourier Cosine Transform, *Journal of the ACM*, **15**, 115–123.

Gaver, D. P. (1966) Observing Stochastic Processes, and Approximate Transform Inversion, *Operation Research*, **14**, 444–459.

Ichikawa, S. and Kishima, A. (1972) Applications of Fourier Series Technique to Inverse Laplace Transform, *Kyoto University Memoires*, **34**, Part I, 53–67.

Stehfest, H. (1970) Numerical Inversion of Laplace Transforms, *Communications of the ACM*, **13**, 47–48.

Nomenclature

A	heat transfer surface area, m^2
A_c	cross-sectional area, m^2
a	thermal diffusivity, m^2/s
B	temperature coefficient of solubility, kg /(m^3 K)
Bi	Biot number, $Bi = hL/k_w$, dimensionless
C	heat capacity, J/K; also conduction wave propagation velocity, m/s
c	concentration, mol/m^3; also specific heat capacity, J/(kg K)
c_p	isobaric specific heat capacity, J/(kg K)
$\mathcal{D}$	apparent conductivity or dispersion coefficient, W/(m K)
e	inner energy, *J/m*^{3}
$F_i(z)$	transformed forms of $f_i(z)$ in the Laplace transform domain, dimensionless
$f_i(z)$	inlet temperature change of fluid i, dimensionless
Fo	Fourier number, $Fo = a_w \tau_{r1}/L^2$, dimensionless
$g_i(x)$, $g_w(x)$	initial dimensionless temperature distributions in fluid i and in the wall, dimensionless
h	heat transfer coefficient, W/(m^{2}K); also specific enthalpy, J/kg
h_r	comprehensive constant of mass transfer coefficient and reaction constant, m/s
h_w	heat transfer coefficient between the fluid and the end surface of the tube, W/(m^{2}K)

I	tubeside flow arrangement, $I = 0$ and $I = 1$ correspond to the tubeside flow arrangements I and II shown in Fig. 3.6, respectively.
K	overall heat transfer coefficient, W/(m^{2}K)
k	heat conductivity, W/(mK)
L	nominal length of a heat exchanger, m
l	space coordinate in the flow direction, m
M	Mach number, $M = u/C$, dimensionless
$\dot{m}$	mass flow rate, kg / s
N	number of tubeside passes
N_R	number of tube rows in a pass
N_T	number of tubes in a row
NTU	number of transfer units, $NTU = \left[\frac{1}{(hA)_1} + \frac{1}{(hA)_2}\right]^{-1} \frac{1}{\dot{W}_1} = \frac{U_1 U_2 R_2}{U_1 + U_2 R_2}$, dimensionless
P	perimeter, m; also period, s
Pe	the dispersive Péclet number, $Pe = \dot{W}L/\mathcal{D}A_c$, dimensionless
Pr	Prandtl number, $Pr = \nu/a$, dimensionless
p	pressure, Pa
$\dot{\mathbf{q}}$	heat flux, W/m^2
Re	Reynolds number, $Re = ud/\nu$, dimensionless
R_1	ratio of the thermal flow rate of fluid 1 to that of fluid 2, $R_1 = \dot{W}_1/\dot{W}_2 = 1/R_2$, dimensionless
R_{c1}	ratio of the fluid heat capacities, $R_{c1} = C_1/C_2$, dimensionless
R_{c2}	ratio of the fluid heat capacities, $R_{c2} = C_2/C_1$, dimensionless
R_w	ratio of the total heat capacity of the tube wall to that of fluids, $R_w = C_w/(C_1 + C_2)$, dimensionless
R_τ	ratio of the residence time of fluid 2 to that of fluid 1, $R_\tau = \dot{W}_1 C_2/\dot{W}_2 C_1$, dimensionless
r	radius of a tube, m

s	the Laplace transform variable with respect to z, dimensionless
$\dot{s}$	heat source, W/m^3
T	transformed form of t in the Laplace transform domain, dimensionless; also the absolute temperature, K
$\widehat{T}$	transformed form of t in the dual Laplace transform domain
t	dimensionless temperature, $t = (\theta - \theta_{r2})/(\theta_{r1} - \theta_{r2})$, dimensionless
U_1, U_{1i}	parameter, $U_1 = (hA)_1/\dot{W}_1$, $U_{1i} = (hA)_{1i}/\dot{W}_1$, dimensionless
U_2, U_{2i}	parameter, $U_2 = (hA)_2/\dot{W}_2$, $U_{2i} = (hA)_{2i}/\dot{W}_2$, dimensionless
u	axial velocity, m/s
$\mathbf{v}$	velocity vector, m/s
$\dot{W}$	thermal flow rate, W/K
x	dimensionless coordinate, $x = l/L$ or $x = l_1/L_1$, dimensionless
y	dimensionless coordinate, $y = l_2/L_2$, dimensionless
z	dimensionless time, $z = (\dot{W}_1/C_1)\tau$, dimensionless

Greek symbols

α_1, α_{1i}	parameter, $\alpha_1 = U_1/(1+R_{c2})$, $\alpha_{1i} = U_{1i}/(1+R_{c2})$, dimensionless
α_2, α_{2i}	parameter, $\alpha_2 = U_2/[R_\tau(1+R_{c1})]$, $\alpha_{2i} = U_{2i}/[R_\tau(1+R_{c1})]$, dimensionless
β	expansion coefficient, 1/K
ε	emissivity; also effectiveness
η_0	overall finned surface efficiency, $\eta_0 = 1-(1-\eta_f)A_f/A$, dimensionless
η_c	$\eta_c = \tau/P_c$ for the cold period in a regenerator

η_f	fin efficiency, dimensionless
η_h	$\eta_h = \tau/P_h$ for the hot period in a regenerator
θ	temperature, K
θ_b	bulk temperature of the main stream, K
θ_i	temperature at the interface between the foulant layer and the fluid, K
θ_r	reference temperature, K
Λ	reduced regenerator length, $\Lambda = \dfrac{hV\sigma}{\dot{W}(1-\varphi)}$, dimensionless
λ	dimensionless heat conductivity, dimensionless
ν	kinematic viscosity, m^2/s
ξ	the dimensionless spatial variable, $\xi = l/L$
Π	reduced periodic time, $\Pi = \dfrac{hP\sigma}{(1-\varphi)\rho_s c_{ps}}$, dimensionless
ρ	density, kg/m^3
σ	step change ratio of the flow velocity, $\sigma = u/u_r$, dimensionless; also specific surface, m^2/m^3; also Stefan–Boltzmann constant, W/m^2K^4
τ	time, s
Φ	viscous dissipation, W/m^3
φ	porosity

Subscripts

a	ambient
c	the cold period in a regenerator
e	outlet
fr	fouling resistance
h	the hot period in a regenerator
i	inner side of a tube or a duct
o	outer side of a tube or a duct
r	reference parameter
s	shell
w	wall

17618